T0335795

"This book is an educational tour de force that presents mathematical thinking as a right-brained activity. Most 'left brain/right brain' education-talk is at best a crude metaphor; but by putting the main focus on the process of (mathematical) abstraction, Eugenia Cheng supplies the reader (whatever their 'brain-type') with the mental tools to make that distinction precise and potentially useful. The book takes the reader along in small steps; but make no mistake, this is a major intellectual journey. Starting not with numbers, but everyday experiences, it develops what is regarded as a very advanced branch of abstract mathematics (category theory, though Cheng really uses this as a proxy for mathematical thinking generally). This is not watered-down math; it's the real thing. And it challenges the reader to think — deeply at times. We 'left-brainers' can learn plenty from it too."

— Keith Devlin, Stanford University (Emeritus), author of *The Joy of Sets*.

"Eugenia Cheng loves mathematics — not the ordinary sort that most people encounter, but the most abstract sort that she calls 'the mathematics of mathematics.' And in this lovely excursion through her abstract world of category theory, she aims to give those who are willing to join her a glimpse of that world. The journey will change how they view mathematics. Cheng is a brilliant writer, with prose that feels like poetry. Her contagious enthusiasm makes her the perfect guide."

— John Ewing, President, Math for America

"Eugenia Cheng's singular contribution is in making abstract mathematics relevant to all through her great ingenuity in developing novel connections between logic and life. Her latest book, *The Joy of Abstraction*, provides a long-awaited fully rigorous yet gentle introduction to the 'mathematics of mathematics,' allowing anyone to experience the joy of learning to think categorically."

— Emily Riehl, Johns Hopkins University,
author of *Category Theory in Context*

"Archimedes is quoted as having once said: 'Mathematics reveals its secrets only to those who approach it with pure love, for its own beauty.' In this fascinating book, Eugenia Cheng approaches the abstract mathematical area of category theory with pure love, to reveal its beauty to anybody interested in learning something about contemporary mathematics."

— Mario Livio, astrophysicist, author of *The Golden Ratio*
and *Brilliant Blunders*

"Eugenia Cheng's latest book will appeal to a remarkably broad and diverse audience, from non-mathematicians who would like to get a sense of what mathematics is really about, to experienced mathematicians who are not category theorists but would like a basic understanding of category theory. Speaking as one of the latter, I found it a real pleasure to be able to read the book without constantly having to stop and puzzle over the details. I have learnt a lot from it already, including what the famous Yoneda lemma is all about, and I look forward to learning more from it in the future."

— Sir Timothy Gowers, Collège de France, Fields Medalist, main editor of *The Princeton Companion to Mathematics*

"At last: a book that makes category theory as simple as it really is. Cheng explains the subject in a clear and friendly way, in detail, not relying on material that only mathematics majors learn. Category theory — indeed, mathematics as a whole — has been waiting for a book like this."

— John Baez, University of California, Riverside

"Many people speak derisively of category theory as the most abstract area of mathematics, but Eugenia Cheng succeeds in redeeming the word 'abstract.' This book is loquacious, conversational and inviting. Reading this book convinced me I could teach category theory as an introductory course, and that is a real marvel, since it is a subject most people leave for experts."

— Francis Su, Harvey Mudd College, author of *Mathematics for Human Flourishing*

"Finally, a book about category theory that doesn't assume you already know category theory! . . . Eugenia Cheng brings the subject to us with insight, wit and a point of view. Her story of finding joy — and advantage — in abstraction will inspire you to find it too."

— Patrick Honner, award-winning high school math teacher, columnist for *Quanta Magazine*, author of *Painless Statistics*

The Joy of Abstraction

Mathematician and popular science author Eugenia Cheng is on a mission to show you that mathematics can be flexible, creative, and visual. This joyful journey through the world of abstract mathematics into category theory will demystify mathematical thought processes and help you develop your own thinking, with no formal mathematical background needed. The book brings abstract mathematical ideas down to earth using examples of social justice, current events, and everyday life — from privilege to COVID-19 to driving routes. The journey begins with the ideas and workings of abstract mathematics, after which you will gently climb toward more technical material, learning everything needed to understand category theory, and then key concepts in category theory like natural transformations, duality, and even a glimpse of ongoing research in higher-dimensional category theory. For fans of *How to Bake Pi*, this will help you dig deeper into mathematical concepts and build your mathematical background.

DR. EUGENIA CHENG is world-renowned as both a researcher in category theory and an expositor of mathematics. She has written several popular mathematics books including *How to Bake Pi* (2015), *The Art of Logic in an Illogical World* (2017), and two children's books. She also writes the "Everyday Math" column for the *Wall Street Journal*. She is Scientist in Residence at the School of the Art Institute of Chicago, where she teaches abstract mathematics to art students. She holds a PhD in category theory from the University of Cambridge, and won tenure in pure mathematics at the University of Sheffield. You can follow her @DrEugeniaCheng.

The Joy of Abstraction

An Exploration of Math, Category Theory, and Life

EUGENIA CHENG

CAMBRIDGE
UNIVERSITY PRESS

CAMBRIDGE
UNIVERSITY PRESS

University Printing House, Cambridge CB2 8BS, United Kingdom

One Liberty Plaza, 20th Floor, New York, NY 10006, USA

477 Williamstown Road, Port Melbourne, VIC 3207, Australia

314–321, 3rd Floor, Plot 3, Splendor Forum, Jasola District Centre, New Delhi – 110025, India

103 Penang Road, #05–06/07, Visioncrest Commercial, Singapore 238467

Cambridge University Press is part of the University of Cambridge.

It furthers the University's mission by disseminating knowledge in the pursuit of education, learning, and research at the highest international levels of excellence.

www.cambridge.org
Information on this title: www.cambridge.org/9781108477222
DOI: 10.1017/9781108769389

First published 2023

A catalogue record for this publication is available from the British Library.

ISBN 978-1-108-47722-2 Hardback

To
Martin Hyland

Contents

ix

Prologue

Abstract mathematics brings me great joy. It is also enlightening, illuminating, applicable, and indeed "useful", but for me that is not its driving force. For me its driving force is joy. The joy it gives me moves me to want to pursue it further and further, immerse myself in it more and more deeply, and engage with the discipline it seems to involve.

This is how I lead all of my life, where I can. I might look disciplined: in the way I do research, write books, practice the piano, follow complex recipes. But I firmly believe that you only need discipline to accomplish something unpleasant. I prefer to find a way to enjoy it. This approach also works better: the enjoyment takes me much further than discipline ever could.

Abstract mathematics is different from what is usually seen in school mathematics. Mathematics in school is typically about numbers, equations and solving problems. Abstract mathematics is not. Mathematics in school has a focus on getting the right answer. Abstract mathematics does not. Mathematics in school unfortunately puts many people off the subject. Abstract mathematics need not.

The aim of this book is to introduce abstract mathematics not usually seen by non-specialists, and to change attitudes about what mathematics is, what it is for, and how it works. The purpose might either be general interest or further study. The specific subject of the book is Category Theory, but along the way we will get a taste of various important mathematical objects including different types of number, shape, surface and space, types of abstract structure, the worlds they form, and some open research questions about them.

This Prologue will provide some background about the book's motivation, style and contents, and guidance about a range of intended audiences. In summary: if you're interested in learning some advanced mathematics that is very different from school math, but find traditional textbooks too dry or requiring too much background, read on.

The status of mathematics

Math has an image problem. Many people are put off it at school and end up as adults either hating it, being afraid of it, or defensively boasting about how bad they are at it or how irrelevant it is anyway. Complaints about math that I hear most commonly from my art students include that it is rigid, uncreative, and requires too much memorization; that the questions have nothing to do with real life and that the answers involve too many rules to be interesting; that it's useful for scientists and engineers but pointless for anyone else.

On the other hand, as an abstract mathematician I revel in how flexible and creative the field is, and how *little* memorization it requires. I am invigorated and continually re-awakened by how the way of thinking is pertinent to all aspects of life. I adore how its richness and insight come from not having to follow anyone else's rules but instead creating different worlds from different rules and seeing what is possible. And I believe that while certain parts are useful for science and engineering, my favorite parts are powerful and illuminating for everyone.

I think there are broadly three reasons math education is important.

1. As a foundation for further study in mathematical fields.
2. For direct usefulness in life.
3. To develop a particular way of thinking.

The first point, further study, is the one that is obviously not relevant to everyone: it doesn't apply if you have absolutely decided you are not going into further study in mathematical (and by extension scientific) fields.

The second point, usefulness, is often emphasized as a reason that math is compulsory for so long at school, but there seems to be a wide range of views about what this actually means. It certainly doesn't seem to justify the endless study of triangles, graph sketching, solving quadratic equations, trigonometric identities and so on. Some people focus on arithmetic and are convinced that math is important so that we don't have to rely on calculators to add up our grocery bill, calculate a tip at a restaurant, or work out how much we'll pay when something is on sale at 20% off. Others argue that the math we teach is not relevant enough and we should teach things like mortgages, interest rates, and how to do your taxes. All of these views are much more utilitarian than the view this book will take.

The third point is about math as a way of thinking, and is the one that drives both my research and my teaching. Abstract mathematics is not just a topic of study. It is a way of thinking that makes connections between diverse situations to help us unify them and think about them more efficiently. It focuses our attention on what is relevant for a particular point of view and temporarily disregards the rest so that we can get to the heart of a structure or an argument. In making these connections and finding these deep structures we package up intractably complex situations into succinct units, enabling us to address yet more complicated situations and use our limited brain power to greater effect. This starts with numbers, where instead of saying "1 + 1" all the time we can call it 2, or we fit squares together and call the result a cube, and then build up to more complex mathematical structures as we'll see throughout this book.

This is what I think the power and importance of abstract mathematics are. The idea that it is relevant to the whole of life and thus illuminating for everyone may be surprising, but is demonstrated by the wide range of examples that I have found where category theory helps, despite the field being considered perhaps the "most abstract" of all mathematics. This includes examples such as privilege, sexism, racism, sexual harassment. These are not the sort of contrived real life examples involving the purchase of 17 watermelons, but are *real* real life questions, things we actually do (or should) think about in our daily lives.

If people are put off math then they are put off these ways of thinking that could really intrigue and help them. The sad part is that they are put off an entirely different kind of math usually involving algorithms, formulae, memorization and rigid rules, which is not what this abstract math is about at all. Math is misunderstood, and the first impression many people get of it is enough to put them off, forever, something that they might have been able to appreciate and benefit from if they saw it in its true light.

Traditional mathematics: subjects

A typical math education is a series of increasingly tall hurdles. If these really were hurdles it would make sense not to try higher ones if you're unable to clear the lower ones.

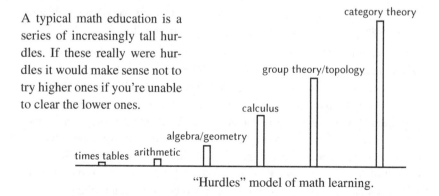

"Hurdles" model of math learning.

However, math is really more like an interconnected web of ideas, perhaps like this; everything is connected to everything else, and thus there are many possible routes around this web depending on what sort of brain you have.

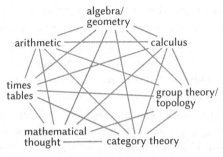

"Interconnectedness" in math learning.

Some people do need to build up gradually through concrete examples towards abstract ideas. But not everyone is like that. For some people, the concrete examples don't make sense until they've grasped the abstract ideas or, worse, the concrete examples are so offputting that they will give up if presented with those first. When I was first introduced to single malt whisky I thought I didn't like it, but I later discovered it was because people were trying to introduce me "gently" via single malts they considered "good for beginners". It turns out I only like the extremely smoky single malts of Islay, not the sweeter, richer ones you might be expected to acclimatize with.

I am somewhat like that with math as well. My route through the web of mathematics was something like this diagram.

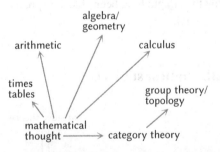

My progress to higher level mathematics did not use my knowledge of mathematical subjects I was taught earlier. In fact after learning category theory I went back and understood everything again and much better.

I have confirmed from several years of teaching abstract mathematics to art students that I am not the only one who prefers to use abstract ideas to illuminate concrete examples rather than the other way round. Many of these art students consider that they're bad at math because they were bad at memorizing times tables, because they're bad at mental arithmetic, and they can't solve equations. But this doesn't mean they're bad at math — it just means they're not very good at times tables, mental arithmetic and equations, an absolutely tiny part of mathematics that hardly counts as abstract at all. It turns out that they do not struggle nearly as much when we get to abstract things such as higher-dimensional spaces, subtle notions of equivalence, and category theory structures. Their blockage on mental arithmetic becomes irrelevant.

It seems to me that we are denying students entry into abstract mathematics when they struggle with non-abstract mathematics, and that this approach is counter-productive. Or perhaps some students self-select out of abstract mathematics if they did not enjoy non-abstract mathematics. This is as if we didn't let people try swimming because they are a slow runner, or if we didn't let them sing until they're good at the piano.

One aim of this book is to present abstract mathematics directly, in a way that does not depend on proficiency with other parts of mathematics. It doesn't have to matter if you didn't make it over some of those earlier hurdles.

Traditional mathematics: methods

When I studied modern languages at school there were four facets tested with different exams: reading, writing, speaking and listening. Of those, writing and speaking are "productive" where reading and listening are "receptive". For full mastery of the language all four are needed of course, but if complete fluency is beyond you it can still be rewarding to be able to do only some of these things. I later studied German for the purposes of understanding the songs of Schubert (and Brahms, Strauss, Schumann, and so on). My productive German is almost non-existent, but I can understand Romantic German poetry at a level including some nuance, and this is rewarding for me and helps me in my life as a collaborative pianist.

I think there is a notion of "productive" and "receptive" mathematics as well. Productive mathematics is about being able to answer questions, say, homework questions or exam questions, and, later on, produce original research. There is a fairly widely held view that the only way to understand math is to work through problems. There is a further view that this is the only way of doing math that is worthwhile. I would like to change that.

I view "receptive" mathematics as being about appreciating math even if you can't solve unseen problems. It's being able to follow an argument even if you wouldn't be able to build it yourself.

I can appreciate German poetry, restaurant food, a violin concerto, a Caravaggio, a tennis match. Imagine if appreciation were only taught by doing. I can even read a medical research paper although I can't practice medicine. The former is still valuable. In math some authors call this "mathematical tourism" with undertones of disdain. But I think tourism is fine — it would be a shame if the only options for traveling were to move somewhere to live there or else stay at home. I actually once spoke to a representative from a health insurance company who thought this was the case, and did not comprehend the concept that I might visit a different state and ask about coverage there.

One particular feature of this book is that I will not demand that the reader does any exercises in order to follow the book. It is standard in math books to exhort the reader to work through exercises, but I believe this is offputting to many non-mathematicians, as well as some mathematicians (including me). I will provide "Things to think about" from time to time, but these will really be questions to ponder rather than exercises of any sort. And one of the main purposes of those questions will be to develop our instincts for the sorts of questions that mathematicians ask. The hope is that as we progress, the reader will think of those questions spontaneously, before I have made them explicit. Thinking of "natural" next questions is one important aspect of mathematical thinking. Where working through them is beneficial to understanding what follows I will include that discussion afterwards.

The content in this book

Category theory was introduced by Eilenberg and Mac Lane in the 1940s and has since become more or less ubiquitous in pure mathematics. In some fields it is at the level of a language, in others it is a framework, in others a tool, in others it is the foundations, in others it is what the whole structure depends on.

Category theory quickly found uses beyond pure math, in theoretical physics and computer science. The view of things at the end of the 20th century might be regarded like this, with the diagram showing applications moving outwards from category theory:

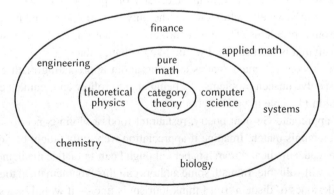

However, since then category theory has become increasingly pervasive, finding direct applications in a much wider range of subjects further from pure mathematics, such as ecological diversity, chemistry, systems control, engineering, air traffic control, linguistics, social justice. The picture now might be thought of as more like this:

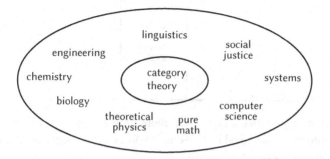

Mac Lane

For some time the only textbook on the subject, from which everyone had to try and learn, was the classic graduate text by Mac Lane, *Categories for the Working Mathematician* (from 1971). The situation was this: there was a huge step up to Mac Lane, which many people, even those highly motivated, failed to be able to reach.

As is the way with these things, what started as a research field had become something that graduate students (tried to) study, and eventually it trickled down into a few undergraduate courses at individual universities around the world that happened to have an expert there who wanted to teach it at that level. This spawned several much more approachable textbooks at the turn of the 21st century, notably those by Lawvere and Schanuel (1997), followed by a sort of second wave with Awodey (2006), Leinster (2014) and Riehl (2016). There was still a gap up to those books, and the gap was still insurmountable for many people who didn't have the background of an undergraduate mathematician, either in terms of experience with the formality of mathematics or background for the examples being used.[†]

In 2015 I wrote *How to Bake π*, a book about category theory for an entirely

---

[†] Lawvere and Schanuel include high school students in their stated target audience but I think they have in mind quite advanced ones. There are also some recent books aimed at specific types of audience, which are less in the vein of standard textbooks; see Further reading.

general audience with no math background whatever. The situation became like this:

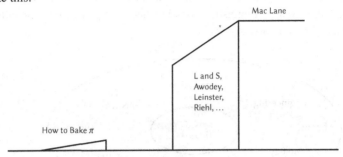

How to Bake π provides a ramp from more or less nothing but does not get very far into the subject and remains mostly informal throughout. The role of the present work is to fill the remaining gap:

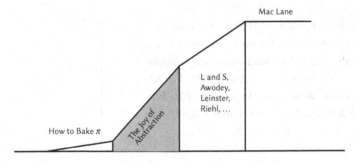

This book will not assume undergraduate level mathematics training, nor even a disposition for it, but it will become technical. The aim, after all, is to bridge the gap up to the undergraduate textbooks. We will build up very gradually towards the rigorous formality of category theory. While there is no particular technical prerequisite for learning category theory, the formality of any mathematical field can be offputting to those who are not used to it. The intention is that you can read this book *and turn the pages*, in contrast with so many math books where you have to sit and think for an hour, week or month about a single paragraph (although those books have their place too). It will be informal and personal in style, including descriptions of my personal feelings about various aspects of the subject. This is not strictly part of mathematics but I believe that building an emotional engagement is an important part of learning anything. There will also be many diagrams to help with visualizing, partly to engage visual people and partly because the subject is very visual. There will be both formal mathematical diagrams, and informal "schematic" diagrams. All these things mean that the book will in some sense be the opposite of terse — long, perhaps, for the technical content contained in it. But I believe this is the key to reaching a wide audience not inclined (or not yet prepared) for reading terse technical mathematics.

Audience

I am aiming this book at anyone who wants to delve into some contemporary mathematics that is not done in school, and is very different from the kind of math done in school. This is regardless of your previous math achievement and your future math goals. I will not assume any knowledge or any recollection of school math, and I will gradually build up the level of formality in case it is the symbols and notation that have previously put you off. Here are some different types of reader I imagine:

- Adults who regret being put off math in the past and think, deep down, that they should be able to understand math if it's presented differently.
- Adults who always liked math and miss it, and feel like having some further mathematical stimulation.
- Anyone who wishes to learn some contemporary mathematics not covered in the standard curriculum, though I hope it might be one day.
- Math teachers who want to extend their knowledge and/or get ideas for how to teach abstract math without much (or any) prerequisite.
- School students who want extra math to stretch themselves, and/or an introduction into higher level math different from what students are usually stretched with at school.
- School students who are unhappy with their math classes and who might benefit from seeing a more profound, less routine type of math.
- Non-mathematicians who want to learn category theory but find the existing textbooks beyond them. Judging from correspondence since *How to Bake π* this might include programmers, engineers, business people, psychologists, linguists, writers, artists and more.
- Undergraduate mathematicians who have heard that category theory is important but are not sure why such abstraction is necessary, or are not sure how to approach it.
- Those with a math degree who would still like to have a gentle companion book to the existing texts, that contains more of the spirit of category theory alongside the technicalities.
- Home-schools and summer camps.

How to Bake π is not exactly a prerequisite but having read it will almost certainly help.

This material is developed from my teaching art students at the School of the Art Institute of Chicago. Most of the students had bad experiences of school math and many of them either can't remember any of it or have deliberately forgotten all of it as they found it so traumatic. This book seeks to be different from all of those types of experiences. It might seem long in terms of pages, but I hope you will quickly find that you can get through the pages much faster than you can for a standard math textbook. If the content here were written in a standard way it might only take 100 pages. I didn't want to make it shorter by explaining things less, so I have made it longer by explaining things more fully. I will gradually introduce formal mathematical language, and have included a glossary at the end for quick reference. I occasionally include the names of related concepts that are beyond the scope of this book, not because I think you need to know them, but in case you are interested and would like to look them up.

One obstacle to non-mathematicians trying to learn category theory is that the examples used are often taken from other parts of pure mathematics. In this book I will be sure to use examples that do not do that, including examples from daily experience such as family relationships, train journeys, freezing and thawing food, and more hard-hitting subjects such as racism and privilege. I have found that this helps non-mathematicians connect with abstract mathematics in ways that mathematical examples do not. Where I do include mathematical examples I will either introduce them fully as new concepts, or point out where they are not essential for continuing but are included for interest for those readers who have heard of them.

In particular, if you think you're bad at mental arithmetic, terrible at algebra, can't solve equations, and shudder at the thought of sketching graphs, that need not be an obstacle for you to read this book. I am not saying that you will find the book easy: abstraction is a way of thinking that takes some building of ability. We will build up through Part One of the book, and definitely take off in Part Two. It should be intellectually stretching, otherwise we wouldn't have achieved anything. But your previous experiences with math need not bar your way in here as they might previously have seemed to do. Most of all, aside from the technicalities of category theory I want to convey the joy I feel in the subject: in learning it, researching it, using it, applying it, thinking about it. More than technical prowess or a huge litany of theorems, I want to share the joy of abstraction.

PART ONE

BUILDING UP TO CATEGORIES

1

Categories: the idea

An overview of what the point of category theory is, without formality.

I like to think of category theory as the mathematics of mathematics.

I admit this phrase sounds a bit self-important, and it comes with another problem, which is the widespread misunderstandings about what mathematics actually is. This problem is multiplied (or possibly raised to the power of infinity) here by the reference of math to itself.

Another problem is that it might make it seem like you need to understand the whole of mathematics before you could possibly understand category theory. Indeed, that is not far from what the prevailing wisdom has been about studying category theory in the past: that you have to, if not understand *all* of math, at least understand a large amount of it, say up to a graduate level, before you can tackle category theory. This is why category theory has traditionally only been taught at a graduate level, and more recently sometimes to upper level undergraduates who already have a solid background in upper level pure mathematics. The received wisdom is that all the motivating examples come from other branches of pure mathematics, so you need to understand those first before you can attempt to understand category theory.

Questioning "received wisdom" is one of my favorite pastimes. I don't advocate just blindly going against it, but the trouble with received wisdom, like "common sense", is that it too often goes unquestioned.

My experience of learning and teaching category theory has been different from that received wisdom. I did first learn category theory in the traditional way, that is, only after many undergraduate courses in pure math. However, those other subjects didn't help me to understand category theory, but the other way round: category theory was much more compelling to me and I loved and understood it in its own right, whereupon it helped me to understand all those other parts of pure math that I had never really understood before.

I eventually decided to start teaching category theory directly as well, to students with essentially no background in pure mathematics. I am convinced that the ideas are interesting in their own right and that examples illustrating

13

those ideas can be found in life, not just in pure math. That's why I'm starting this book with a chapter about those ideas.

I think we can sometimes unintentionally fall into an educational scheme of believing that we need to learn and teach math in the order in which it was developed historically, because surely that is the logical order in which ideas develop. This idea is summed up in the phrase "ontogeny recapitulates phylogeny", although that is really talking about biological development rather than learning.[†] I think this has merit at some levels. The way in which children grasp concepts of numbers probably does follow the history of how numbers developed, starting with the counting numbers 1, 2, 3, and so on, then zero, then negative numbers and fractions (maybe the other way round) and eventually irrational numbers. However, some parts of math developed because of a lack of technology, and are now somewhat redundant. It is no longer important to know how to use a slide rule. I know very few ruler and compass constructions, but this has not hindered my ability to do category theory, just like my poor skills in horse riding have not hindered my ability to drive a car. Of course, horse riding can be enjoyable, and even crucial in some walks of life, and by the same token there are some specific situations in which mental arithmetic and long division might be useful. Indeed some people simply enjoy multiplying large numbers together. However, none of those things is truly a prerequisite for meeting and benefiting from category theory.

Crucially, I think we can benefit from the ideas and techniques of category theory even outside research math and aside from direct technical applications. Mathematics is among other things a field of research, a language, and a set of specific tools for solving specific problems. But it is also a way of thinking. Category theory is a way of thinking about mathematics, thus it is a way of thinking about thinking. Thinking about how we think might sound a bit like convoluted navel-gazing, but I believe it's a good way of working out how to think better. And in a world of fake news, catchy but contentless memes, and short attention spans, I think it's rather important for those who do want to think better to find better and better ways of doing it, and share them as widely as possible rather than keeping people under a mistaken belief that you have to learn a huge quantity of pure math first.

I have gradually realized that I use the ideas and principles of category theory in all my thinking about the world, far beyond my research, and in areas that probably wouldn't be officially considered to be applications. It is these ideas and principles that I want to describe in this first chapter, before starting to delve into how category theory implements those ideas and how it trains us

---

[†] Also the phrase was coined by Ernst Haeckel who had some repugnant views on race and eugenics, so I'm reluctant to quote him, but technically obliged to credit him for this phrase.

in the discipline of using them all the time. This chapter is in a way an informal overview of the entire book; it might seem a little vague but I hope the ideas will become clearer as the book progresses.

We are going to build up to the definitions very gradually, so if you're feeling impatient you might want to glance at Chapter 8 in advance, but I urge you to read the chapters of build-up to get into the spirit of this way of thinking. In the Epilogue I will come back to the ideas and spirit of category theory, but from a more technical point of view after we have met and explored the formalism.

1.1 Abstraction and analogies

Mathematics relies heavily on abstraction to get it going. Its arguments are all based on rigorous logic, and rigorous logic only works properly in abstract settings. We can try and use rigorous logic in less abstract settings, but we will probably always[†] run into problems of ambiguity: ambiguity of definitions, ambiguity of interpretations, ambiguity of behavior, and so on.

In normal life situations there is always the possibility that something will get in the way of logic working perfectly. We might think that, logically, one plus one is always two, but in real life some aspect of the objects in question might get in the way. If someone gives you one cookie and then another cookie you might have two cookies but it depends if you ate them. If you had one flower and you buy another then you might have two, but perhaps you bought another because the first one died.

Abstraction is the process of deciding on some details to ignore in order to ensure that our logic does work perfectly. In the situations above this might consist of specifying that we don't eat the cookies, or that the flowers don't die (or reproduce). This is an important part of the process of doing mathematics because one of the aims is to eliminate ambiguity from our arguments. This doesn't mean that ambiguity is bad; indeed ambiguity is one of the things that can make human life rich and beautiful. However, it can also make arguments frustrating and unproductive. Math is a world in which one of the aims is to make arguments unambiguous in order to reach agreement on something. We will go into detail about how abstraction works and what its advantages and disadvantages are in the next chapter. The idea is that abstraction itself has the potential to be ambiguous, and category theory provides a secure framework for performing abstractions.

[†] I am tempted to say "always" but my precise mathematical brain prevents me from making absolute statements without some sort of qualification such as "probably" or "I believe" or "it is almost certainly true that".

1.2 Connections and unification

One of the aims and advantages of abstraction is to make connections between different situations that might previously have seemed very different. It might seem that abstraction takes us further away from "real" situations. This is superficially true, but at the same time abstraction enables us to make connections between situations that are further apart from one another. This is one of the ways in which math helps us understand more things in a more powerful way, by making connections between different situations so that we can study them all at once instead of having to do the work over and over again. Once we've understood that one plus one is two (abstractly) we don't have to keep asking ourselves that question for different objects. Spotting similarities in our thought processes enables us to make more efficient use of our brain power.

One way in which this arises is in pattern spotting. A pattern can arise as a connection within a single situation, such as when we use a repeating pattern to tile a floor or wall. Or it can arise as a connection between different situations, such as when we see a pattern of certain types of people dominating conversations or belittling others, whether it's at work or in our personal lives, in "real life" or online.

Making connections between different situations is a step in the direction of unification. In math this doesn't mean making everything the same, but it is more about making an abstract theory that can encompass and illuminate many different things. Category theory is a unifying theory that can simultaneously encompass a broad range of topics and also a broad range of scales by zooming in and out, as we'll see. Chapter 3 will be about patterns, and how this gives us a start at recognizing abstract structures.

1.3 Context

One of the starting points of category theory is the idea that we should always study things in context rather than in isolation. It's a bit like always setting a frame of reference first. This is one crucial way to eliminate ambiguity right from the start, because things can take on very different meanings and different characteristics in different contexts. Our example of one plus one giving different results was really a case of context getting in the way of our logical outcomes. One plus one does always equal two provided we are in a context of things behaving like ordinary numbers and not like some other kind of number. But there are plenty of contexts in which things behave differently, as we'll see in Chapter 4. One of the disciplines and driving principles of category theory is to make sure we are always aware of and specific about what context we're considering. This is relevant in all aspects of life as well. For example, the context of someone's life situation, how they grew up, what is going on for them in their personal life, and so on, has a big effect on how they behave, and what their achievements represent. The same achievement is much more impressive to me when someone has struggled against many obstructions in life, because of race, gender, gender expression, sexual orientation, poverty, family circumstance, or any number of other struggles. Sometimes this is controversially referred to as "positive discrimination" but I prefer to think of it as contextual evaluation.

1.4 Relationships

One of the crucial ways in which category theory specifies and defines context is via relationships. It takes the view that what is important in a given context is the ways in which things are related to one another, not their intrinsic characteristics. The types of relationship we consider are often key to determining what context we're in or should be in. For example, in some contexts it matters how old people are relative to one another, but in other contexts it matters what their family relationships are, or how much they earn. But if we're thinking about, say, how good different people will be at running a country, then it might not seem relevant how much money they have relative to one another. Except that in some political systems (notably the US) being very rich seems quite important in getting elected to political office.

There can also be different types of relationship between the same things in mathematics, and we might only want to focus on certain types of relationship at any given moment. It doesn't mean that the others are useless, it just means that we don't think they are relevant to the situation at hand. Or perhaps we want to study something else for now, in something a bit like a controlled experiment. Numbers themselves have various types of relationship with each other. The most obvious relationship between numbers is about size, and so we put numbers on a number line in order of size. But we could put numbers in a different diagram by indicating which numbers are divisible by others. In category theory those are two different ways of putting a category structure on the same set of numbers, by using a different type of relationship. We will go into more detail about this in Chapter 5.

The relationships used in category theory can essentially be anything, as long as they satisfy some basic principles ensuring that they can be organized in a mildly tractable way. This will guide us to the formal definition of a category. To build up to that we will look at the idea of formalism in Chapter 6, to ease into this aspect of mathematics that can sometimes be so offputting. In Chapter 7 we'll look at a particular type of relationship called equivalence relations, which satisfy many good properties making them exceedingly tractable. In fact, they satisfy too many good properties, so they are too restrictive to be broadly expressive in the way that category theory seeks.

We will see that category theory is a framework that achieves a remarkable trade-off between good behavior and expressive possibilities. If a framework demands too much good behavior then expressivity is limited, as in a totalitarian state with very strict laws. On the other hand if there are too *few* demands, then there is great potential for expressivity, but also for chaos and anarchy. Category theory achieves a productive balance between those, in the way it specifies what type of relationship it is going to study.

Part One of the book will build up to the formal definition of a category. We will then take an Interlude which will be a tour of mathematics, presenting various mathematical structures as examples of categories. The usual way of doing this is to assume that a student of category theory is already familiar with these examples and that this will help them feel comfortable with the definition of category theory. I will not do that, but will introduce those examples from scratch, taking the ideas of category theory as a starting point for introducing these mathematical topics instead. In Part Two of the book we will then look more deeply into the sorts of things we do with category theory.

1.5 Sameness

One of the main principles and aims of category theory is to have more nuanced ways of describing sameness. Sameness is a key concept in math and at a basic level this arises as equality, along with the concept of equations. Indeed, many people get the impression that math is *all* about numbers and equations. This is very far from true, especially for a category theorist. First of all, while numbers are an example of something that can be organized into a category, the whole point is to be able to study a much, much broader range of things than numbers. Secondly, category theory specifically does not deal in equations because equality is much too strong a notion of sameness in category theory.

The point is, many things that we write with an equals sign in basic math aren't really equal deep down. For example when we say $\boxed{5 + 1 = 1 + 5}$ we really mean that the outcomes are the same, not that the two sides of the equation are actually completely the same. Indeed, if the two sides were completely the same there would be no point writing down the equation. The whole point is that there is a sense in which the two sides are different and a sense in which the two sides are the same, and we use the sense in which they're the same to pivot between the senses in which they're different in order to make progress and build up more complex thoughts. We will go into this in Chapter 14.

Numbers and equations go together because numbers are quite straightforward concepts,[†] so equality is an appropriate notion of "sameness" for them. However, when we study ideas that are more complex than numbers, much more subtle notions of sameness are possible. To take a very far opposite extreme, if we are thinking about people then the notion of "equality" becomes rather complicated. When we talk about equality of people we don't mean that any two people are actually the same person (which would make no sense) but we mean something more subtle about how they should be treated, or what opportunities they deserve, or how much say they should have in our democracy. Arguments often become heated around what different people mean by "equality" for people, as there are so many possible interpretations.

Math is about trying to iron out ambiguity and have more sensible arguments. Category theory seeks to study notions of sameness that are more subtle and complex than direct equality, but still unambiguous enough to be discussed in rigorous logical arguments. Sometimes a much better question isn't to ask whether two things are equal or not, but *in what ways* they are and aren't equal, and furthermore, if we look at some way in which they're not equal, how much and in what ways do they fail to be equal? This is a level of subtlety provided by category theory which we sorely need in life too.

1.6 Characterizing things by the role they play

Category theory seeks to characterize things by the role they play in context rather than by some intrinsic characteristics. This is related to the idea of context and relationships being so important. Once we understand that objects take on very different characteristics in different contexts it becomes clearer that the whole idea of intrinsic characteristics is rather shaky.

I think this applies to people as well. I don't think I have an intrinsic personality because I behave very differently depending on what sort of situation I'm in. In some situations I'm confident and talkative, and in other situations I'm nervous and shy. Even mathematical objects do something similar, although in that case the characteristics we're thinking about aren't personality traits, but mathematical behaviors.

[†] Actually they're very profound, but once they're defined there's not much nuance to them.

For example, we might think the number 5 is prime "because it's only divisible by 1 and itself", but we really ought to point out that the context we're thinking of here is the whole numbers, because if we allow fractions then 5 is divisible by everything really (except 0).[†]

In normal life we often mix up when we're characterizing things by role and by property in the way that we use language. For example "pumpkin spice" is named after the role that this spice combination plays in classic American pumpkin pie, but it has now come to be used as a flavoring in its own right in any number of things that are not actually pumpkin pie, but it's still called pumpkin spice, which is quite confusing for non-Americans. Conversely "pound cake" is named after the fact that it's a recipe consisting of a pound each of basic cake ingredients. So it's named after an intrinsic property, and it's still called pound cake even if you change the quantity that you use. I, personally, have never made such an enormous cake.

One of the advantages of characterizing things by the role they play in context is that you can then make comparisons across different contexts, by finding things that play analogous roles in other contexts. We will talk about this when we discuss universal properties in Chapter 16. This might sound like the opposite of what I just described, as it sounds a bit like properties that are universal regardless of context, but what it actually refers to is the property of being somehow extreme or canonical within a context. This can tell us something about the objects with that property, but it can also tell us something about the context itself. If we go round looking at the highest and lowest paid employees in different companies, that tells us something about those companies, not just about the employees. It is only one piece of information (as opposed to a whole distribution of salaries across the company) but it still tells us something.

1.7 Zooming in and out

One of the powerful aspects of category theory's level of abstraction is that it enables us to zoom in and out and look at large and small scale mathematical structures in a similar light. It's like a theory that unifies the sub-atomic level with the level of galaxies. This is one of my favorite aspects of category theory.

If we study birds then we might need to make a theory of birds in order to make our study rigorous. However, that theory of birds is not itself a bird — it's one level more abstract. On the other hand if we study mathematical objects then we similarly might need a theory of them. I find it enormously

---

[†] Also this is more of a characterization than a definition.

satisfying that that theory is itself also a mathematical object, which we can then study using the same theory. Category theory is a theory of mathematics, but is itself a piece of mathematics, and so it can be used to study itself. This sounds self-referential, but what ends up happening is that although we are still in category theory we find ourselves in a slightly higher dimension of category theory. Dimensions in this case refer to levels of relationship. In basic category theory our insight begins by saying we should study relationships between objects, not just the objects themselves. But what about the relationships? If we consider those to be new mathematical objects, shouldn't we also study relationships between those? This gives us one more dimension.

Then, of course, why stop there? What about relationships between relationships between relationships? This gives us a third dimension. And really there is no logically determined place to stop, so we might keep going and end up with infinite dimensions. This is essentially where my research is, in the field of higher-dimensional category theory, and we will see a glimpse of this to finish the book. To me this is the ultimate "fixed point" of theories. If category theory is a theory of mathematics, then higher-dimensional category theory is a theory of categories. But a theory of higher-dimensional category theory is still higher-dimensional category theory.

This is not just about abstraction for the sake of it, although I do find abstraction fun in its own right. It is about subtlety. Category theory is about having more subtle ways of expressing things while still maintaining rigor, and every extra dimension gives us another layer of possible subtlety.

Subtlety and nuance are aspects of thinking that I find myself missing and longing for in daily life. So much of our discourse has become black-and-white in futile attempts to be decisive, or to grab attention, or to make devastating arguments, or to shout down the opposition. Higher-dimensional category theory trains us in balancing nuance with rigor so that we don't need to resort to black-and-white, and so that we don't *want* to either.

I think mathematics is a spectacular controlled environment in which to practice this kind of thinking. The aim is that even if the theory is not directly applicable in the rest of our lives, the thinking becomes second nature. This is how I have found category theory to help me in everyday life, surprising though it may sound.

1.8 Framework and techniques

As I have described it so far, category theory might sound like a philosophy more than anything else. But the point is that it is only *guided* by these vari-

ous philosophies. It is still entirely rigorous technical mathematics. It sets up
a framework for implementing these philosophies and pursuing these goals
rigorously. The framework consists of a formal definition of a category as an
algebraic structure, and then techniques for studying these structures and for
constructing and investigating particular features that might arise in them.

To this end, a certain amount of formal mathematics is needed if we are ever
going to get very far into the theory itself, rather than poetically exploring the
ideas behind it. This is one of the things that can be offputting about mathematics, and I do advocate the idea of seeing and appreciating the ideas of math
even if you can't or don't want to follow the formality. However that is not
the aim of this book. (It was, in a way, the aim of my book *How to Bake π*.)
I do think that way of appreciating mathematics is under-rated. It is a bit like
going to visit a country without learning to speak the language. I think it would
be culturally limiting for us to decide we should never visit a country without
learning the language first. However, I also think that if we can learn at least
some of the language then even if we're not fluent we will get much more out
of a visit. This is what this book is for.

Mathematics is sometimes taught as if the only valid interaction with it is to
be able to do it. As I said in the Prologue, languages are taught with a "productive" and a "receptive" component (as well as a cultural component, in my experience not examinable). When we talk about basic education we sometimes
talk about "reading, writing and arithmetic". Aside from the over-emphasis
on boring arithmetic (for which we basically all have phone calculators now),
there is again the idea that for language the skills of reading and writing are
separate, but math is just math.

In this book I'm not going to expect readers to become fluent in all aspects of
category theory. My aim isn't to be able to get you to be able to do research in
category theory, but mainly to be able to read and appreciate it, and have some
build-up into the formality of it in case you do want to go further. Tourism is
sometimes used as a derogatory word, tourists thought of as superficial visitors who take selfies and then leave. But well-informed and curious tourists
are a valuable part of cultural exchange. I have always appreciated living in
places that are interesting enough to attract tourists from around the world.
And tourists do sometimes become long-term visitors, permanent residents, or
even citizens. One way to learn a language is to be deposited in a foreign country where nobody speaks your native tongue, but I want to do something more
gentle than that. The next few chapters will build up to the formal language
gradually.

2

Abstraction

An overview of the abstract side of mathematics, to put us in a good state of mind for category theory. It's important not to be thinking of math in the common narrow way as numbers, equations and problem solving. This chapter will still have little formality.

2.1 What is math?

In the previous chapter I described category theory as "the mathematics of mathematics". So I'd better start by describing in more depth how I think of mathematics. If you take too narrow a view of what math is, then the phrase "the mathematics of mathematics" doesn't make any sense.

Math is not just the study of numbers and equations, and it's not all about solving problems. Those are some aspects of math, and they are often the aspects that are emphasized in school math, and in math for non-mathematicians. Math in school tends to start with numbers and arithmetic, moves on to equations, and then maybe deals with things like trigonometry and a bit of geometry. Trigonometry and geometry aren't really about numbers but about shapes; however they still involve a lot of numbers and things like calculating angles and lengths, expressed in numbers.

Math as used in the world does involve quite a bit of solving problems using numbers and equations. There are usually some measurable quantities measured in numbers, and some equations relating them, and the task is to calculate the ones we don't know, using the ones we can measure.

At least, this is the most visible and obvious way in which math is used in the world, so I don't blame anyone for thinking that's all there is. But there is more going on behind that: the *theory* of how those things work. That theory is what enables us to make sure they work, to refine them, and to develop new versions to deal with more complex and nuanced situations. This is like the foundations of a building: you can't see them, but without them the building would not stand up.

Every academic discipline provides a way of reaching truths of some form. Each discipline is seeking a particular type of truth, and develops a method or framework for deciding what counts as true. In this era of information excess (and indeed general excess) I think understanding those methods and frameworks is far more important than knowing the truths themselves. The important thing is to know *how* to decide what should count as true — how to build good foundations on which to base our understanding. I strongly believe that this understanding of process and framework is what is most transferrable about studying any subject, especially math.

2.2 The twin disciplines of logic and abstraction

The framework of mathematics involves the twin disciplines of logic and abstraction. Math is not unique in its use of either of these things, but I regard it as being more or less defined by its use of these in combination.

I would say that philosophy uses logic, but applies it to real questions about life experiences. Art uses abstraction, but does not primarily build on its abstractions by logic. Math uses logic and abstraction together. It uses logic to build rigorous arguments, and uses abstraction to ensure that we are working in a world where logic can be made rigorous.

This might make it sound like we can never be talking about the "real" (or rather, concrete[†]) world as we will always be working in an abstract world. While this is in some sense true, it is also reductive. Abstractions are facets of the concrete world, or views from a particular angle. While they will never give us the full explanation of the concrete world, it is still valuable to get a very full understanding of particular aspects of the concrete world. As long as we are clear that each one is only a partial view, we can then move flexibly between those different views to build up a clearer picture.

There is a subtle difference between this and the approach of studying the concrete world directly. In the direct approach we typically get only a partial understanding, because the concrete world is too messy for logic. The following diagram illustrates the difference.

---

† What is real anyway?

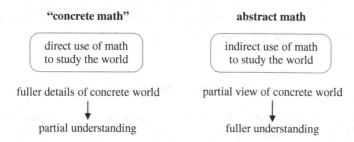

Thus abstract math still studies the world, just in a less direct way. Its starting point is abstraction, and the starting point for abstraction is to forget some details about a situation.

2.3 Forgetting details

Abstraction is about digging deep into a situation to find out what is at its core making it tick. Another way to think of it is about stripping away irrelevant details, or rather, stripping away details that are irrelevant to what we're thinking about now. Those details might well be relevant to something else, but we decide we don't need to think about them for the time being. Crucially, it's a careful and controlled forgetting of details, not a slapdash ignoring of details out of laziness or a desire to skew an argument in a certain direction.

If someone says "women are worse at math than men" then they are omitting crucial details and opening up ambiguities, or deliberately using data in a misleading way. This inflammatory statement has some truth *in some sense*, which is that fewer women are currently employed as math professors than men, not that there is any evidence that women are *innately* worse at math than men. It's a pedantically correct expression of the fact that women are currently doing worse in the field of mathematics than men are.[†]

Whereas if we observe that one apple together with another one apple makes two, and that one banana together with another banana makes two, and we say that one thing together with another thing makes two, then we are ignoring the detail of applehood and bananahood as that is genuinely irrelevant to the idea of one thing and another thing making two things. That is abstraction, and is how numbers come into being. Numbers are one of the first abstract concepts we come across, but we don't always think of them as being abstract. That is a good thing, as it shows that we have raised the baseline level of abstraction that we're comfortable with in our heads. This is like the fact that things can

---

[†] In *The Art of Logic* I wrote about pedantry being precision without illumination. In this case it's even worse: it's precision with active obfuscation.

seem hard at first, but later seem so easy they're second nature. It is all a sign that we've progressed.

2.4 Pros and cons

Before I go into more detail about how we perform abstraction, I want to talk about the pros and cons. It might be tempting just to talk about all the benefits of doing something, but that can be counter-productive if other people see disadvantages and think you're being dishonest or misleading. Instead, I think it's important to see the pros and cons of doing something, and weigh them up. There are rarely exclusively positives or negatives to doing something.

The advantages of abstraction, as I see them, are broadly that we unify many different situations in a sort of inclusivity of examples; this then enables us to transfer insight across different situations, and thus to gain efficiency in our thought processes by studying many things at once. The world is a complicated place and we need to simplify it in order to be able to understand it with our poor little brains. One popular way to simplify it is to ignore some of the detail, but I think that's a dangerous way of simplifying it. Another way is to make connections inside it so that our brains can deal with more of it at once. I think this is a better way. The best way overall is to become more intelligent so that the world becomes simpler relative to our brains.

So much for the advantages; I will now acknowledge some disadvantages of abstraction. One is that it does take some effort, but I do think this is about front-loading effort in order to reap rewards from it later. I think that it's an investment, and the extra effort *early* means that we can understand more things more deeply with less effort *later*.

Another disadvantage is that we lose details. However, I think this just means we shouldn't remain *exclusively* in the abstract world, but should always bear in mind that at some point some details will need to go back in. Losing the details temporarily is an important part of finding connections between situations, so again I think that this is a net positive.

Another disadvantage is that this takes us further from the normal everyday world that we're used to, which can be scary. It can be scary because it can seem like we don't have our feet on the ground any more. It can be scary because we can't touch things or feel things or see things, and we can't use a lifetime of intuition any more. However, intuition is itself a double-edged sword.[†] Intuition both helps and hinders us, whether in math or in life. It helps

---

[†] I've always found the metaphor of a double-edged sword a bit strange, as it doesn't seem to me that the two edges of a double-edged sword work in opposition to one another.

us in situations where we do not have enough information or time to use logic. It helps us by drawing on our experience quickly. But it is thus also limited by our experience, and if it is used instead of logic it can be dangerously misleading. For example, it's unavoidable to have a gut instinct when we meet a new person, but it's wrong to hold onto that instead of actually responding to the person as we get to know them, especially when our gut instinct is skewed by implicit bias as it (by definition) always is.

Likewise, in math it's not wrong to have intuitions about things, and indeed this is how much research gets started, by a vague idea coming from inside a mathematician's figurative gut somewhere. But the key is then to investigate it using logic and not rely too much on that intuition.

One crucial point is that the framework of building arguments by rigorous logic in math can take us much further than our intuition can. It can take us into places where we have no intuition, such as infinite-dimensional space, or worlds of numbers that have no concrete interpretation in the normal world. For example, one of the points of calculus is to understand what gives rise to interesting features in graphs, like gaps, spikes, places where the graph changes direction. This means we can seek those features even when the graph itself is much too complicated to draw, so that we can't just look for them visually.

One possible objection to this "advantage" is that you might think you'll never find yourself in a place that's so far beyond your intuition. That may well be true. But it might still be good for your brain to be stretched into those places, so that your intuition can develop. I am convinced that my years of stretching my brain in those abstract ways have enabled me to think more clearly about the world around me, and more easily make connections between situations, connections that others don't see. Often when I give my abstract explanations of arguments around social issues such as sexual harassment, sexism, racism, privilege, power imbalance and so on, people ask me how I thought of it. The answer is that a lifetime of developing my abstract mathematical brain makes these things come to me quite smoothly. It's good to train yourself to be able to do more than you think you'll need, so that the things you do need to do feel easier.

The last advantage I want to give for abstraction is that it can be fun. Fun can seem a little frivolous in trying political times, but if we only stress the utility of something it might start sounding awfully boring. I find it enormously satisfying to strip away outer layers of a situation to find its core. It appeals to my general aesthetic, which is typically that I am not so interested in superficial appearances, but care about what is going on in the heart of things, deep down, far below the surface. Abstraction in its own right really does bring me joy.

2.5 Making analogies into actual things

Abstraction comes from seeing analogies between different situations. This is a particular form of detail-forgetting, based in finding connections between different situations, rather than just arbitrarily ignoring details.

If we say that one situation "is analogous to" another situation, essentially what we are saying is that if we ignore some surface-level details in each situation then the two are really the same. In math, unlike in normal life, we don't just say that the situations are "analogous", but we make very precise what feature is the same in both situations, which is causing the analogy we want to consider. Some of what follows I have also written about in *The Art of Logic*.

If we think about two apples and two bananas, we can consider them to be analogous because they're both examples of two things. But we could also consider them to be analogous because they're both examples of two fruits. Neither of those is "right" or "wrong", neither is "better" or "worse". What we can say, however, is that "two things" is a further level of abstraction than "two fruits" because it forgets more of the details of the situation; conversely, saying "two fruits" is less abstract and leaves us closer to the actual situation.

The more abstract version takes us further away from "reality" (whatever that is), and one major upside is that it enables us to include more distant examples in our analogy. In this example it means we could include two chairs, or two monkeys, or two planets.

In a way, abstraction is like looking deeper into a situation, but it is also like taking a step back and seeing more of the big picture rather than getting lost in the details. The fact that we can find different abstractions of the same thing makes it sound ambiguous but is in fact key. I find it helpful to draw diagrams of different levels of abstraction.

Here is a diagram showing that two apples and two bananas are analogous because they're both two fruits, but also that if we go up further to the level of two things then we get to include two chairs as well.

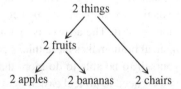

The key in math is that we don't just say that things are *analogous*. Rather, we precisely specify our level of analogy, and then go a step further and regard that as an object in its own right, and study it. That is how we move into abstract worlds, and it is in those abstract worlds that we "do" math. In the above example that's the level of "2 things": the world of numbers.

Pinning down what is causing the analogy, rather than just saying things are analogous, is like the difference between telling someone there is a path through a field, and actually marking out where the path is. I remember going on hiking expeditions in the rolling hills of the South Downs in the south-east of England when I was growing up. We would climb over a stile[†] into a field, and a little wooden signpost would tell us that a path was that way. But it didn't actually tell us where the path was, and invariably we would get to the other side of the field and no stile would be in sight because we had strayed from the supposed "path".

Pinning down structure rather than just observing that it exists is an idea that we will come back to, especially when we are doing more nuanced category theory later on. But for now the main point is that pinning things down helps us iron out some ambiguity, especially as there is always the possibility of different abstractions of the same situation.

2.6 Different abstractions of the same thing

It might seem confusing that there can be different abstractions of the same thing, but in a way that's the whole point. Abstraction is not a well-defined process, which is to say that it can produce different results depending on how you do it. (In math "well-defined" means "unambiguously determined"; it doesn't just vaguely mean that a definition has been done well.) This is the cause of many acrimonious arguments in normal life, because if we just declare something to be "analogous to" something else, it leaves open an ambiguity of what level we're talking about. Then someone else is likely to get angry and either say "It's not the same, because. . ." and then point out some way in which your two things are different, or they will say "Well that's the same as saying. . ." and then invoke some ludicrous level of abstraction and implicitly accuse you of using that level. In the above example the first case would be like someone saying "2 apples aren't the same as 2 bananas because you can't eat the skin of bananas", and the second case would be like someone saying "Well if you think 2 apples are the same as 2 chairs you've obviously never tried sitting on an apple!"

Here is a much less trivial example. In 2018 in Colorado there was a court case involving some Christian bakers who refused to bake a cake for a same-sex wedding. I saw someone online (no less) who tried to argue that saying Christians should have to bake a cake for a same-sex wedding is like saying Jews should have to bake a cake for a Nazi wedding. I sincerely hope that your gut

---

[†] A stile is a traditional form of wooden step to enable people to climb over a fence into or out of a field, without letting livestock out. They're quite common in the UK where there are often "rights of way" giving the public the legal right to walk through a field.

reaction is "That's not the same!" but I actually think it's important to be able to acknowledge the *sense in which* those things are analogous, even if this sense disgusts us. I really think that just retorting "That's not the same" is ineffective, and that using some more careful logic is more productive. Here is a level of abstraction that does yield that analogy, and a different level that differentiates between those two conclusions:

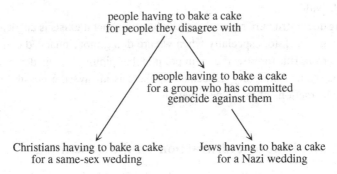

My wonderful PhD supervisor, Martin Hyland, taught me that when you're doing abstract mathematics, the aim shouldn't be to find the most abstract possible level, but to find a good level of abstraction for what you're trying to do. "Horses for courses" one might say.[†] He also instilled in me the idea of starting sentences with "There is a sense in which. . ." because math isn't about right and wrong, it's not about absolute truth; it's about different contexts in which different things can be true, and about different senses in which different things can be valid. Abstraction in mathematics is about making precise which sense we mean, so that instead of having divisive arguments, whether it's about abstract theories or about homophobic bakers, we can investigate more effectively what is causing certain outcomes to arise.

In the end, we could always make absolutely anything the same by forgetting just about all of the details, like when I take my glasses off and everything looks equally blurry. We do this in life as well: sometimes it is tempting to declare that we're all human, so we're all the same really. This might sound happily unifying, but it also negates various people's struggles against oppression, prejudice, poverty, illness and any number of other things. There is the opposite tendency as well, to fragment into so many different identities and combinations of identities, to emphasize our unique experiences. Pushing too far to either extreme is probably unhelpful. What is helpful is to be able to see what all the levels are, and maintain enough flexibility to be able to move between them and draw different insights from different levels.

---

† One also might not, if one has never heard of this saying.

Things To Think About

T 2.1   What are some senses in which addition and multiplication are "the same"? What are some senses in which they are "different"?

Addition and multiplication are both *binary operations*: they take two inputs and produce one answer at the end. In the first instance they are binary operations on numbers, but as math progresses through different levels of abstraction we find ways of defining things like addition and multiplication on other types of mathematical object as well.

As binary operations, they have some features in common, including that the order in which we put the numbers doesn't matter (which is *commutativity*) and nor does how we parenthesize (which is *associativity*). However, addition and multiplication behave differently in various ways. For example, addition can always be "undone" by subtraction, but multiplication can only sometimes be "undone" by division: multiplication by 0 can't be undone by division. This is sometimes thought of as "you can't divide by 0" but we'll come up with some better abstract accounts of this later.

2.7 Abstraction journey through levels of math

As mathematics progresses, aspects of it become more and more abstract. There is a sort of progression where we move through levels of abstraction gradually, through the following steps:

1. see an analogy between some different things,
2. specify what we are regarding as causing the analogy,
3. regard that thing as a new, more abstract, concept in its own right,
4. become comfortable with those new abstract concepts and not really think of them as being that abstract any more,
5. see an analogy between some of those new concepts,
6. iterate...

One of the advantages of taking abstract concepts seriously as new objects is that we can then build on them in this way. Here's an example of an initial process of abstraction in basic math.

This is the infamous process of "turning numbers into letters", which is the stage of math many people tell me is where they hit their limit.

(In fact there was a level below that, where we went from objects like apples and bananas to numbers in the first place.) Why have the numbers turned into letters? It's so that we can see things that are true abstractly, regardless of what exact numbers we're using. For example:

$$1 + 2 = 2 + 1$$
$$1 + 3 = 3 + 1$$
$$2 + 3 = 3 + 2$$
$$5 + 7 = 7 + 5$$
$$\vdots$$

Something analogous is going on in all of these situations, and it would be impossible for us to list all the combinations of numbers for which this is true as there are infinitely many of them. We could describe this in words as "if we add two numbers together it doesn't matter what order we put them", but this is a bit long-winded.

The concise abstract way of saying it is: given any numbers a and b,

$$a + b = b + a.$$

We have "turned the numbers into letters" so that we can make a statement about tons of different numbers at once, and make precise what pattern it is that is causing the analogy that we see. Not only is this more concise and thus quicker to write down (and mathematicians are very lazy about writing down long-winded things repeatedly), but the abstract formulation can help us to go a step further and pin down similarities with other situations.

But there's a level more abstract as well, in the direction we were going at the end of the previous section. If we think about similarities between addition and multiplication we see that they have some things in common. For a start, they are both processes that take two numbers (at a basic level) and use them to produce an answer. The processes also have some properties that we noticed, such as commutativity and associativity.

When we call them a "process that takes two numbers and produces an answer" that is a further level of abstraction. It's an analogy between addition and multiplication.

Here is a diagram showing that new level, with the symbol ⊙ representing a binary operation that could be +, × or something else. There is a journey of abstraction up through levels of this diagram that is a bit like the journey through math education.

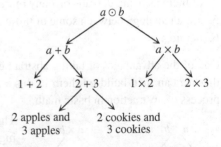

At the very bottom level we have the sorts of things you might do in pre-school or kindergarten where you play around with familiar objects and get nudged in the direction of thinking about numbers. At the next level up we have arithmetic as done in elementary school, perhaps, and then we move into algebra as done a little bit further on at school. The top level here, with the abstract binary operation, is the kind of thing we study in "abstract algebra" if we take some higher level math at university. Binary operations are studied in group theory, for example, and this is one of the topics we'll come back to. Incidentally I always find the term "abstract algebra" quite strange, because all algebra is abstract and, as I've described, what we even consider to count as "abstract" changes as we get more used to more abstraction.

There is indeed a further level of abstraction, which one might call "very abstract algebra". At this level we can think about more subtle ways of combining two things, where instead of taking just any two things and producing an answer, we can only take two things that fit together like in a jigsaw puzzle. For example, we can think about train journeys, where we can take one train journey and then another, to make a longer train journey — but this only makes sense if the second journey starts where the first one ends, so that you can actually change train there. This means you can't combine *any old* train journeys to make longer ones, but only those that meet up suitably where one ends and the other begins.

This is the sort of way we'll be combining things in category theory. Binary operations are still an example of this, but as with all our higher levels of abstraction, we will now be able to include many more examples of things that are more subtle than binary operations. This includes almost every branch of math, as they almost all (or maybe even all) involve some form of this way of combining things. Here is a diagram showing that, including the names of some of the mathematical topics we'll be exploring later in this book.

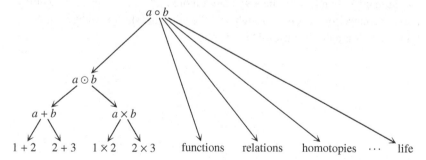

I've included "life" in the examples here, to emphasize that the higher level of abstraction may seem further away from normal life, but at the same time, the higher level is what enables us to unify a much wider range of examples, including examples from normal life that are not usually included in abstract mathematics. I think this is like swinging from a rope — if you hang the rope from a higher pivot then you can swing much further, provided you have a long rope. It's also like shining a light from above — if you raise it higher then the light will become less bright but you will illuminate a wider area. The result is perhaps more of an overview and an idea of context, rather than a close-up of the details. However, you won't lose the details forever as long as you retain the ability to move the light up and down. Furthermore, if you can find a way to make the light itself brighter, then you can see more detail and more context at the same time. I think this is an important aspect of becoming more intelligent, and that abstract mathematics can help us with that.

"More abstract" doesn't necessarily mean less relevant and it doesn't necessarily mean harder either. Too often we assume that things get harder and harder as we move up through those levels of abstraction, and thus that we shouldn't try to move up until we've mastered the previous level. However I think this is one of the things that can hold some people back or keep them excluded from mathematics. Actually the higher levels might be easier for some people, either because, like me, they enjoy abstraction and find it more satisfying, or because it encompasses more examples and those might be more motivating than the examples included at the lower levels. If you're stuck at the level of $a + b$ and the only examples involve adding numbers together then you may well feel that the whole thing is tedious and not much help to you either. After all, some things are fun, some things are useful, and some things are both, but the things that are neither are really the pits.

I think we should stop using the lower levels of abstraction as a prerequisite for the higher ones. If at the higher levels you get to deal with examples from life that you care about more than numbers, perhaps examples involving people and humanity, then it could be a whole lot more motivating. Plus, if you enjoy making connections, seeing through superficiality, and shining light, then those higher levels are not just useful but also fun.

3

Patterns

We can make something abstract, then find patterns, then ask if those patterns are caused by some abstract structure. This chapter still has little formality.

3.1 Mathematics as pattern spotting

Patterns are aesthetically pleasing, but they are also about efficiency. They are a way to use a small amount of information to generate a larger amount of information, or a small amount of brain power to understand a large amount of information.

Humans have used patterns across history and across most, if not all, cultures. Patterns are used for designs on fabric, on floors, on walls.

One carefully designed tile can be used to generate a large and intricate pattern covering a very large surface, such as in this tiled wall at the Presidential Palace of Panama. If you look closely you can see that the pattern is actually made from just one tile, rotated and placed at different orientations.[†]

Using one tile like this requires much less "information" than drawing an entire mural from scratch. But aside from this sort of efficiency, patterns can be very satisfying. They give our brains something to latch onto, so that they don't get too overwhelmed. It's like when a chorus or refrain comes back in between each verse of a song.

Mathematics often involves spotting patterns. This can help us understand what is going on in general, so that we can use less of our brain power to understand more things.

[†] It might be hard to see in black-and-white, but you can see a color version at this link: eugeniacheng.com/tile/.

T 3.1   Patterns in the way we write numbers can help us quickly understand
some characteristics of individual numbers. What patterns are there in multiples of 10? What about multiples of 5? Multiples of 2? Why are these patterns
more noticeable than the patterns in multiples of 3 or 7?
└───┘

Multiples of 10 all end in 0: 10, 20, 30, 40, and so on. This is something we
learn when we are quite small, typically, and it helps us to multiply things by 10
quickly, by sticking 0 on the end. Counting in 10's is something that children
are usually encouraged to learn how to do, but if they just do it by "rote" then it
might not be as powerful as understanding the pattern, and understanding *why*
multiplication by 10 works in this way.

Multiples of 5 end in 5 or 0, and these alternate. Multiples of 2 end in any
of the possible even digits: $0, 2, 4, 6, 8$. These patterns are much more obvious
than the pattern for multiples of 3, which can end in any digit, and go in this
order: $3, 6, 9, 2, 5, 8, 1, 4, 7, 0$. Likewise multiples of 7, but in a different order.

These patterns don't exactly tell us anything about the numbers $10, 5, 2, 3$,
and 7 by themselves, but rather, they tell us about those numbers *in the context*
of their relationship with the number 10. This is because the way we normally
write numbers is based on powers of 10: called base 10, or the decimal number
system. This means that the last digit is just 1's (that is, 10 to the power of 0),
the second to last digit is 10's, then 100's, then 1000's and so on. So a four-digit
number $abcd$ is evaluated as $\boxed{1000a + 100b + 10c + d.}$

The fact that multiples of 10 end in 0 is nothing to do with any inherent
property of the number 10, but is because we have chosen to write numbers in
base 10. If we wrote numbers in base 9 then it would be all the multiples of 9
that end in 0, and multiplying by 9 is what would be done by sticking a 0 on
the end. Similarly the fact that multiples of 5 and 2 have such a tidy pattern
in their last digit is nothing to do with any inherent property of 5 and 2, but
because of their *relationship* with 10: they are factors.

──────────────────────── Technicalities ────────────────────────
In base 9, the last digit would still indicate the 1's but then the second to last
digit would be 9's, and the third to last would be 81's, and then 729's, and so on.
If we write this down in general, then a four-digit base 9 number $abcd$ would
"translate" back into an ordinary base 10 number as: $\boxed{729a + 81b + 9c + d.}$

The multiples of 3 in base 9 look like: $0, 3, 6, 10, 13, 16, 20, 23, 26, \ldots$ The
last digit has a repeating pattern because of how 3 is related to 9 (it is a factor).

T 3.2 Here is a table for how we add up hours on a 12-hour clock. 1 hour past 12 o'clock is 1 o'clock. 4 hours past 10 o'clock is 2 o'clock. See if you can fill in the rest of the table and then observe some patterns. What would happen on a 24-hour clock? It would take a long time to fill in and would quickly become repetitive, so can you describe the principles governing it instead?

	1	2	3	4	5	6	7	8	9	10	11	12
1	2	3	4	5	6	7	8	9	10	11	12	1
2												
3												
4									2			
5												
6												
7												
8												
9												
10												
11												
12												

For the patterns on the 12-hour clock, here is the rest of the table filled in. The numbers form diagonal stripes, with the whole row of numbers "shifting" over to the left as we move down the table, and the number that falls off on the left-hand side re-appears at the right in the next row down.

	1	2	3	4	5	6	7	8	9	10	11	12
1	2	3	4	5	6	7	8	9	10	11	12	1
2	3	4	5	6	7	8	9	10	11	12	1	2
3	4	5	6	7	8	9	10	11	12	1	2	3
4	5	6	7	8	9	10	11	12	1	2	3	4
5	6	7	8	9	10	11	12	1	2	3	4	5
6	7	8	9	10	11	12	1	2	3	4	5	6
7	8	9	10	11	12	1	2	3	4	5	6	7
8	9	10	11	12	1	2	3	4	5	6	7	8
9	10	11	12	1	2	3	4	5	6	7	8	9
10	11	12	1	2	3	4	5	6	7	8	9	10
11	12	1	2	3	4	5	6	7	8	9	10	11
12	1	2	3	4	5	6	7	8	9	10	11	12

This will be similar on the 24-hour clock so we really don't need to draw it all out. In fact, although we don't usually consider other numbers of hours per day, we could *imagine* clocks with other numbers of hours, say, a 10-hour clock, or a 4-hour clock, or indeed an n-hour clock for any whole number n. Regardless of how many hours there are, the pattern would still be in some sense "the same" on the table, with those diagonals made by numbers shifting over to the left row by row. That principle is quite a deep mathematical structure. We will come back to exactly what the structure is, but for now I want to talk more about the idea of what patterns are and what they tell us.

3.2 Patterns as analogies

Patterns are really analogies between different situations, which is what category theory is going to be all about. At a visual level, we could talk about "stripes", for example, and understand that we mean alternating lines in different colors. We might not know the specific details – how wide the lines are, what the colors are, what direction they're pointing – but there is something analogous going on between all different situations involving stripes, and the abstract concept behind it is the concept of a "stripe". Patterns for clothes are also analogies, this time between different items of clothing. A dress pattern is like an abstract version of the dress.

In our previous examples of numbers we were looking at analogies between numbers. With the repeating patterns of final digits on multiples of numbers it was analogies between multiples. The numbers 10, 20, 30, 40, 50, and so on are all analogous via the fact that they consist of two digits: some number n followed by 0.

With the 12-hour clock the pattern on the table was an analogy between the different rows: in each row, the numbers go up one by one to the right (and the number 1 counts as "one up" from the number 12). The only difference between the rows is then what number they start at.

The analogy between clocks with different numbers of hours is then a sort of meta-pattern — an analogy between different patterns. The thing the different tables have in common is the "shifting diagonal" pattern. We have, in a sense, gone up another level of abstraction to spot a pattern among patterns.[†]

──────────────── **Things To Think About** ────────────────

T 3.3   What patterns can you think of in other parts of life? In what sense could they be thought of as analogies between different situations? You could think about patterns in music, social behavior, politics, history, virus spread, language (in terms of vocabulary and grammar)...

3.3 Patterns as signs of structure

Spotting patterns in math is often a starting point for developing a theory. We take the pattern as a sign of some sort of abstract structure, and we investigate what abstract structure is causing that pattern.

[†] At this point in writing I went into lockdown for the COVID-19 crisis and finished the draft without leaving the house again. I feel the need to mark it here.

──────── Things To Think About ────────

T 3.4 Here is an addition table for the numbers 1 to 4. Can you find a line of symmetry on this table, that is, a line where we could fold the grid in half and the two sides would match up. Why is that line of symmetry there? What about in a multiplication table for the same numbers?

+	1	2	3	4
1	2	3	4	5
2	3	4	5	6
3	4	5	6	7
4	5	6	7	8

Here's the addition table for the numbers 1 to 4 with a line of symmetry marked. I have also highlighted an example of a pair of numbers that correspond to each other according to this line of reflection. The one on the lower left is the entry for 3 + 1 whereas the one on the upper right is the entry for 1 + 3.

+	1	2	3	4
1	2	3	4	5
2	3	4	5	6
3	4	5	6	7
4	5	6	7	8

We can *see* that these entries are the same, but the *reason* they are the same is that $3 + 1 = 1 + 3$, which we might know as the commutativity of addition. If you check any other pair of numbers that correspond under the symmetry, you will find that they are all pairs of the form $a + b$ and $b + a$. The entries on the diagonal where the line is actually drawn are all examples of $x + x$ so switching the order doesn't change the entry. That is a form of symmetry in itself: in the expression $a + b = b + a$, if a and b are both x then the left and right become the same. We say the equation is symmetric in a and b.

An analogous phenomenon happens in the multiplication table, with the line of symmetry now being a visual sign of commutativity of multiplication. Often when we spot visual patterns we ask ourselves what abstract or algebraic structure is giving rise to that visual pattern.

──────── Things To Think About ────────

T 3.5 Here is a grid of the numbers 0 to 99. We have already talked about the patterns for multiples of 2, 5, and 10. If we mark in all the multiples of 3, what pattern arises and why? What about multiples of 9?

0	1	2	3	4	5	6	7	8	9
10	11	12	13	14	15	16	17	18	19
20	21	22	23	24	25	26	27	28	29
30	31	32	33	34	35	36	37	38	39
40	41	42	43	44	45	46	47	48	49
50	51	52	53	54	55	56	57	58	59
60	61	62	63	64	65	66	67	68	69
70	71	72	73	74	75	76	77	78	79
80	81	82	83	84	85	86	87	88	89
90	91	92	93	94	95	96	97	98	99

Here is a picture of the multiples of 3 on the number grid. When we only listed their last digits it was less obvious how much of a pattern there was, because it seemed a bit random: 0, 3, 6, 9, 2, 5, 8, 1, 4, 7. However, when we draw them on this square it's visually quite striking that the multiples of 3 go in diagonal stripes, a bit like the diagonal stripe pattern we saw on the addition table above.

0	1	2	3	4	5	6	7	8	9
10	11	12	13	14	15	16	17	18	19
20	21	22	23	24	25	26	27	28	29
30	31	32	33	34	35	36	37	38	39
40	41	42	43	44	45	46	47	48	49
50	51	52	53	54	55	56	57	58	59
60	61	62	63	64	65	66	67	68	69
70	71	72	73	74	75	76	77	78	79
80	81	82	83	84	85	86	87	88	89
90	91	92	93	94	95	96	97	98	99

This is because the last multiple of 3 under 10 is one less than it, and so when we move down a row in this table, the pattern shifts one to the left. If the last multiple were 2 less than it, then the pattern would shift two to the left and be less striking. In this picture of multiples of 9 we see something similar happening, just with fewer stripes.

This pattern is essentially where we get that cute trick for the 9 times table where you hold down one finger at a time and read the number off the remaining fingers. So for 4×9 you can hold up all 10 fingers, then put down the fourth one from the left, and read off 3 from the left of that and 6 from the right to get 36. Tricks like that can be a way to bypass understanding or a way to deepen understanding.

0	1	2	3	4	5	6	7	8	9
10	11	12	13	14	15	16	17	18	19
20	21	22	23	24	25	26	27	28	29
30	31	32	33	34	35	36	37	38	39
40	41	42	43	44	45	46	47	48	49
50	51	52	53	54	55	56	57	58	59
60	61	62	63	64	65	66	67	68	69
70	71	72	73	74	75	76	77	78	79
80	81	82	83	84	85	86	87	88	89
90	91	92	93	94	95	96	97	98	99

──────── Things To Think About ────────

T 3.6   How does that trick generalize for other multiples? We will have to change what base we're working in.

A general principle of pattern spotting is that *visual* patterns might be easy to spot in simple examples, but the *abstract* structure might be easier to reason with, use or even verify, in more complex situations. It would be harder to draw the table for multiples of a much larger number, and if the patterns in the table were less obvious then it would be harder to see them. In situations of higher dimensions it is then even harder.

It is fairly easy to see all sorts of patterns of different sized triangles in this grid. However if this
were a 3-dimensional space filled with triangular
pyramids it would be rather harder to see, and in
higher-dimensional space we can't even fit it into our
physical world. But those are very helpful structures
in many fields of research; it just requires more abstract ways of expressing them.

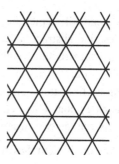

3.4 Abstract structure as a type of pattern

If category theory is the mathematics of mathematics, then categorical structure is about patterns in patterns. It's about seeing the same pattern in different
places, and about making analogies between patterns.

Here's an example. We might talk about a "mother–daughter" relationship
abstractly, as opposed to thinking about a specific mother's relationship with
their daughter. Now we could think about a relationship between someone's
grandmother and mother. This is another type of "mother–daughter" relationship; it's just a particular type where another generation also does exist. The
difference this makes to our considerations depends on the context.

Here's a tiny family tree for that structure. In a family
tree we're not taking into account any context other than
parent–child relationships. In the diagram there is no difference
in the abstract structure depicted between the grandmother and
her daughter, or the mother and her daughter.

Alex
|
Billie
|
Cat

However, if we're writing a book about sociology, or the psychology of family units, or about motherhood, then we might well want to think about how
the relationship between a mother and her daughter changes when the daughter has her own daughter in turn. However, the abstract similarity between the
grandmother–mother and the mother–daughter relationships is still an aspect
of what frames this question.

Another example is if we think about power dynamics between different
groups of people in society. White people hold structural power over non-white
people, and male people hold structural power over non-male people. This is
about overall structures, not individuals — it doesn't mean that every white
person holds power over every individual non-white person, or that every male
person holds power over each individual non-male person. It's about the way
the structures of society are set up. In any case even if you dispute this fact

we can still depict the abstract structure that I am describing, because we can describe abstract structure as a separate issue from the question of how the abstract structure manifests itself in "real life".

We could depict these power structures like this, emphasizing the
analogy that I am claiming exists
between the two structures.

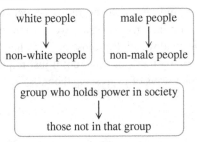

We could emphasize it further
by going one level more abstract
to this, which immediately unifies
many situations.

This now includes all sorts of other examples such as straight people over non-straight people, cisgender[†] people over trans people, rich people over non-rich people, educated people over non-educated people, employed over unemployed, people with homes over people experiencing homelessness, and so on.

3.5 Abstraction helps us see patterns

Finding the abstract structure in situations and expressing it in some way, often as a diagram to make it less abstract, can help us see the patterns and relationships between situations. It can help clear our mind of clutter and distraction and emotions. Distraction and emotions aren't bad per se, they can just get in the way of us seeing the actual structure of an argument rather than the window-dressing. Sleight of hand and flattering clothing can be fun, but if we're at the doctor's getting diagnosed for something it would be much more productive to show the whole truth and not be afraid of getting naked.

One of my favorite examples is the way I have been drawing diagrams of analogies and levels of abstraction. This way of specifying their structure has helped me pin down much more clearly where disagreements around analogies are coming from. I described it in *The Art of Logic* in terms of disagreements largely taking two possible forms.

It starts by someone invoking an analogy in this form, between two ideas A and B, but crucially without specifying what abstract level X they're referring to.

---

† Cisgender people are those whose gender identity matches the one they were assigned at birth.

Someone then objects, for either of these two reasons:

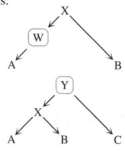

1. They see a more *specific* principle W at work that is really the reason behind A, so they do not consider A and B to be analogous:

2 They see a more *general* principle Y that they think the first person is invoking. This makes some other thing C analogous, and they object to that.

Things To Think About

T 3.7   Try modeling these two arguments with the above abstract structures:

1. Some people say COVID-19 is just like the normal seasonal flu. However medical professionals and scientists (and I) say it is not.
2. Some people (including me) believe that same-sex marriage is just like opposite-sex marriage, but others erroneously think that this means we also accept incest and pedophilia.

For COVID-19, we have the first case: in the diagram A is COVID-19 and B is seasonal flu. Some people believe X, that they are both contagious viruses that cause respiratory problems. This is true. However others recognize W, which is that COVID-19 is a new virus so there is only a new vaccine, no herd immunity, and less understanding of how to treat it (at time of writing). This is aside from it being more contagious and more deadly across the population.

For the second example, A is opposite-sex marriage and B is same-sex marriage. Some of us believe X, that marriage is between any two consenting unrelated adults. But others hallucinate that we believe a further level Y of just "adults" or even "people" which would then include examples C such as incest or pedophilia. But X and Y are very different levels.

These structures are abstract, and the abstraction allows me to see the pattern in common between essentially all arguments in which analogies are disputed. It also allows me to see how to make these arguments better: by being clearer about what the principles at work really are, and to explore the sense in which the different cases are and aren't analogous, rather than just declaring that something is or is not analogous.

If we aren't clear about different levels of abstraction then it causes angry disagreement. This is particularly a shame because different levels of abstraction are inevitable and indeed important. Different patterns might be relevant in different situations, and so we really *need* to be able to invoke different abstractions in different contexts. In the next chapter we'll look more at the sorts of different patterns that can arise depending on context.

4

Context

Introduction to the idea that math is all relative to context as things behave differently in different contexts. A little more formality.

Mathematics is often thought of as being rigid and fixed, having "absolute truth", clear right and wrong, and unbreakable laws. In this chapter we are going to examine a sense in which this is an unnecessarily and unrealistically narrow view of mathematics. Like many stereotypes it has a kernel of truth to it, but that small truth has been blown out of proportion.

We are going to look at how mathematical objects behave very differently in different contexts; thus they have no fixed characteristics, just different characteristics in different contexts. Thus truth is not absolute but is contextual, and so we should always be clear about the context we're considering. This idea is central to category theory.

Pedantically one might declare that the "truth-in-context" is then absolute, but I think this amounts to saying that truth is *relative* to context. Your preferred wording is a matter of choice, but I have made my choice because I think it is important to focus our attention on the context in which we are working, and not to regard anything as fixed.

Moreover, I don't think "truth-in-context is absolute" is what anyone typically means by the "rigidity" of mathematics; usually they're not thinking about different things being true in different contexts. For example, people who aren't mathematicians often declare that one plus one just *is* two. But we are going to see that even this basic "truth" is relative to context.

There *is* a certain rigidity to mathematics, but there's also a crucial fluidity, and the two work together. Fluidity comes from the vast universe of mathematical worlds that we can move between. Rigidity or constraint comes from a decision to move between worlds only in a way that makes sense, like pouring water into a glass without spilling it, or going to the moon without destroying your space ship. This is a different sort of constraint from the kind where you don't go anywhere at all and, moreover, you assume there's nowhere to go.

The popular concept of math is that it is a dead tree, completely fixed and dry. I think a more accurate view is that it's a living tree, whose base is fixed but it can still sway and its branches and roots still grow. An impatient explorer might say "Trees are boring. They don't move." And yes, perhaps compared with lions and elephants they don't move. But if you take time to stare at them long enough they might become fascinating.

4.1 Distance

We're going to look at a world in which some familiar things behave differently from usual. Although actually it's a very common world in "real life", it just behaves differently from some common mathematical worlds that might be regarded as "fixed", which is not really *its* fault.

We're going to think about being in a city with streets laid out on a grid — so probably not a European city. Of course, even American cities aren't on a *perfect* grid — there are usually some diagonal streets somewhere, but we're going to imagine a perfectly regular grid with evenly spaced parallel roads.

Now, when traveling from A to B we can't take the diagonal because there isn't a street there. Instead, the distance we'd actually have to go along the streets will be something like in this diagram. We have to go 3 blocks east and 4 blocks south making a total of 7 blocks.

Calculating the distance "as the crow flies", that is, the direct distance through the air in a straight line regardless of obstacles, is rather academic as we can't usually make use of that route (unless we're pinpointing a location by sound or something, as in the film "Taken 2").

—————————— **Things To Think About** ——————————

T 4.1 We can use Pythagoras to calculate the distance as the crow flies which in this case will be 5 blocks. Is the road distance always more than the crow distance?

Is the road distance between two places always the same even if we turn at different points, for example as in this diagram?

This type of distance is sometimes called "taxi-cab" distance (as if we all travel around in taxis all the time).

The taxi distance can be the same as the crow distance if *A* and *B* are on the same street, but the taxi distance can never be *smaller* than the crow distance.

Also it doesn't matter where you turn as long as you don't go back on yourself, because wherever you turn you still have to cover the same number of blocks east and the same number of blocks south, making the same total. By contrast the path shown on the right covers more blocks because it does go back on itself.

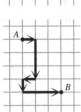

It might seem like I'm trying to say something complicated, but it really is just like counting the blocks you actually walk when you walk around a city laid out on a grid. If your brain is flashing "Math class!" warning lights then you might be trying to read too much into this. No hypotenuse is involved.

We can now think about what a circle looks like in this world. You might think a circle is just a familiar shape, but there is a *reason* that shape looks like that, and in math we are interested in reasons, or ways we can characterize things precisely.

How could you describe a circle to someone down the phone? One way is to say that you pick a center and a distance and draw all the points that are this distance away from that center. That chosen distance is called the *radius*.

But in the taxi world we can't do distance along diagonals, so what will "all the points the same distance from a chosen center" look like? Let's try a circle of radius 4 blocks.

| Here are the most obvious points that are a distance of 4 blocks from *A*. | Here is something we can't do. | So we definitely can't get all these points — most aren't on the grid at all. |

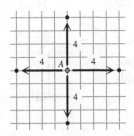 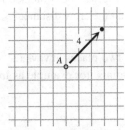 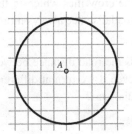

However we can go 4 blocks with a turn in the middle.	If we do this in all possible directions we get these points.	If we also do 3 blocks and 1 block to make 4 we get all these points.
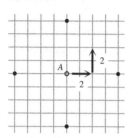	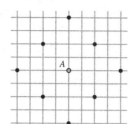	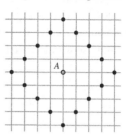

The last picture is what a circle looks like in this taxi world. After filling in the points for going "2 blocks and 2 blocks" you might have seen the pattern to help you realize you could also do 3 and 1, which is good.

However, you might also have been tempted to join the dots like this picture. You are welcome to do so on paper but it won't mean anything in the taxi world because those lines are not lines we can travel on. They include points we can't get to in the taxi world — the taxi world circle really doesn't include those lines.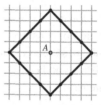

So a circle in this taxi world is just a collection of disconnected dots in a diamond shape. How does that make you feel? Do you feel uncomfortable, as if this somehow violates natural geometry? Or do you feel tickled that a circle can look so funny? Both reactions are valid. The main thing is to appreciate that even some of the most basic things we think we know are only true in a particular context, and things can look very different in another context. It is important to be

a) clear what context we're considering at any given moment (which we usually aren't in basic math lessons), and

b) open to shifting context and finding different things that can happen.

—————————————————— Technicalities ——————————————————

What we have done here is find a different scheme for measuring distance between points in space. In fact there are many different possible ways of measuring distance and these are called *metrics*. Not every scheme for measuring distance will be a reasonable one, and in order to study this and any abstract concept rigorously we decide on criteria for what should count as reasonable. A metric space is then a set of points endowed with a metric. The idea of a metric space is to focus our attention on not just the points we're thinking about, but the *type of distance* we're thinking about.

The usual way of measuring distance "as the crow flies" is called the Euclidean metric. The taxi-cab way really is called the taxi-cab metric or the rectilinear metric. More formally it is called the L_1 metric and is the first in an infinite series of L_n metrics. The next one, L_2, is in fact the Euclidean metric.

4.2 Worlds of numbers

Many people say to me "Well, one plus one just *does* equal two." I reply, "In some worlds it's zero."

It's true that $1 + 1 = 2$ in ordinary numbers. But that's because it's how "ordinary numbers" are defined. Most people who think $1 + 1$ just *does* equal 2 are not considering that this is only true in some contexts and not others, as they're so used to one particular context. This is a bit like people who've never visited another country, and don't realize that people drive on the other side of the road in other places. Some people who haven't traveled don't understand that some ways of doing things are highly cultural and possibly arbitrary. For example:

- "it's math not maths" (not in the UK);
- "steering wheels are on the right of a car" (in the UK).

Some people can't imagine not having a car and others can't imagine having one.

We have seen that distance is contextual and thus "circles" are also dependent on context. We will now see a way that the behavior of numbers is also dependent on context. So all the arithmetic we are forced to learn in school is contextual, not fixed; it's not an absolute truth of the universe, unless we take its context as part of that truth. The context is the integers, that is, all the whole numbers: positive, negative and zero. The set of integers is often written as $\mathbb{Z}$. Here is a diagram showing those different points of view.

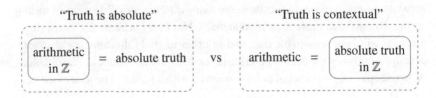

"Truth is absolute"   "Truth is contextual"

arithmetic in $\mathbb{Z}$ = absolute truth   vs   arithmetic = absolute truth in $\mathbb{Z}$

Arithmetic might seem like absolute truth if you think the integers are the only possible context. But in fact most people *do* know other contexts, they just don't come to mind when thinking about arithmetic. For example if you dump a pile of sand onto a pile of sand you still just get one pile of sand; it's just bigger.

─────────────── **Things To Think About** ───────────────

T 4.2   What contexts can you think of in which $1 + 1$ is something other than 2? Can you think of other contexts in which it's 1? What about 0, or 3 or more? What about other ways in which arithmetic sometimes works differently?

Here are some places where arithmetic works differently.

Telling the time. 2 hours later than 11 o'clock isn't 13 o'clock unless you're using a 24-hour clock. On a 12-hour clock it's 1 o'clock, that is $11 + 2 = 1$. On a 24-hour clock 2 hours later than 23 o'clock is not 25 o'clock, it's 1 o'clock, that is $23 + 2 = 1$. (While we don't usually say "23 o'clock" out loud in English, it does happen in French.)

I'm not not hungry. Particular kinds of children find it amusing to say things like "I'm not not hungry" to mean "I am hungry". If we count the instances of "not" we get $1 + 1 = 0$.

Rotations. If you rotate on the spot by one quarter-turn four times in the same direction you get back to where you started, as if you had done zero quarter-turns. So if we count the quarter-turns, $1 + 1 + 1 + 1 = 0$. We could generalize this to any n by rotating n times by $\frac{1}{n}$ of a turn each time.

Mixing paint. If you add one color paint to another you do not get two colors, you get one color.[†] Likewise a pile of sand or drop of water. So $1 + 1 = 1$.

Pairs. If one pair of tennis players meets up with another pair for an afternoon of tennis, there are 6 potential pairs of tennis players among them, if everyone is happy to partner with any of the others.[‡]

The first three of these situations all have something in common and we are now going to express exactly what that analogy might be.

───────────────

[†] This example was brought to my attention by my art students at SAIC.

[‡] This example was also brought to my attention by my art students at SAIC, though in a less child-friendly formulation.

In each of those first three cases there is one fixed number
n that is like 0, in the sense that once you get to *n* you start
counting again from 0 like going round a clock. In the first
example *n* is 12 (or 24), in the next one it is 2, and in the
third it is 4 and then *n*. We could depict this as a generalized
clock with *n* hours. Here is a 6-hour clock, which would be
relevant if, say, you had to take a medicine every 6 hours.

On this clock we have various relationships as in the table below, but we
could also draw them on a spiral as if
we had taken an ordinary number line
and wrapped it around itself:

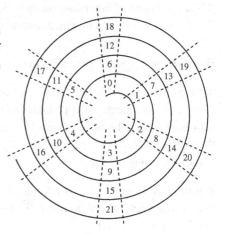

    7   is the same as   1

    8   is the same as   2

    9   is the same as   3

             ⋮

Each of the six positions on the
clock has a whole tower of numbers that "live" there.

We could make an analogous clock for any fixed number
n. Here is a 2-hour clock:

These pictures are vivid but not very practical for working anything out. Imagine trying to work out where 100 sits on the 6-hour clock. You could count all
the way round the clock to 100 but it would be slow, tedious and boring.

──────────────── **Things To Think About** ────────────────

T 4.3   Can you see a relationship between all the numbers living in the same
place? This would give us a *relative* characterization of all the numbers that
count as the same. Can you also think of a *direct* characterization?

This is just a glimpse of the world of *n*-hour clocks, which are technically
called the "integers modulo *n*". In Chapter 6 will discuss why we might want
to represent this world more formally, and how we do it. The idea is that it
makes it easier to work out exactly what is going on and be sure we are right.

Abstract math looks at more and more abstract accounts of these worlds that make connections with more, apparently unrelated, structures.[†]

For now, however, we just need to appreciate that this is a different context in which numbers interact and behave differently from usual.

4.3 The zero world

When I was little, I was allergic to artificial food color. In those days all candy had it, which meant that no matter how many sweets I was given, I effectively had 0. I lived in the zero world of candy, in which everything equals 0. This is the world you end up in if you decide to try and declare $1 + 1 = 1$, but still want some other usual rules of arithmetic to hold: in that case we could subtract 1 from both sides and get $1 = 0$. If that's true then *everything* will be 0. (You might like to try proving that.)

By contrast, in the world of paint we have $1 + 1 = 1$ without landing in the zero world — if we add 1 color to 1 color we get 1 color, and it is not the same as having 0 colors. The difference here is that colors do not obey the other usual rules of arithmetic. In particular you can't subtract colors and so you can't "subtract 1 color from both sides" as we did in the argument involving numbers. This is how that situation avoids being in the zero world.

The zero world might not seem very interesting — you can't really do anything in it. Or rather, you can do *anything* in it and it's all equally valid. It turns out that a world in which everything is equally valid is not very interesting, and is more or less the same as a world in which nothing is possible.

In fact, while this world is not very interesting, it is very important, like some people who are important but not very interesting. We will later see that this world is a useful one to have in the universe. So inside the context of the zero world itself it's not that interesting, but if we zoom out to the context of the universe, the zero world is important and useful. It's another case where context makes a difference.

The zero world is important in the universe because of its extreme relationships with everything else in the universe. In the next chapter we're going to see how different contexts can arise from considering different types of relationship.

[†] They are examples of quotient groups in group theory, which are examples of coequalizers in category theory. I say this just in case you feel like looking them up.

5

Relationships

The idea of studying things via their relationships with other things. Revisiting some of the concepts we've already met, and reframing them as types of relationship, to start getting used to the idea of relationships as something quite general.

In the last chapter we saw that objects take on very different qualities in different contexts. Now we're going to see that different contexts can be provided by looking at different types of relationship. For example, if we relate people by age we get a different context from if we relate them by wealth or power.

One example we investigated was taxi-cab distance. Distance can be viewed as a relationship between locations, and the taxi-cab relationship gives us a different context from the "crow distance" relationship. There are also other possibilities — we could take one-way streets into account, or we could use walking distance, which might be different, since cars and pedestrians often have access to different routes.

In the case of the n-hour clocks we saw a different type of relationship between numbers, in which, for example, $1+1$ can equal 0. This equation is really a relationship between the numbers 1 and 0. We do not have this relationship in the ordinary numbers, where we only have $1 - 1 = 0$ (and $-1 + 1 = 0$). So the existence of the relationship $1 + 1 = 0$ tells us we are in the context of the 2-hour clock, technically called the integers modulo 2.

In the case of the zero world, everything is related by being the same. Of course, everything isn't *actually* the same, but is *considered* to be the same in that world. This is an important distinction that we will keep coming back to. When we focus on context, we are looking at how things appear in *that* context. Things can appear the same in one context but not another, just like when I take my glasses off and everyone looks more or less the same to me.

In this chapter we will develop a way of dealing with relationships that heads towards the way in which category theory deals with them. We will also start drawing diagrams in the way they are drawn in category theory.

5.1 Family relationships

We sometimes depict family relationships in a family tree like this:

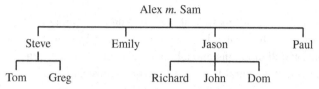

There are only three types of relationship directly depicted here: marriage, parent–child, and siblings. In fact we could view "parent–child" and "sibling" as part of the same depiction, in which case we are depicting only two types of relationship. In any case, we don't need to depict grandparents directly, because we can deduce those relationships from two consecutive parent–child relationships.

Here are two ways to depict this. The first one is a little more rigid as it depends on positions on the page.

$$A$$
$$|$$
$$B$$
$$|$$
$$C$$

$$A \xrightarrow{\text{is a parent of}} B \xrightarrow{\text{is a parent of}} C$$
$$\underbrace{\qquad\qquad\qquad}_{\text{is a grandparent of}}$$

If we represent a relationship using arrows rather than physical positioning on the page, we can draw things any of these ways up (and others) without affecting the relationship we're expressing:

$$A \longrightarrow B \qquad \begin{matrix} A \\ \downarrow \\ B \end{matrix} \quad \begin{matrix} B \\ \uparrow \\ A \end{matrix}$$

That is the point of the arrowhead. Different choices might help us visually, so the flexibility is beneficial (as flexibility typically is). The arrows also encourage us to "travel" along them to deduce other relationships such as:

$$A \xrightarrow{\text{sister of}} B \xrightarrow{\text{mother of}} C \qquad A \xrightarrow{\text{sister of}} B \xrightarrow{\text{mother of}} C \xrightarrow{\text{mother of}} D$$
$$\underbrace{\qquad\qquad}_{\text{is an aunt of}} \qquad\qquad \underbrace{\qquad\qquad\qquad\qquad}_{\text{great-aunt of}}$$

Things To Think About

T 5.1   Can you think of any situations where we can travel along two arrows and the resulting relationship could be several different things?

Note that we all have a relationship with ourself: $A \xrightarrow{\text{self}} A$ for any person A. For this among other reasons things might be ambiguous as in these examples.

$$A \xrightarrow{\text{sister of}} B \xrightarrow{\text{sister of}} C$$
$$\underbrace{\qquad\qquad}_{\text{sister or self}}$$

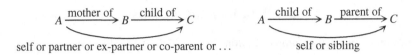

We are going to see that this way of depicting and compiling relationships is remarkably fruitful. It is general and flexible enough to be usable in a vast range of situations, and illuminating enough to have become a widespread technique in modern math and central to category theory. However we will see that we do need to impose some conditions to make sure we don't have ambiguities.

5.2 Symmetry

Some things might not initially seem like a type of relationship, but can be viewed like that by shifting our point of view slightly. We are going to see that categories are built from a very general type of relationship, so if we can view something as a relationship we have a chance of being able to study it with category theory. This is often how we find new examples of existing mathematical structures — it's not exactly that the example is new, but we look at it in a new way so that we see the sense in which it is an example. It's a bit like the fact that if we consider traffic like a fluid then we can understand its flow better using the math of fluid dynamics, leading to effective (and perhaps counter-intuitive) methods for easing congestion.

Symmetry is something that we might think of as a property, but we can alternatively think of it as a relationship between an object and itself. For example a square has four types of rotational symmetry: rotation by 0°, 90°, 180° or 270°. (It doesn't matter which direction we pick as long as we're consistent.)

The symmetry can be seen as this property of a square: if you rotate it by any of these angles it goes back to looking like itself.

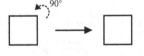

You can't tell the difference unless I put something on it.

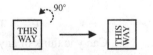

Now, the *fact* that we can do this is a property. It's a property that a rectangle, for example, doesn't possess.

A rectangle in general looks different after we rotate it 90° even without anything written on it.

In abstract math we are moving away from facts and moving towards processes. The *process* of turning a shape around is a relationship.

In the case of the rectangle it's a relationship between these two pictures.

In the case of the square it's a relationship between these two pictures: the square and itself.

Note that now we're thinking of symmetry as a process or relationship we can ask what happens if we do one process and then another.

For example if we do these two processes

the end result is the same as doing this all in one go.

We can see this by checking that with the words written on it looks like this:

─── **Things To Think About** ───

T 5.2   What happens if we do 90° and then 270°? Remember the end result should be one of our rotations: 0°, 90°, 180°, 270°.

If we rotate by 90° and then 270° that's the same as rotating by 360°, but this isn't in our list of rotations because it's "the same" as doing 0°.

Here "the same" means the *result* of rotating by 360° is the same as the *result* of rotating by 0° as shown here.

We could put all that information in this single diagram, which has the added benefit of looking just like the diagrams we drew for family relations previously. (The symbol looking like two short vertical lines is a rotated equals sign.)

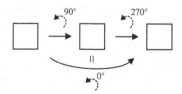

We could depict all these relationships in a table like our previous addition and multiplication tables, to help us keep track of what is going on. So far we have these relationships.

	0	90	180	270
0				
90				
180		270		
270		0		

┌─────────────── **Things To Think About** ───────────────┐

T 5.3   You might like to try filling in the rest of this table yourself and seeing
if you can see some patterns in the table. We'll come back to it in Chapter 11.

└───┘

5.3 Arithmetic

We can also depict ordinary arithmetic as relationships. By "ordinary arithmetic" I mean the familiar kind of arithmetic with any familiar numbers such
as whole numbers $0, 1, 2, 3, \ldots$ which are technically called the *natural numbers*. Let us start by thinking about addition, and depict it as a relationship.

We can express addition as arrows like this.

$$1 \xrightarrow{\;+1\;} 2$$

We can then do two in succession, and see what
single relationship that amounts to, as before.

$$1 \xrightarrow{\;+1\;} 2 \xrightarrow{\;+1\;} 3$$
$$\underset{+2}{\overset{\parallel}{\frown}}$$

We could try it with a different starting point, such
as this:

$$2 \xrightarrow{\;+1\;} 3 \xrightarrow{\;+1\;} 4$$
$$\underset{+2}{\overset{\parallel}{\frown}}$$

In fact we know that $1 + 1 = 2$ in the natural numbers, so adding 1 twice will always be the same
as adding 2, no matter where we start. We could
depict this generally like this.

$$? \xrightarrow{\;+1\;} ? \xrightarrow{\;+1\;} ?$$
$$\underset{+2}{\overset{\parallel}{\frown}}$$

It now looks a lot like our depiction of rotations,
and we can go ahead and put these in a table; it's
rather large (well, infinite) but has some patterns
in it that help us tell what is going to happen later
without us having to fill in the whole table. Learning arithmetic is really a process of becoming familiar with these patterns, just — alas — usually
without seeing the patterns drawn out.

+	0	1	2	3	4	$\cdots$
0	0	1	2	3	4	$\cdots$
1	1	2	3	4	5	$\cdots$
2	2	3	4	5	6	$\cdots$
3	3	4	5	6	$\cdots$	
4	4	5	6	$\cdots$		
$\vdots$	$\vdots$	$\vdots$	$\vdots$			

We started looking at the importance of patterns in Chapter 3.

5.4 Modular arithmetic

The above grid for addition of natural numbers was rather large as there is an
infinite number of those numbers, but a grid for *modular* arithmetic (that is,

the n-hour clock) will be smaller. We only have a finite quantity of numbers on the n-hour clock — the numbers $1, \ldots, n$ (or $0, \ldots, n-1$). We regard n as being the same as 0, a bit like when we regarded $360°$ as being the same as $0°$ for rotations of a square.

Now we have different relationships. Suppose we do a 4-hour clock, so 4 is the same as 0.

We'll now get things like these:

$$1 \xrightarrow{+3} 0 \qquad 2 \xrightarrow{+3} 1$$

along with combinations of arrows like this.

$$1 \xrightarrow{+3} 0 \xrightarrow{+2} 2$$
$$\parallel$$
$$+1$$

As with ordinary arithmetic it doesn't matter where we start: the compiled or "composite" relation is the same.

$$\bullet \xrightarrow{+3} \bullet \xrightarrow{+2} \bullet$$
$$\parallel$$
$$+1$$

Here the gray blobs represent places for a number to be put, like the question marks previously. In fact, as it is the processes we're interested in rather than the specific answers, we will eventually see that those gray blobs don't need to be there at all — we're really just interested in the arrows.

The arrows in this last example encapsulate the same information as simply saying $3+2 = 1$ on the 4-hour clock, and we could put all the relationships in a table as we have done for many relationships already. So far we have this:

	+0	+1	+2	+3
+0				
+1				
+2				
+3		+1		

Things To Think About

T 5.4   Try filling in the rest of the table.

1. What patterns do you see? How are they similar to the ones for rotations of a square? *Why* are they similar?
2. What symmetry can you see? What is the symmetry (visual feature) telling us about the behavior of the numbers (non-visual feature)?

We'll come back to this in Chapter 11.

5.5 Quadrilaterals

There is a meme that goes round every tax season saying "I sure am glad I learned about quadrilaterals every time quadrilateral season comes round".

This is possibly amusing, but misses the point about why we study anything: sometimes it's because the thing itself is important, but sometimes it's because it's a good arena for practicing some sort of disciplined thought process. Quadrilaterals themselves might not be particularly important for life, but I've put them here because they are a handy place to explore different types of relationship and ways of depicting them. One reason it is handy is because quadrilaterals are things we can draw and look at. Once we've got the idea, we can try this in more momentous situations or indeed anywhere we like. It's the principles that are useful, not the quadrilaterals themselves. We are going to explore "special case" relationships between types of quadrilateral.

For example, a square is a special case of a rectangle, one that happens to have all its sides of equal length. This is sometimes depicted as a Venn* diagram and this is a reasonable visualization for simple purposes.

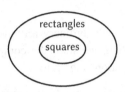

However we are going to depict it like this: square $\xrightarrow[\text{case of}]{\text{special}}$ rectangle
Here are some reasons for doing this.

1. We can depict more complex interactions this way, as we'll see.
2. This abstraction looks like the way we have depicted other relationships in this chapter. In fact this is how we depict relationships in category theory and it turns out to be very fruitful for reasons including (1).

Another example of a quadrilateral relationship involves parallelograms.

Remember that parallelograms have opposite pairs of sides the same length and parallel. Thus opposite angles are the same.

A rectangle is a special case of a parallelogram as it satisfies those conditions plus more: *all* its angles are the same (right angles) not just opposite pairs.

So we have the following compilation of relationships, and this diagram of arrows that looks just like some other situations we've seen already:

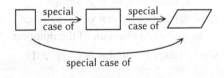

The content of the "special case" diagram is that if A is a special case of B and B is a special case of C then we can immediately deduce that A is a special case of C. (One might say it is a *very* special case.) When we compile relationships in this way it is called *composition* and we say we are composing the arrows.

---

* Technically an Euler diagram. This will be the case every time I refer to a Venn* diagram.

T 5.5   What other types of quadrilateral are there and what are their defining characteristics? Can you draw a diagram of all the special case relationships?

Special cases are often found by imposing extra conditions on a situation, and generalizations are found by relaxing conditions. Thinking about the conditions making a quadrilateral a square, or a trapezoid,[†] or some other special case can help us see how to relax them gradually to make more general cases.

For example a rhombus has all sides the same, and opposite angles the same, whereas a kite has adjacent pairs of edges the same, and only one pair of opposite angles the same.

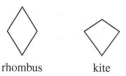

rhombus   kite

For quadrilaterals the conditions involve the lengths of sides, the angles, and whether any sides are parallel. Combining all relationships between quadrilaterals we get this diagram. Note that there are two paths from "square" to "parallelogram": it doesn't matter in what order we relax the condition on the angles and on the sides, we get a parallelogram either way.

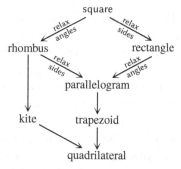

T 5.6   A trapezoid is a special case of a rhombus and of a rectangle; why don't we need to draw those arrows? What other arrows haven't we drawn? What would happen if we tried to draw this as a Venn* diagram?

We don't have to draw the extra arrows as we can deduce those relationships by composing the other ones; they are the type of thing I called "very special case" earlier. On the right are a few examples.

square ⟶ parallelogram
square ⟶ kite
rectangle ⟶ trapezoid
parallelogram ⟶ quadrilateral

---

† It is a point of some contention whether the following sorts of things should be allowed to count as a trapezoid (or "trapezium" in British English): ⟋⟍ ⟍⟋ I prefer to say yes as it fits better with the principle of gradually relaxing conditions. In order to disallow parallelograms from being trapezoids we would have to *add* a condition, not just relax a condition. In abstract mathematics we tend to prefer generalizing so that previous ideas are special cases rather than thrown out.

If we try to put all the types of quadrilateral into a Venn* diagram it's hard. We can get as far as the diagram below but then we have a problem.

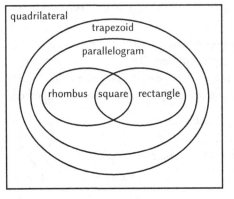

- All rhombuses are kites, so the kite circle needs to contain the entire rhombus circle.
- A kite that is not a rhombus is also not a parallelogram or a trapezoid, so the kite circle needs to stick out into the random quadrilateral part without overlapping the parallelogram or trapezoid circles.

This means we would either have to have a disconnected "kite circle", or have some empty intersections. Either option breaks the connection between the actual situation and the visual representation of it. The point of a visual representation is that it is supposed to clarify the relationships in question. This is a sense in which the arrow depiction has more possibilities for helping us understand more complex interactions.

In fact, in the diagram with arrows I implicitly used the height on the page as well as the arrows, as the heights correspond to how special the objects are, with "square" at the top, and things becoming progressively more general down the page. Next we will see another example where the heights on the page can be invoked to help us organize our diagrams.

5.6 Lattices of factors

We have already looked at relating numbers to each other via addition, with diagrams like this:

$$a \xrightarrow{\;+2\;} b$$

We could try this with multiplication too. This gives us a very different context. For example, if we relate numbers by addition we can build all the natural numbers $(1, 2, 3, 4, \ldots)$ just starting with the number 1. But if we relate them by multiplication we can't get anywhere with the number 1; if we keep multiplying by 1 we just get 1. In fact there is no single number that we can use to get all the natural numbers by multiplication. In this section we're going to investigate the relationships between numbers by multiplication, using diagrams of arrows to help us visualize structures.

Consider the number 30. If we think about all its factors, those are the numbers a related to it as shown.

$a \xrightarrow{\times \text{ something}} 30$

The factors of 30 are all the numbers that "go into" 30. We can write them out in a list: $1, 2, 3, 5, 6, 10, 15, 30$. This has no indication of any relationships between them; we have taken them entirely out of context.

Instead we can draw in all the relationships where anything is a factor of anything else, such as this.

$2 \xrightarrow{\times 3} 6$

As usual, we don't need to draw arrows that we can deduce from composite arrows like this one.

$2 \xrightarrow{\times 3} 6 \xrightarrow{\times 5} 30$
$\times 15$

Here is the whole diagram. As well as omitting composite arrows I have omitted arrows showing that each number is a factor of itself. Those omissions make the diagram tidier without changing the information. In fact we could also omit the labels on the arrows, as we can deduce what each one represents by looking at the endpoints.

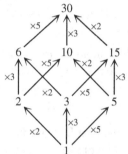

This diagram is now a cube shape, and I find that much more interesting than a list of factors in a row.

─── **Things To Think About** ───

T 5.7 1. What does each face of the cube represent? What do you notice about the labels on the edges?
2. Why did this turn out to be a cube? What other numbers will produce a cube? What shapes will other numbers produce if not cubes? You could try $42, 24, 12, 16, 60$ and anything else you like.
3. How can you tell what shape will be produced before drawing it?

We could try this with 42. This has eight factors just like 30 does, so it has a chance of also being a cube. You might be tempted to arrange the factors in size order as in the case of 30, like this:

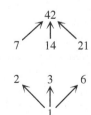

The top and bottom are fine but we have some problems in the middle.

- Nothing goes into 7 except 1 (and itself) as it is prime.
- We have factors on the same level: 7 goes into 14 and 21, and 2 and 3 go into 6. This didn't happen before, where all arrows went up a level.

The first point gives us a clue to how we can fix this: nothing goes into 7 except 1 (and 7) so we should move it down a level. Some things go into 6 other than 1 so we should move it up. This produces a cube diagram just as 30 did.

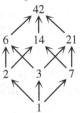

┌──────────────── **Things To Think About** ────────────────┐

T 5.8   What do the numbers at each level have in common with each other? Can you build on that idea to find a level of abstraction that makes this "the same" situation as the cube for 30?
└──┘

The bottom number is always going to be 1 as nothing goes into it. Directly above that we'll have all the numbers (except 1) divisible by nothing except 1 and themselves — that is, the prime numbers. Note that 1 is not prime and also is not at this level; we will come back to this point.

At the next level we have the numbers that are a product of two of the primes from the level below.

$$6 = 2 \times 3$$
$$14 = 2 \times 7$$
$$21 = 3 \times 7$$

At the top we have 42, a product of three primes.

$$42 = 2 \times 3 \times 7$$

This is the key to "why" this is a cube: 42 has three distinct prime factors.[†] In this depiction each one gives a dimension, so we have a 3-dimensional object. Notice that the parallel edges are all the same type of relationship — either ×2, ×3 or ×7.

The square faces tell us things like the fact that multiplying by 2 and then 3 is the same as multiplying by 3 and then 2, i.e. that multiplication is commutative.

Commutativity of multiplication (together with associativity) tells us that each way of building 42 by multiplication is "the same", and in this diagram all those different ways are represented by different paths from 1 to 42. We are

---

[†] We might say it has three "different" prime factors but usually in mathematics the technical word is "distinct" as it is a little more precise in broader contexts. For example we can speak of three "distinct" points in the sense that they are separate from each other, even if as points they don't look different. Differentness is subjective; it depends what criteria we use.

going to see that in category theory this is called commutativity of a diagram, whatever the diagram represents.

The upshot is that both 30 and 42 are a product of three distinct primes, which we could call a, b, c so that both cubes are instances of the one shown here, in which a, b, and c give us a dimension each:

I think this type of diagram vividly depicts the idea of building numbers by multiplying primes together, and gives us another way to understand why 1 is excluded from being prime: it is 0-dimensional here so we can't build any shapes from it at all.

What about a number that is not of the form $a \times b \times c$ for distinct primes a, b, c? We will now try $24 = 2 \times 2 \times 2 \times 3$. The factors are $1, 2, 3, 4, 6, 8, 12, 24$. There are eight factors so it might be tempting to try and make it a cube. But we know that a cube has three dimensions produced by three distinct prime factors, whereas here we only have two distinct prime factors, 2 and 3, with 2 being repeated several times. We are only going to put 2 in the diagram once to avoid redundancy; we are going to see that the repeatedness of the factor appears in a different, more geometric way.

The key is to remember the idea about the levels: we should put 1 at the bottom, then the primes, then the products of two primes, then three, and so on. Note that for these levels we are not thinking about *distinct* primes, so for example $4 = 2 \times 2$ and this counts as a "product of two primes", whereas $12 = 2 \times 2 \times 3$ which is a "product of three primes". Here is the whole diagram with the levels marked in.

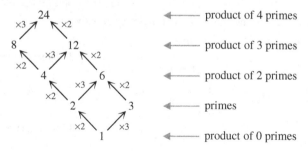

Things To Think About

T 5.9 How many dimensions does this shape have, and why? What visual feature comes from the fact that 2 is a repeated factor? What other numbers will have the same shape? What is the abstract version of this diagram, like when we expressed the cube using a, b, c?

This shape is 2-dimensional: it consists of three (2-dimensional) squares side by side. The two dimensions are given by ×2 and ×3 — they arise from the fact that 24 has only two *distinct* prime factors. The fact that 2 is a repeated factor (three times) appears visually in the fact that there are three squares in a row. This is because we can travel three times in the "×2" direction.

Any other number will have the same shape if it has two distinct prime factors, one repeated three times, i.e. $a^3 \times b$, $a \neq b$, such as:

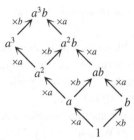

$$3 \times 3 \times 3 \times 2 \ = \ 54$$
$$2 \times 2 \times 2 \times 5 \ = \ 40$$

Here the symbol $\neq$ means "not equal to".

Things To Think About

T 5.10   Can you now deduce what shape will arise from *any* number, based on its prime factorization? In particular, what numbers will be 1-dimensional? Can you try drawing a 4-dimensional example, e.g. $210 = 2 \times 3 \times 5 \times 7$?

This might all seem like little more than a cute game for turning numbers into shapes, but we're now going to see a powerful consequence of pursuing the further levels of abstraction.

Once we have reached the level of abstraction of a, b, c it doesn't really make any difference if we multiply the numbers together or not. Consider the diagram of the abstraction of 30 and 42. It might as well be a diagram of subsets of $\{a, b, c\}$. This denotes a set of three things or "elements", a, b, c, and a subset of this is then any set containing some combination of those elements: possibly all of them, none of them, or something in between.

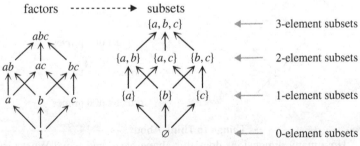

The symbol $\varnothing$ denotes the empty subset, the one with no elements.

An arrow like the one on the right now shows the relationship "A is a subset of B".

$$A \longrightarrow B$$

To make a tidier diagram, we only need to draw arrows for situations where exactly one element is "added" to the subset (thrown in, not actually +). When two elements are added in we can break the process down into adding one element at a time.

We can depict this in a familiar diagram.

$$\{a\} \xrightarrow{\text{add } b} \{a, b\} \xrightarrow{\text{add } c} \{a, b, c\}$$

add b and c

Now for the point of the abstraction: at this level of abstraction we can have a, b, c being *any* three things, not necessarily prime numbers, and not necessarily numbers at all because we are not trying to multiply them.

For example they could be three types of privilege, such as rich, white, male. We get this diagram:[†]

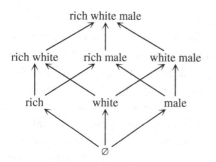

Each arrow depicts hypothetically gaining one type of privilege while everything else about you stayed the same: the theory of privilege says you would be expected to be better off in society as a result of that one change.

This depicts some important things about privilege that are often misunderstood: white privilege does not mean that *all* white people are better off than *all* non-white people, nor even that they are better off on average, nor that they are better off in every way. It is just about the abstract idea of keeping everything else fixed and moving up an arrow.

For example some white men who did not grow up rich think they do not have white privilege. However there is an arrow

non-rich non-white men $\longrightarrow$ non-rich white men

representing the gain in privilege of whiteness. This means that non-rich men who are *not* white are even worse off. However there might well be (and indeed there are) rich non-white men who are better off. But there is no arrow from rich non-white men to non-rich white men, so the theory of privilege does not say anything about the relative situations of these two groups.

An abstract explanation is that some people are confusing the hierarchy of *overall privilege* with the hierarchy of *number of types of privilege*. This is

---

[†] This previously appeared in *The Art of Logic*.

analogous to thinking we could make the cube of factors of 42 by putting the
smallest numbers at the bottom level above 1. That didn't work because $6 < 7$
so 6 is lower in the hierarchy of sheer size, but 6 has more prime factors than
7 so is higher up in the hierarchy of number of factors.

Later we will see that these are questions of

1. totally ordered sets vs partially ordered sets, and
2. order-preserving vs non-order-preserving maps.

For now it is enough to appreciate that the formalism of drawing these diagrams of arrows can help us gain clarity about different interactions between
things, and, as I've aimed to demonstrate with the examples in this chapter, that
the formalism of these arrows is very flexible and can be used in a remarkably
wide range of situations.

It may seem unusual to discuss issues of privilege and racism in relation
to abstract mathematics. In fact I am sometimes asked if I'm not worried that
talking about social justice issues in math will put some people off math, those
who don't agree with me about social justice issues. The thing is that when I
didn't talk about social justice issues in math, nobody asked me if I was worried
about putting off those who *do* care about those issues. I think there are plenty
of existing math books and classes that don't touch on those issues, and that
talking about them includes and motivates many people who have previously
not felt that math is relevant to them. I would like to stress that you don't have
to agree with my stances in order to see the logical structures I am presenting
in them. I think it's important for us all to be able to see the logical structures in
everyone's points of view, whether we agree with them or not. I think abstract
mathematics helps with that.

In the next chapter we will discuss the formalism of mathematics and why
it can be helpful for progressing through the theory while also being offputting
and obstructive when you are new to it.

6

Formalism

Easing us from informal ideas into formal mathematics. This chapter will motivate that move, develop more formal approaches to the structures we've already seen, acclimatize us to the formalism, and see what we get out of it.

6.1 Types of tourism

It is possible to visit a country whose language you don't speak, and still have a culturally rewarding trip, in which you expand your mind and don't just behave like an obnoxious tourist demanding that everyone speak your language to you. Perhaps it is less accepted that it is possible to "visit" the world of math in this way, without having any formal understanding and without knowing any of the technical language, but still appreciating the culture and being enriched by it. This sort of visit to category theory is more or less the idea of my book *How to Bake π*; perhaps I would push the analogy a little further and say that that book also contained some basic phrases as an introduction to the language.

In the present book, however, we're aiming to take things a little further. I want to provide a way in to the true language, a route that is approachable but is still a decent beginning that doesn't just consist of memorizing how to say "Good morning", "Thank you" and "My name is Eugenia".

I have already been gradually ramping up the level of formalism, by gradually introducing technical mathematical notation, much of which is actually category theory notation. But I want to take a little pause and talk directly about formalism and notation because I know that this is one of the offputting aspects of learning math. Mathematicians get very fluent and take it for granted, which makes them forget to explain it sometimes. It's a bit like how it can be very hard to get a native speaker of a language to explain the grammar, if they know it too instinctively.

If you're already comfortable with formal mathematical notation you can probably skip or skim this chapter. However, the formal mathematical notation you've seen might not be used in quite the same way as in abstract mathematics, so I'd suggest reading this chapter quickly anyway.

6.2 Why we express things formally

One of the important issues around formalism is why we do it. If you don't find something inherently interesting and you also don't see the point of it then it will almost certainly be offputting. If you find it fun then it doesn't really matter whether or not you think it has a point, but if you find it offputting then understanding why it's helpful is at least a starting point.

The formalism of mathematics is like a language specifically designed for what we are trying to do. Normal language is geared towards expressing things about our daily human experience, and less towards making rigorous arguments about abstract concepts. So our normal language falls short when we're trying to do math.

In fact, even among normal languages, different ones express different things in different ways. People who speak several languages often find themselves missing words and expressions in one language when they're speaking another. In English I miss the concise and pithy four-character idioms that my family often uses in Cantonese; one of my favorites, which somehow doesn't translate quite so pithily into multi-syllabic English, is something like "rubbish words covering the sky" to indicate that someone is spouting utter nonsense.

Formal mathematical language and notation are partly to do with efficiency, and partly to do with abstraction (and those two concepts are themselves related). The efficiency is to do with when we keep talking about the same thing over and over again, in which case it's really more efficient to have a quick way to refer to it. If you're only going to mention it once then it doesn't matter so much. If you come to visit Chicago for a week you might talk about driving up the highway that runs alongside the Lake. However, if you live here it might get tedious to keep saying "the highway that runs alongside the Lake" so it helps to know it's called Lakeshore Drive, and then if you really talk about it a lot you might just start calling it "Lakeshore" or "the Drive" or (somewhat amusingly) LSD.

Mathematical language and notation happens similarly. If you eat two cookies and then another three cookies and then never do anything similar again in your life then you might never need any more language or notation. But if you ever plan to develop thoughts about that sort of thing any further then it can help to write the numbers 2 and 3 rather than the words "two" and "three". It's quicker, takes up less space, and is faster for our brains to process. We can then also move from writing "2 and 3" to $2 + 3$ for similar reasons. I would then argue that it makes it easier to see further analogies such as the analogy between $2 + 3$ and 2×3.

This is where the part about abstraction comes in. Concise notation can help us to package up multiple concepts into a single unit that our brains can treat as a new object, which is one of the crucial aspects of abstraction.

For example the expression on the right is fine and precise and rigorous, but it is long-winded.

$$2 + 2 + 2 + 2 + 2$$

So we make the following notation and get the idea that multiplication is repeated addition.

$$5 \times 2$$

Note that this is not the *only* way to think of multiplication. Some people think this means we should never talk about multiplication as repeated addition, but I think it's more productive to see the sense in which it *can* be thought of in this way, while also being open to how we can generalize if we don't *insist* on it being thought of in this way.

Anyway, we have now made new notation for *iteration* of an operation, and this created a new operation. We could now repeat that process, that is, we could apply that entire thought process to the new operation ×, treating it as analogous to the original operation +. This might not have been so obvious if we hadn't used this formal notation.

If we iterate multiplication we get things like this:

$$5 \times 5 \times 5$$

Here is the notation that mathematicians created for that, making a new operation called exponentiation.

$$5^3$$

At time of writing, we have just all gone into lockdown over the COVID-19 virus. Viruses spread by exponentiation, because each infected person infects a certain number of people on average. Thus the number of infected people is multiplied by that factor repeatedly, producing exponentiation. Without that precise language and notation it would be much harder to reason with the concept and understand it.

As it is, mathematicians understand that exponentiation makes numbers grow extremely rapidly, and it's not just that they grow rapidly, but the growth itself gets faster and faster as it goes along, as shown here.

This is very different from repeated addition, where things might grow fast but they never get any faster, as shown by the straight line graph.

This understanding is why scientists and mathematicians know that advance precautions are critical when exponential growth is involved, even if things seem to be moving very slowly at that particular point in time. The formalism of math helps us to model the exponential in question even when it's early on in the slow part of the growth, and predict what will happen in the future if we don't take steps to slow it down. To people who don't understand exponentials this has unfortunately seemed like panic and over-reaction.

When we're doing research and we come up with a new concept, one of the first things we do is give it a name and some notation, so that we can move the concept around in our heads more easily. It's a bit like when a baby is born and one of the first things we do is give it a name.

I'm now going to take a few of the examples from the previous chapters and introduce the formality that mathematicians would use to discuss them.

6.3 Example: metric spaces

In Section 4.1 we thought about a type of distance measured "as the taxi drives" (in a grid system, like in many American cities) as opposed to "as the crow flies". Different types of distance in math are called *metrics*. This more abstract technical-sounding word may sound alien but it serves the purpose of

a) reminding us that we're not thinking of straightforward distance, and
b) opening our minds to the possibility of generalizations which have some things in common with distance but aren't actually distance.

Here is how mathematicians write that down formally. We start by thinking about what "distance" really means, right down to its bare bones. At the most basic, it is some process where you take two locations, and produce a number.

In order to think about this more carefully we're going to use some "place-holders" to refer to those things. This is a bit like in normal life when we're talking about two random people, and we might call them person A and person B. This is only necessary if things get complicated, say we're discussing an interaction and we need to keep clear about who's doing what. So we might say "one person makes the coffee and the other person gets the cookies and gives them to the person who made the coffee and the person who made the coffee takes the coffee round while the person who makes the cookies cleans up". That is long-winded, and we might sum it up more clearly as "A makes the coffee, B gets the cookies and gives them to A, then A takes the coffee round while B cleans up". We could make it even more efficient with a table:

A	B
makes coffee	gets cookies and gives to A
takes coffee round	cleans up

Now for our case of distance we're talking about *locations A and B*. If we were just going to talk about features of them individually that would be enough, but we want to talk about the distance between them, which is a number involving both of them. It's a bit like a subtraction, but not quite. If A and B were just numbers on a number line

we could find the distance between them just by doing a subtraction $B - A$ (or $A - B$ if A is bigger). But if they're positions in 2-dimensional space or higher then it's a bit more complicated. Anyway we want to separate out the notation for a type of distance from the instructions for how to calculate it. $B - A$ is more of an instruction or a method of calculating the distance between them. So we could say this:

> distance between A and B on number line $= B - A$

Now we get to the efficiency/abstraction part. If we write "distance between A and B on number line" a lot, we might get bored of doing it. Also it's a lot of stuff for our eyes to take in.

So we might shorten it to this, for example: $\qquad d(A, B) = B - A$

Here the d stands for distance and the parentheses don't indicate any kind of multiplication; they are just a sort of box in which to place the two things whose distance apart we're taking.

─────────────── **Technicalities** ───────────────

We'll later see that basically what we're doing here is defining a *function* and that this notation using parentheses is a standard way to write down what happens when you apply a function to an input. In this case the function happens to have two inputs, A and B. Other functions like sin just take one input and then the output is written as $\sin(x)$.

────────────────

We can now *generalize* this by a sort of leap of faith: even if we have yet to work out how we're going to calculate a distance, we can still use the notation $d(A, B)$ for the distance between A and B. In formal mathematical notation we make a declaration saying that's what we're going to do. It's a bit like saying "I hereby declare that I am going to use the notation $d(A, B)$ to mean the taxi-cab distance between A and B". This is still a bit imprecise because we haven't said what A and B are allowed to represent. Are they numbers or are they people or are they locations? So the full declaration might look like this:

> *Let A and B be locations on a grid. We will write d(A, B) for the taxi-cab*
> *distance between A and B according to the grid.*

Now we might want to get even more abstract and talk about new types of distance that we haven't even defined yet, and then we want to say in what circumstances something deserves to count as a type of "distance".

That's like what happens if some random person comes along and says "Hey, I've discovered a treatment for COVID-19" — we'd want (I hope) to do some tests on it to see if it really deserves to be called a treatment, and the thing we'd have to do before doing those tests is decide what the criteria are for being counted as a treatment.

So if some random person comes along and says "Hey, I've discovered a new type of distance", mathematicians want to have some criteria for deciding if it's going to count or not. Here are some criteria.

Criterion 1: distance is a number

The first criterion is that distance should be a way of taking two locations and producing a number. So given any locations A and B, $d(A, B)$ is a number. There are different kinds of number though, so if we're being really precise we should say what kind of number. We probably want distances to go on a continuum, so we don't just want whole numbers or fractions, but *real numbers*: the real numbers include the rational numbers (fractions) and irrational numbers, and sit on a line with no gaps in it. So, for any locations A and B, $d(A, B)$ is a real number.

Criterion 2: distance isn't negative

We typically think of distance as having size but not direction — it's just a measure of the gap between two places. This means that we shouldn't get negative numbers. So in fact for any locations A and B, $d(A, B) \geq 0$.

Criterion 3: zero

Can the distance between places ever be 0? Well yes, the distance from A to A should be 0. In our formal notation: $d(A, A) = 0$. But is there any other way that distance can be 0? Usually not, but in fact there are more unusual forms of distance where it might be possible. We don't usually include those in *basic* notions of distance though. So now what we're saying is "the distance from A to B can only be 0 if $A = B$".

———————————————— Note on A and B being the same ————————————————

It might be confusing to have locations A and B that turn out to be the same one. Let's go back for a second to the example of the people A and B who were sorting out the cookies and the coffee. Usually in normal life if we talk about person A and person B we really are talking about two entirely different people. But if the group needing cookies and coffee was short-staffed that day then it's possible one person could take on both roles, in which case person A and person B would both be the same person.

In math even if we use two different letters for things there is always the possibility that they could represent the same thing. For example if you write x for the number of sisters you have and you write y for the number of brothers you have it's quite possible that those are the same number. Writing those variables as x and y means that the numbers *could be* different but it still allows for the possibility of them being the same. Whereas if you write them both as x then you have declared in advance that they're the same and ruled out the possibility of them being different.

If we really want to rule out the possibility of them being the same in math we call them "distinct", so instead of saying "Let A and B be locations" we would have to say "Let A and B be distinct locations". That is equivalent to adding in an extra condition saying "$A \neq B$" and mathematicians don't like extra conditions if we can avoid them, so it's usually preferable or more satisfying to see if we can include the possibility of A and B being equal.

In summary: for distance we have decided that the distance from a location to itself is 0, and that this is the only way we can have a 0 distance. In formal language we say: $\boxed{d(A, B) = 0 \text{ if and only if } A = B.}$

I will discuss more about what "if and only if" means in the next section.

Criterion 4: symmetry

We said in Criterion 2 that we were just going to measure the gap between places, without direction being involved. We have specified that this means the answer is always a non-negative number, but that's not enough: we need to make sure that the "distance from A to B" doesn't come out being different from the "distance from B to A".

That is, for any locations A and B we have this: $\boxed{d(A, B) = d(B, A).}$

This is called *symmetry* because we have flipped the roles of A and B. Later we'll look at symmetry more and more abstractly. Symmetry starts off as being about folding things in half and the two sides matching up, but eventually we generalize it to being about doing a transformation on a situation and it still looking the same afterwards. That's what symmetry is about here.

Criterion 5: detours

The last criterion is about taking detours. This is the one that you might not immediately think up if you're just sitting around thinking about what should count as a sensible notion of distance, but here it is: if you stop off some-where along the way it should not make your distance any *less*. At best it won't change the distance, but it might well make the journey longer. Now let's try putting that into formal mathematics.

It might help to draw a diagram: we're trying to go from A to B but we go via some other place X like in this triangle.

The direct journey has this distance: $d(A, B)$

whereas the detour has this distance: $d(A, X) + d(X, B)$

Now the detour distance should be "no less" than the direct distance: $d(A, X) + d(X, B) \geq d(A, B)$

Note that "no less" is logically the same as "greater than or equal to", so in the formal version we use the sign $\geq$.

This condition is often referred to as the *triangle inequality* because it relates to the triangle picture above, and produces a specific inequality relationship. In math, when we refer to things as an inequality we usually don't just mean that two things are unequal; after all, most things are unequal. Usually we are specifically referring to a relationship with $\geq$ or $>$ in. The point is that even if we can't be sure two things are equal, sometimes it helps to know for sure that one is bigger, or definitely not bigger. See, math isn't just about equations: sometimes it's about inequations.

That's the end of the formal criteria for being a type of distance in math. It is quite typical that we now give it a formal name, both to emphasize that we've made a formal definition, and to remind ourselves that some things that aren't physical distance can count as this more abstract type of distance. In this case, we call it a *metric* in math. We have carefully put some conditions on something to say when it deserves to be called a metric. Crow distance and taxi distance are both examples. Sometimes the money or energy spent getting from one place to another could be examples too.

Once we have the definition with a set of criteria, aside from working with it rigorously and finding more examples, we can think about generalizing it by relaxing some criteria. We might think there are some interesting examples that *almost* satisfy all the criteria but not quite, and maybe they deserve to be studied too. Rather than just throw them out or neglect them, mathematicians prefer to make a slightly more relaxed notion that will include them, and then study that. It often makes the theory more difficult because things aren't so rigidly defined, but it often also makes it more interesting, and anyway, it's more inclusive.

┌─────────────── **Things To Think About** ───────────────┐

T 6.1 Can you think of some measures that are a bit like distance as we defined it above but that fail some of these criteria? See if you can think of some that fail each criterion in turn, except maybe the first one.

└──┘

6.4 Basic logic

While we are talking about the formalism of mathematics it's just as well to make sure we're clear about the basic logic underpinning it. As we are going to be building arguments using rigorous logic, the way in which basic logic flows is going to be critically important, although a deep background in formal logic is definitely not needed. Really what matters is following the logic of the arguments I make as we go along, so even if you don't feel you understand all of this section out of context, you could keep going and just come back to it if you want more clarification about a later argument.

Logic is based on *logical implication*. Given any statements A and B we can make a new statement "A implies B" which means "if A is true then B has to be true", or, to put it more succinctly "if A then B". The fact that it's a *logical* implication means that the truth of B follows by sheer logic, not by evidence, threats or causation. This usually means it is sort of inherent from some definitions.

We sometimes write "implies" as a double arrow like this: $\boxed{A \Rightarrow B.}$ This means we have fewer words to look at, and also helps emphasize the directionality of logic. One source of confusion is that in normal language we use the following types of phrases interchangeably (in vertical pairs):

if A is true then B is true	if you are a human then you are a mammal
B is true if A is true	you are a mammal if you are a human

This looks like we've changed the order around, but we haven't changed the direction of the *logic*, we've just changed the order in which we *said* it: the statements are logically equivalent. Note that some statements are *logically* equivalent without being *emotionally* equivalent, sort of like the fact that Obamacare and The Affordable Care Act are the same thing by definition, but induce very different emotions in people who hate Obama so much that they immediately viscerally object to anything that has his name in it. This means that in normal life we need to choose our words carefully.

In math we are interested in things that are logically equivalent, but we still

choose our words carefully because sometimes there are different ways of stating the same thing that make it easier or harder to understand, or shed different
light on it that can help us make progress.

In summary we have these several ways of stating the same implication, and
they sort of go in pairs (horizontally here) corresponding to ways of stating the
thing in the opposite order without changing the direction of the logic:

$$
\begin{array}{cc}
A \text{ implies } B & B \text{ is implied by } A \\
\text{if } A \text{ then } B & B \text{ if } A \\
A \Rightarrow B & B \Leftarrow A
\end{array}
$$

I particularly like using the arrows here because to my brain it's the clearest
way of seeing that the left- and right-hand versions are the same, without really
having to think. Or rather, I don't have to use logical thinking, I can just use a
geometrical intuition.

We are going to use a lot of notation involving arrows as that is one of the
characteristic notational features of category theory.

The arrow notation also makes it geometrically clear (at least to
me) that if we change the way the arrow is pointing at the actual
letters then something different is going on, as in this pair.

$$
\begin{array}{c}
A \Rightarrow B \\
B \Rightarrow A
\end{array}
$$

This reversing of the flow of logic produces the *converse* of the original
statement. The converse is logically independent of the original statement. This
means that the two statements might both be true, or both false, or one could
be true and the other false. Knowing one of them does not help us know the
other. Here are some examples of all those cases.

Both $A \Rightarrow B$ and $B \Rightarrow A$ are true

A = you are a legal permanent resident of the US
B = you are a green card holder in the US

Both $A \Rightarrow B$ and $B \Rightarrow A$ are false

A = you are an immigrant in the US
B = you are undocumented[†]

One is true but the other is false

A = you are a citizen of the US
B = you can legally work in the US

As a statement and its converse are logically independent we have to be care-

---

[†] You can be born in the US but not have any documents about it.

ful about not confusing them. But if they are both true then we say they are logically equivalent (even if they aren't emotionally equivalent) and say

> A is true *if and only if* B is true.

Mathematicians like abbreviating "if and only if" to "iff" in writing sometimes, but this is hard to distinguish from "if" in speaking out loud so it should still be pronounced "if and only if" out loud. The phrase "if and only if" encapsulates both the implication and its converse:

> | A if B | means | $A \Leftarrow B$ |
> | A only if B | means | $A \Rightarrow B$ |

The last version has a tendency to cause confusion so if it confuses you it might be worth pondering it for a while. My thought process goes something like this: *A* implies *B* means that whenever *A* is true *B* has to be true, which means that *A* can only be true if *B* is true.

We have this full table of equivalent ways of saying the same statement, including a double-headed arrow for "if and only if":

original		converse	
$A \Rightarrow B$	$B \Leftarrow A$	$B \Rightarrow A$	$A \Leftarrow B$
A implies B	B is implied by A	B implies A	A is implied by B
if A then B	B if A	if B then A	A if B
A only if B	only if B then A	B only if A	only if A then B

logical equivalence

$$A \Longleftrightarrow B$$
A if and only if B
A iff B

I think that the notation with the arrows is by far the least confusing one and so I prefer to use it.

6.5 Example: modular arithmetic

In Section 4.2 we looked at some worlds of numbers in which arithmetic works a bit differently. One of the examples was the "6-hour clock".

We started making a list of some of the numbers that are "the same" as each other on this clock, such as these.

7	is the same as	1
8	is the same as	2
9	is the same as	3

Now the thing is that we can't just sit and make a list of *all* possible numbers that are the same as each other, because there are infinitely many. In situations where it is impossible to write everything down one by one it can be extremely helpful to have a formal and/or abstract way of writing things, so that we can give a sort of recipe or make a machine for working every case out, without actually having to write out every case one by one.

We then drew this spiral diagram showing what happens if you keep counting round and round the clock and writing in the numbers that live on the "same hour" as each other. If you pick one position maybe you can spot a relationship between all the numbers in that position. In the 0 position it's perhaps the most obvious — all the numbers in that position are divisible by 6.

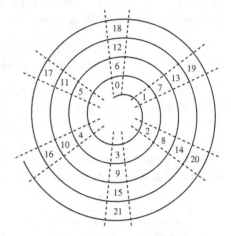

If you look at the 1 position perhaps you can see that the numbers jump by 6: $1 \longrightarrow 7 \longrightarrow 13 \longrightarrow 19 \longrightarrow \cdots$. This jumping by 6 happens in each position on the clock. This makes sense as we're "wrapping" an ordinary number line around a circle — each time it goes round it wraps six numbers around, so we will get back to the same place six later. What we've informally found is this: *Two numbers live in the same position provided they differ by a multiple of* 6.

Now let's try and express that in formal mathematics. The first step is to refer to our two numbers as something, say, a and b. As with the locations in the previous section, we leave open the possibility that a and b are actually the same number. Of course, if they're actually the same number then they obviously are going to live in the same position on the clock, and hopefully this will still make sense when we've put our conditions on.

Next we want to see if they "differ by a multiple of 6", which is a condition on their difference $b - a$. Like with distance we're really thinking about the gap and ignoring whether it's positive or negative, so we could also say $a - b$ but for various (slightly subtle) reasons mathematicians prefer using $b - a$.

Now how are we going to say "is a multiple of 6"? Well, we don't know what multiple it's going to be, so we could use another random letter, say k, to represent a whole number, and then a multiple of 6 is anything of the form $6k$. However this extra letter is a bit annoying, so mathematicians have invented this notation instead: $6 \mid m$ means "6 divides m", or "6 goes into m", which is the same as saying m is a multiple of 6, just the other way round.

So we have the following more formal characterization:

> *a and b live in the same position provided* $6 \mid b - a$.

We have been referring to this as a and b "living in the same position" on the 6-hour clock" but in mathematics this is called "congruent modulo 6" or "congruent mod 6" for short, and we write it like this: $a \equiv b$ (mod 6).

The idea is that we're using notation that is reminiscent of an equals sign, because this is in fact a bit like an equation, it's just an equation in a different world — the world of the 6-hour clock. This is technically called the "integers modulo 6" and is written $\mathbb{Z}_6$. We can now try writing out an entire definition of how congruence works in $\mathbb{Z}_6$. One more piece of notation will help us: we write "$a \in \mathbb{Z}$" as a shorthand for "a is an element of $\mathbb{Z}$", that is, a is an integer. Then we have this:

Definition 6.1 Let $a, b \in \mathbb{Z}$. Then $a \equiv b$ (mod 6) if and only if $6 \mid b - a$.

You might notice that I've used the words "if and only if" rather than the symbol $\Longleftrightarrow$ here, despite me having said that I like the clarity of the symbols. The difference is a subtle point: here we're making a definition of the thing on the left via the thing on the right, rather than declaring two pre-existing things to be logically equivalent. This feels slightly different and is less symmetric as a statement, and under those circumstances I prefer the words to the symbol. These are often the sorts of subtleties that mathematicians don't say out loud but just *do* and hope that people will pick them up.

One of the things that I think becomes clearer by this formulation (although it took us a while to get here) is how to generalize this to other clocks, that is, to the integers modulo n for other values of n; all we have to do is replace every instance of 6 with n. Here n can be any positive integer, and I might write the positive integers[†] as $\mathbb{Z}_+$.

Definition 6.2 Let $n \in \mathbb{Z}_+$ and $a, b \in \mathbb{Z}$. Then $a \equiv b$ (mod n) iff $n \mid b - a$.

We'll now do a small proof to see what it looks like when we use this formalism. It's good to try and do it yourself first even if you don't succeed.

[†] Some people refer to the positive integers as the natural numbers, written $\mathbb{N}$, but there is disagreement about whether the natural numbers include 0 or not. While some people have a very strong opinion about this I think it's futile to insist on one way or the other as everyone disagrees. I think there are valid reasons for both ways and so the important thing is to be clear which one you're using at any given moment. Some people like worrying about things like this; I'd rather save my worry for social inequality, climate change and COVID-19.

┌──────────────── **Things To Think About** ────────────────┐

T 6.2   Can you show that if $a \equiv b$ and $b \equiv c$ then $a \equiv c$ (mod n)?

└───┘

To work out how to prove something it is often helpful to think about the intuitive idea first, before turning that idea into something rigorous that we can write down formally. The idea here is that if the gap between a and b is a multiple of n and the gap between b and c is also a multiple of n, then the gap between a and c is essentially the sum of those two gaps, so is still a multiple of n. Nuances of what happens when c is actually in between are taken care of by the formal definition of "gap". This is the informal idea behind this formal proof. Proofs are not always in point form but I think this one is clearer like that. Note that the box at the end means "the end".

Proof   We assume $a \equiv b$ and $b \equiv c$, and aim to show that $a \equiv c$ (mod n).

- Let $a \equiv b$ (mod n), so $n \mid b - a$.
- Let $b \equiv c$ (mod n), so $n \mid c - b$.
- Now if $n \mid x$ and $n \mid y$ then $n \mid x + y$.
- Thus it follows that $n \mid (c - b) + (b - a)$, which is $c - a$, so $n \mid c - a$.
- That is $a \equiv c$ as required.   □

┌──────────────── **Things To Think About** ────────────────┐

T 6.3   On our n-hour clock we just drew the numbers 0 to $n - 1$. How do we know that every integer is congruent to exactly one of these?

└───┘

Here is the formal statement of that idea. (A lemma is a small result that we prove, smaller than a proposition, which is in turn smaller than a theorem.)

Lemma 6.3   *Given $a \in \mathbb{Z}$, there exists b with $0 \leq b < n$ and $a \equiv b$ (mod n).*

The idea is that we are "wrapping" the integers round and round the clock. Every multiple of n lands on 0 and then the next $n - 1$ integers are a "leftover" part, which wraps around the numbers 1 to $n - 1$. This is behind the idea of "division with remainder", where you try and divide an integer by n, and then if it doesn't go exactly you just say what is left over at the end instead of making a fraction. This sounds complicated but is really what children do *before* they hear about fractions, and it's what we do any time we're dealing with indivisible objects like chairs: you put the leftovers to one side.

I think this is an example where the idea is very clear but the technicality of writing down the proof is a little tedious and unilluminating, so I'm not going to do it. But it's worthwhile to try it for yourself to have a go at working with formality; at the same time it's not a big deal if you don't feel like it.

Often when we get comfortable with abstract ideas we work at the level of ideas rather than the level of details. This sounds unrigorous but we can put in technical work early on to make that level rigorous. This is what category theory is often about, which is one of the reasons I love it so much: I feel like I'm spending much more time working at the level of ideas, which is my favorite level, rather than the level of technicalities.

This is a slightly hazy way in which mathematicians working at a higher level might seem unrigorous to those who are not so advanced, as they are more comfortable taking large steps in a proof knowing that they *could* fill them in rigorously. It's like a tall person being able to deal with stepping stones that are much further apart.

6.6 Example: lattices of factors

In Chapter 5 we drew some diagrams of lattices of factors of a particular number, using arrow notation for the factor relationships. The arrow notation is a lot like the vertical bar notation we used in the previous section, it's just a bit more general, a bit more flexible, and a bit more geometric. In the previous section we wrote a divides b as $a \mid b$ as that's how it's typically done if we're not thinking about category theory. Writing it $a \longrightarrow b$ instead is only superficially visually different from the vertical bar version, but that visual difference opens the door to drawing geometric diagrams, which would have been impossible or impractical and confusing using the vertical bar.

Different formal presentations are possible even when we're using the same kind of abstraction, and in this sense although formality and abstraction are related, they're not quite the same issue. We are going to see that category theory does both of these things in a productive way:

- it comes up with a really enlightening form of abstraction using the idea of abstract relationships, and also
- it comes up with a really enlightening formal presentation of that abstraction, using diagrams of arrows.

To warm up into the first part of this we're going to spend the next chapter investigating a particular type of relationship that will lead us into the definition of category in the chapter after that. From this point on we are going to start using the more formal notation and language that we've discussed in this chapter, so you might find yourself needing to go more slowly if you're unfamiliar with that kind of formality. Mathematics really does depend on that formality and so I think we need it in order to get further into category theory than just skimming the surface.

7

Equivalence relations

A first look at relations more abstractly. Here we will look at properties some types of relationship have, and observe that not many things satisfy those properties. This is to motivate the more relaxed axioms for a category. We will now start using more formal notation.

Category theory is based on the idea of relationships. We build contexts in which to study things. We build them out of objects, and relationships between objects.

Although this is quite general, we do still have to be a *bit* specific about the relationships, because if we allow any old type of relationship we might get complete chaos. So we're going to put a few mild conditions on the type of relationship we study. In this chapter we're going to explore the idea of putting conditions on relationships, and then in the next chapter we'll look at what conditions we actually use in category theory.

In math, and in life, we often start out with conditions that are rather stringent, either to be on the safe side or because we're conservative, or unimaginative, or exclusive. For example, at some egregious point in history, marriage had to be between two people of the same race. Now that condition has mostly been dropped, and in some places we've agreed that the two people don't have to be different genders any more. I believe that thinking clearly, from first principles, leads to less reliance on fearful boundaries, and greater inclusivity.

In this chapter we're going to explore a type of relationship that has rather stringent conditions on it, before seeing how we can relax the conditions but still retain organized structure when we define categories. The idea is to do with generalizing the notion of equality.

7.1 Exploring equality

If we are going to use a more relaxed and inclusive notion of equality, what sorts of things should we look for that won't make our arguments and deductions fall apart too much?

Let's think about how we use equality in building arguments. One of the key ways we use it is in building a long
string of small steps as shown on the right.

$$a = b$$
$$b = c$$
$$c = d$$
$$d = e$$
$$\text{so } a = e$$

Suppose now we say that one person is "about the same
age as" another if they were born within a month. We could
write this as $A \overset{\text{month}}{\approx} B$ if the people involved are our usual
"person A" and "person B". Now we can try making a long
string of these, as shown. But we get a problem.

$$A \overset{\text{month}}{\approx} B$$
$$B \overset{\text{month}}{\approx} C$$
$$C \overset{\text{month}}{\approx} D$$
$$D \overset{\text{month}}{\approx} E$$

Here's the issue: can we now deduce that $A \overset{\text{month}}{\approx} E$? We can't. It *might* be true,
but it doesn't follow by logic, because A and E could be almost four months
apart. If we keep going with enough people, we could get the people at the far
ends to be any number of years apart. That makes for an unwieldy situation we
might not want to deal with.

──────────────── **Things To Think About** ────────────────

T 7.1   Can you think of a way of saying two numbers are "more or less the
same" which avoids this problem? Does it result in some other problems? We
will come back to this in Section 7.7.

Another property of equality we habitually use is that it doesn't matter which
way round we write it: $a = b$ and $b = a$ mean the same thing. This is not true
for inequalities like $a < b$, so manipulating those takes more concentration as
we really have to pay attention to which side is which.

For example if $a = b$ then we can immediately deduce that $a - b = 0$ and
also $b - a = 0$. However, if instead we start with $a < b$ then we get $a - b < 0$ but
$b - a > 0$. It takes some thought to get those the right way round.

Finally it might never have occurred to you to think about the fact that everything equals itself. But if anything failed to equal itself then all sorts of weird
things would happen.[†]

7.2   The idea of abstract relations

We're now going to look at how we express conditions on a relationship without knowing what the relationship is yet. This involves going a level more

───────────────────────

[†] In fact everything would probably equal 0 in which case everything *would* equal itself, we'd
have a contradiction, and the world would implode.

abstract, not just on the objects (which we might call a, b, c or A, B, C) but on the relationship itself.

If we're talking about numbers we might think about inequality relationships such as $1 < 5$ and $6 < 100$. We might then refer abstractly to $a < b$, without specifying particular numbers a and b. When we take it a step further we might be thinking about various different relationships, like $a < b$ or $a = b$, and want to refer to a general relationship without specifying the relationship or the numbers. We might write it as "$a R b$" or $a \,\square\, b$. I like the R because it is asymmetrical, reminding us that the left- and the right-hand sides might be playing different roles here, as with $<$. But I like the notation with the box because it reminds us that we can put something in the box. So I will make the box asymmetrical like this: $a \,\square\, b$.

Technically we could call this a *binary* relation because it's a relation that involves only two things at a time. But I will typically say "relation" and understand it to be binary; if we ever considered relations between more than two objects we could make that clear at that point.

Then the idea is we start with some world that we're going to study, say the world of people, or numbers, or days of the week, or shapes. In the above example about people's ages I sort of implicitly only considered applying the relationship to people. When we're talking about abstract relationships in math we should explicitly declare what world we're considering, so that we don't try and consider the relationship beyond the context intended. So we might say: consider the set of real numbers, and the relation $\square$ defined on it by

$$a \,\square\, b \quad \text{whenever} \quad a < b$$

It's important to note that "$a \,\square\, b$" is a statement (with a verb in it) which might turn out to be true or might turn out to be false. This is very different from when we put a binary operation in between two things. For example $a + b$ is not a statement — it has no verb. If a and b are numbers then $a + b$ is just another number.

We are now going to examine what properties relations might have. In all of what follows we start by fixing a world, that is, a set of objects S, and a relation $\square$ that we are considering in that world.

7.3 Reflexivity

A relation $\square$ is called *reflexive* if it sees everything to be related to itself. That is, every object a satisfies $a \,\square\, a$. It's important to be clear that this is a property of the relation $\square$, not a property of any particular object a. The condition of reflexivity only holds for the relation $\square$ if the condition $a \, \square \, a$ holds for all objects a in S.

At this point I want to introduce another piece of notation because I'm already bored of typing "for all". Mathematicians use this upside-down A to mean "for all": $\forall$. We will also continue to use the symbol $\in$ to indicate that something is in a set. So here is the full definition of reflexivity:

Definition 7.1 A relation $\square$ on a set S is called *reflexive* if

$$\forall \, a \in S, \, a \, \square \, a.$$

Translation: "for all a in S, a is related to a".

Example 7.2 Let $S = \mathbb{R}$ (the real numbers) and let $\square$ be the relation $=$. Now, given $a \in \mathbb{R}$ certainly $a = a$, so $\square$ is indeed reflexive.

Note that when we're proving "for all" statements we don't have to sit and test the conditions on every single object in turn (which would not be possible in an infinite set anyway). What we can do instead is pick one arbitrary member of the set (so we use some sort of letter to represent it) and then test the conditions only using properties that are known to be shared by every member of the set. So if it's the set of all numbers, we pick a random one and make sure we don't use anything about it except that it's a number. If it's the set of all even numbers we can use the fact that it's divisible by 2. It might look like we've only tested one number, but this is one of the powerful aspects of abstraction and why we "turn numbers into letters" — we've only tested one number abstractly, but effectively we've tested every example.

─────── **Things To Think About** ───────
T 7.2 Is the relation $<$ on $\mathbb{R}$ reflexive?

Let $a \in \mathbb{R}$. Then it is *not* true that $a < a$, so $\square$ is *not* reflexive. This is an extreme example of non-reflexivity, in which *no* object is related to itself. However, in general in order to fail reflexivity we don't have to be that extreme. A "for all" statement fails even if it only fails in one place; it doesn't have to fail everywhere. The next example is less "extreme".

─────── **Things To Think About** ───────
T 7.3 Let $S =$ the set of all people in the world, and let $\square$ be the relation "is in love with". Is $\square$ reflexive?

The question we have to ask ourselves is: if we take any random person a in the world, are they necessarily in love with themselves? Well, we can't tell for sure. Some people are in love with themselves but some people aren't. This means that $\square$ is *not* reflexive because there exists at least one $a \in S$ such that $a \not\square a$. I'll use this notation of a slanted line through the box to mean that the relation does not hold in this case. Mathematicians also use this backwards E to say "there exists": ∃. "For all" and "there exists" are called *quantifiers* in logic, because they quantify (specify) the scope of a logical statement, that is, they tell us something about where the statement holds.

Anyway we can formally write down what we've found like this:

$$\exists\, a \in S \ \text{ such that } \ a \not\square a$$

Translation: "there exists a in S such that a is not related to a".

This shows that this particular $\square$ is not reflexive. If there's one case that contradicts a "for all" statement it's called a *counterexample*. There might well be more than one counterexample, but you only need one in order to show that a "for all" statement isn't true.

─────────────── **Things To Think About** ───────────────

T 7.4　Are these relations reflexive? In each case take the set S to be the set of all people in the world.

1. $\square$ = "is the same height as"
2. $\square$ = "is taller than"
3. $\square$ = "is at school with"

If it confuses you how to use "is the same height as" as a relation, imagine putting it inside the box $\square$ everywhere there is a box in the definition. To investigate reflexivity we need to see if this is true:

$$\forall a \in S \quad a \,\boxed{\text{is the same height as}}\, a$$

Translation: "for all people a in the world, a is the same height as a".

This is true, so the relation is indeed reflexive.

For "is taller than" we need to see if this is true:

$$\forall a \in S \quad a \,\boxed{\text{is taller than}}\, a$$

This is another "extreme" failure: given any person a, a is definitely not taller than a (you can't be taller than yourself), so "is taller than" is not reflexive.

The last example, "is at school with" is more subtle. We need to check this:

$$\forall a \in S \quad a \,\boxed{\text{is at school with}}\, a$$

Now, if a person a is at school at all, then they are at school with a (they are at school with themselves). However, some people aren't at school. In math we usually like being completely clear by picking a specific example: say me. I am a person, and I am not at school, therefore I am not at school with myself. I am a counterexample showing that this relation is not reflexive.

We'll now go on to a slightly more complicated condition on relations.

7.4 Symmetry

A relation is called *symmetric* if it's "the same both ways round", that is, it doesn't matter which way round you write it. That doesn't mean it's always true both ways round, it just means that its truth doesn't change if you write it the other way. Here's the formal definition.

Definition 7.3 A relation $\square$ on a set S is called *symmetric* if

$$\forall\, a, b, \in S, \text{ if } a\,\square\,b \text{ then } b\,\square\,a$$

> Translation: "for all a and b in S, if a is related to b then b is related to a".

Note that the definition says $\qquad$ $a\,\square\,b \implies b\,\square\,a$

but if we switch the roles of a and b we get $\quad b\,\square\,a \implies a\,\square\,b$

and so putting those together we have $\qquad a\,\square\,b \iff b\,\square\,a$

So for a symmetric relation the statements $a\,\square\,b$ and $b\,\square\,a$ are logically equivalent. The "symmetry" here is the fact that if we switch the positions of a and b then the statement doesn't change, logically speaking; it just looks different.

Let's try some of the same examples as before.

Example 7.4 Let $S = \mathbb{R}$ and let $\square$ be the relation $=$. Now let $a, b \in \mathbb{R}$. If $a = b$ then it is also true that $b = a$, so this relation is symmetric.

Note that we are not saying a actually is equal to b, we are *supposing* $a = b$ and seeing what we could deduce *if* that were true.

Example 7.5 Let $S = \mathbb{R}$ and let $\square$ be the relation $<$. Now let $a, b \in \mathbb{R}$. If $a < b$ then it is definitely not true that $b < a$, so this relation is not symmetric.

Note that we have again got ourselves a situation where we don't just have one counterexample, we have something that is *always* false, an "extreme" non-symmetric relation.[†] Here is an "in between" example.

Example 7.6 Let $S = \mathbb{Z}$ and let $\square$ be the relation "divides" (and recall

---

[†] In fact these are called "anti-symmetric".

that we write this as a vertical line). Now let $a, b \in \mathbb{Z}$. If $a \mid b$ then it is *possible*
that $b \mid a$ (if they're equal) but it is not *necessarily* true: for example $2 \mid 4$ but
$4 \nmid 2$.[†] So the relation is not symmetric.

In that example, there's a special case: when $a = b$ we do have $a \mid b$ and
$b \mid a$. It might be tempting to think the relation "sometimes symmetric and
sometimes not" but this is not how a "for all" condition works in math. If
something is "sometimes true and sometimes not" then it's not *always* true, so
the "for all" statement is false.

Things To Think About

T 7.5   Are these relations symmetric?   1. ☐ = "is the same height as"
In each case take the set S to be the set   2. ☐ = "is the mother of"
of people in the world.   3. ☐ = "is a sister of"

In each case we need to check that the relevant thing is true *for all* a and b. We
can start by letting a and b be any people, and then we put the given relation
in the box ☐ every time it appears in the definition of symmetry, and see
whether the condition is necessarily true for all people.

The first relation is symmetric: if a is the same height as b then b must be
the same height as a.

However the second one is not symmetric: if a is the mother of b then b
definitely can't be the mother of a. Again, it has failed in an extreme way —
it's not just that it's *sometimes* false the other way round, it's that it's *always*
false the other way round.

The last one is another "in between" case, though it might seem symmetric
at first sight. However if a is a sister of b then b *might* be a sister of a but might
be a brother. This is perhaps a bit sneaky, but it means that this relation is not
symmetric. If we said "sibling" then it would in fact be symmetric.

Note that reflexivity involved considering one object at a time, and symmetry involved considering two. Our last condition involves considering three.

7.5 Transitivity

Our last condition is about whether we can $a = b$
build up long chains of relations or not, a bit $b = c$
like when we built up long chains of equali- $c = d$
ties like the one shown on the right. $d = e$
                                                      so $a = e$

[†]   We generally put a slanted line like this through things to express the fact that they're not true.

However, we only need to demand the condition for chains of length 2 (involving 3 objects), because from there we can build up longer ones if we want.

Definition 7.7   A relation $\square$ on a set S is called *transitive* if

$$\forall a,b,c \in S: \quad \left\{\; a\,\square\, b \text{ and } b\,\square\, c \;\right\} \implies a\,\square\, c.$$

> Translation: "for all a,b,c in S,
> if a is related to b and b is related to c then a is related to c".

I've put those big curly brackets in to try and make it a bit clearer where the logic is, because it can be a bit confusing. Perhaps it would be clearer to do it more geometrically

like this: or maybe like this:

$a\,\square\, b$ which I hope reminds you of this:
$\quad b\,\square\, c$ $a\,\square\, b\,\square\, c$
$a\,\overline{\square}\, c$ $a\,\overline{\square}\, c$ $a \longrightarrow b \longrightarrow c$

Let's try some examples, starting with our usual motivating one, equality.

Example 7.8   Let $S = \mathbb{R}$ and let $\square$ be the relation $=$. Now let $a,b,c \in \mathbb{R}$. If $a = b$ and $b = c$ then it follows that $a = c$, so this relation is transitive.

Again, remember we're not saying that a, b and c actually *are* equal, we are just supposing that $a = b$ and also that $b = c$ to see what we can deduce.

Example 7.9   Let $S = \mathbb{R}$ and let $\square$ be the relation $<$. Now let $a,b,c \in \mathbb{R}$. If $a < b$ and $b < c$ then it follows that $a < c$, so this relation is transitive.

Now let's try some non-examples.

Example 7.10   If a is the mother of b and b is the mother of c then a is definitely *not* the mother of c, so the relation "is the mother of" is not transitive.

That was the extreme failure, where the outcome can never be true. Here's a subtly "in between" one.

—————— **Things To Think About** ——————

T 7.6   Suppose a, b, c are people, and a is a sister of b and b is a sister of c. Can you work out a slightly sneaky way in which a might not be a sister of c? Remember that a, b and c aren't necessarily *different* people.

Here is a counterexample. Consider this a, b and c:

a	=	me	So $a \,\square\, b$
b	=	my sister	and $b \,\square\, c$
c	=	me	but $a \,\not\!\square\, c$ as I am not my own sister.

This might make you feel a bit tricked, and I admit it is a little sneaky. But it's logically completely sound, and this is the sort of thing we need to be very careful about in math. In some cases the relation itself precludes the possibility of a and c being the same person, once we've declared that $a \,\square\, b$ and $b \,\square\, c$ (for example if the relation is "is the mother of") but that's because of the definition of $\square$ in that particular case.

Note that where reflexivity involves only one object at a time, symmetry involves two objects at a time, and transitivity involves three. And let me just re-emphasize that these are not necessarily three *distinct* objects.

┌─────────────────── **Things To Think About** ───────────────────┐

T 7.7   Are these relations transitive? In   1. $\square$ = "is the same height as"
each case take the set S to be the set of   2. $\square$ = "is married to"
people in the world.   3. $\square$ = "is in a choir with"

└──┘

In each case we start by letting a, b and c be any three people (not necessarily distinct), and then we put the given relation in the box $\square$ every time it appears in the definition of transitivity, and see whether the statement is necessarily true for *all* people.

If a is the same height as b and b is the same height as c then yes, a is necessarily the same height as c, so this relation is transitive.

Now suppose a is married to b and b is married to c. According to the laws of most countries, the only way for this to be legal is if a and c are the same person, because otherwise b would be committing bigamy. In any case it is certainly *possible* for $a \,\square\, c$ to fail: take a and b to be any married couple, and then take $c = a$. Then a is not married to c.[†]

The third example shows us what would have happened in the previous example if bigamy were allowed, because it is in fact possible to be in a choir with more than one person (in fact I think by definition it's probably not a choir otherwise). However a could be in a choir with b, and b could be in a different choir with c, and so it doesn't necessarily follow that a and c are in a choir together. So this fails transitivity for a slightly different reason.

[†] Actually the very precise mathematician brain in me observes that some people have started marrying themselves as a radical act of self-love (or something) but there are plenty of people who haven't, whom we can pick for our counterexample.

Note that we can build up longer strings of relations using transitivity even though we only
specified it for sticking two relations together.
For example if we have the chain of individual
relations on the right, we can use transitivity one
step at a time to build up to $a \,\square\, e$.

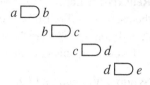

This is why we don't need to specify conditions on any more than two steps of
the relation — that is, three objects — at a time.

7.6 Equivalence relations

We have looked at three conditions on relations that make them well-behaved
in a similar way to the familiar relation =. If a type of relation satisfies all three
then it is particularly well-behaved, and we name it in the following definition.

Definition 7.11   If a relation $\square$ on a set S is reflexive, symmetric and transitive then we call it an *equivalence relation*.

Things To Think About

T 7.8   Are the following relations
equivalence relations? Here again S
is the set of all people in the world.

1. $\square$ = "has the same birthday as"
2. $\square$ = "is older than"
3. $\square$ = "is a friend of"

You might wonder whether I mean "same birthday" as in the day of the year,
or including the actual year of birth as well. In fact it doesn't matter, as long
as you're consistent: as long as you pick one version and stick to it, this is an
equivalence relation.

The big clue here is the words "the same". This indicates that the relation is
based on =, it's just that we've used a process to turn our objects into something
where = makes sense in an unambiguous way. Once we've done it, we know
that = is an equivalence relation, and so our relation will also be one. It's a
form of transferring structure that we'll see more of later.

By contrast "is older than" is not an equivalence relation. It is not reflexive,
as nobody is older than themselves, and it is not symmetric, because if a is
older than b then it's definitely not true the other way round. But it is transitive.

For the case of friendship things become a bit more philosophical, or perhaps, depressing.

Reflexivity: Let *a* be any person. Is it necessarily true that *a* is friends with themselves? I don't think so. Many people are not kind to themselves. Although some people are friends with themselves, there exist people who aren't, so I would say this relation is not reflexive.

Symmetry: Let *a* and *b* be any people. If *a* is a friend of *b*, is it true that *b* is a friend of *a*? I would say not: *a* might be a very good friend to *b* who treats them horribly in return.

Transitivity: This is probably the most obvious and least depressing. If *a* is a friend of *b* and *b* is a friend of *c* then *a* and *c* might not even know each other, so they certainly need not be friends, so this is not transitive.

This is obviously not a rigorous mathematical situation and is somewhat down to interpretation. However, the unambiguous part is how we translate the definitions of reflexive, symmetric and transitive into statements we can check in the real life situations; the ambiguity arises as the answers might not be clear-cut. Here is an example that isn't clear-cut but for different reasons.

─────────────── **Things To Think About** ───────────────

T 7.9   This time let *S* be the set of locations and ▭ be the relation "is east of". Is this reflexive, symmetric, transitive?

Here it really depends whether we allow going all the way around the world as an option. Usually we'd think that if *a* is east of *b* then *b* has to be west of *a* and can't be east of it, but if we're on a globe perhaps everything is east of everything else? Unfortunately we often use rather silly terminology left over from an ignominious Eurocentric past in which there was "The East" and "The West", as if we still think the earth is flat.

Of course, if we did count going all the way round the world then everything would be east of everything, and this would reduce to being a very uninteresting relation in which everything is related to everything else. Instead we could restrict our "world" *S* somewhat, instead of considering all locations everywhere. We could restrict it to, say, one particular city, in which case it will be transitive but not reflexive or symmetric.

This shows that relations take on different characteristics depending on what context we're thinking about. Sometimes if something is behaving in an unwieldy way we put more conditions on it, but sometimes we can restrict our context instead. This is one of the many reasons it's important to be clear what context we're in at any point.

7.7 Examples from math

Let's look at some more mathematical examples now.

Things To Think About

T 7.10 Are the following relations equivalence relations? Here S is $\mathbb{Z}$, the set of integers.

1. $\leq$ ("less than or equal to")
2. divides
3. is congruent mod n (for some fixed n)

The first one $\leq$ is crucially different from just $<$ because of the "or equals" part. This means that it is now reflexive, because $a \leq a$. However this does not make it symmetric: for example $1 \leq 2$ but it is not true the other way round. The relation is transitive.

The relation "divides" is reflexive since everything divides itself, and we've already seen that it isn't symmetric. It is transitive because if $a \mid b$ and $b \mid c$ then it definitely follows that $a \mid c$. (You could try proving that, starting by making a formal definition of "divides".)

For congruence, remember that two integers a and b are called "congruent mod n" if $n \mid b - a$. I hope you *feel* that this relation ought to be an equivalence relation as we're talking about things being "the same" on an n-hour clock. A feeling isn't enough, but I think it's an important start.

Reflexivity: Given $a \in \mathbb{Z}$, $n \mid a - a$ as this is 0. So the relation is reflexive.

Symmetry: We need to show that if $a \equiv b$ then $b \equiv a$. Well, if $n \mid b - a$ then $n \mid a - b$ so this is also true.

Transitivity: We proved that this is true in Section 6.5.

So far the only equivalence relations we've seen have involved some sort of sameness. It's worth wondering how far we can push that.

Things To Think About

T 7.11 Can you make a definition of "more or less the same as" that is an equivalence relation? Why would it cause problems if it's not?

Earlier on we saw a version of "more or less the same as" that is not transitive, when we looked at people born within a month of each other. Although we did it for people and ages, that is essentially the same as for numbers. We could make "more or less the same as" mean "within ε of each other"[†] for any value of $\varepsilon > 0$ and it still wouldn't be transitive, no matter how small we made ε. (It would, however, be reflexive and symmetric.)

[†] Mathematicians traditionally use the Greek letter ε (epsilon) to represent unspecified small positive numbers.

One way of saying "more or less the same as" that is transitive is to round everything to the nearest whole number, or to a certain number of decimal places. This is how we take approximations of numbers, especially in physics, other experimental sciences, and in real life.

However, this results in other problems with things like 1.49 rounding to 1 but 1.5 rounding to 2 despite being almost the same as 1.49. It also means you can add two things that round to 0 and get something that does not round to 0, like when you somehow get full from eating a large quantity of negligible bites of something. We sometimes have to pick which problem to live with though, and usually the problem of anomalies with rounding is a bit irritating, where the problem of non-transitivity is catastrophic.

7.8 Interesting failures

There is plenty more to say about equivalence relations, but the last thing I want to draw attention to is how few things actually turn out to be equivalence relations. Moreover, many of the things that fail to be equivalence relations are mathematically interesting concepts that we may well still want to study. In math (and in life) if something doesn't fit our rigid constraints we can either continue to exclude them, or relax our constraints to let them in. In both math and life I believe in recognizing when our constraints have been unreasonably rigid, excluding things and people just because of preconceptions, out-dated practices, or fear. I believe in inclusivity in math and in life.

Abstract math has a reputation of being rigid and rule-based. It is true that any given context has its rules, but flexibility comes from continually finding ways to shift context or expand contexts in order to include concepts that previously didn't fit.

Equivalence relations are particularly well-behaved relations. But they are so well-behaved that they are, in a way, not that interesting. In fact all the ones we've seen are just some version of sameness where we map our world into a simpler one and apply sameness there, like having the same birthday, being the same height, being the same on an n-hour clock. Equivalence relations turn out to be logically the same as "partitions", where you just partition the world into sections and call two objects related if they're in the same section. Many interesting relations don't work like that, and in category theory we don't want to just exclude them. So we relax our conditions instead. This is what we're going to do in the next chapter.

8

Categories: the definition

We are finally ready for the formal definition of a category, but we'll discuss each part of the definition first.

We're going to define a framework for thinking about relationships. In order to make sure we are thinking about *slightly* reasonable types of relationships, we will stipulate that they at least behave slightly well, so that we can do various things with them. These stipulations will be inspired by some of the things we've already seen, but will be less stringent than the stipulations on equivalence relations. We will talk through all the parts of the definition first, and then present it formally in Section 8.4.

We are building up to the definition of a category, and we'll call it $\mathcal{C}$ (a curly C). Just like when we talk about person A or a number x this is not a particular category, but a general category that we have named with a letter for convenience.

Our definition takes a typical form for abstract mathematical structures, in three parts:

- Data: basic building blocks.
- Structure: things we can do with those building blocks.
- Properties: rules about how those things must fit together.

8.1 Data: objects and relationships

Our basic data in a category are objects and relationships. The objects can be anything, we just have to be clear what we're including. This is a bit like when we defined relations starting with the "world" S of objects we were going to consider. For a category $\mathcal{C}$ we might call the world of objects ob $\mathcal{C}$. In a particular category these could be numbers, people, shapes, mathematical structures of a particular kind, points in a space — really anything. They could also be just unspecified abstract objects. We'll look at plenty of examples soon.

The relationships might be things like "less than" or "is the mother of", but they also might not be anything like relationships, so we more generally call them arrows, or "morphisms", inspired by examples where they transform or "morph" one thing into another. (However in this book we might still informally refer to them as relationships). We draw them like this: $a \rightarrow b$, as we have done for many relationships already in this book. As the notation indicates, each arrow in a category starts at an object and ends at an object. The beginning object is called the "source" and the end object is called the "target". Sometimes these are called domain and codomain, but personally I find source and target much more vivid; in the spur of the moment I can't actually remember which is which out of the domain and codomain.

Unlike the relations we looked at in the previous chapter, the ones in a category aren't just true or false. Given two objects a and b, we're not just asking if they're related or not; we ask a more nuanced question (as we should in life as well) which is: *in what ways* are these objects related?

This means that there could be more than one arrow from a to b and we need to give them names, perhaps like this.

$$a \underset{g}{\overset{f}{\rightrightarrows}} b$$

When there's more than one arrow with the same source and target, as above, we call them *parallel*, because they look parallel.

Another way to have multiple arrows between the same objects is pointing in opposite directions like this.

$$a \overset{f}{\underset{g}{\rightleftarrows}} b$$

There is no limit to the number of arrows going from any object to another. There could be infinitely many and there could be none.

We often focus on one pair of objects a and b and write the collection of arrows from a to b as $\mathcal{C}(a, b)$. Some people[†] write it as $\hom_{\mathcal{C}}(a, b)$, which comes from the idea of "homomorphisms" which we'll see quite a bit of later on. The set of morphisms $\mathcal{C}(a, b)$ is sometimes called a "homset". Homsets are directional: if an arrow starts at b and ends at a then it will be in $\mathcal{C}(b, a)$ and not in $\mathcal{C}(a, b)$.

8.2 Structure: things we can do with the data

Next we declare that there are a couple of things we can definitely do with these relationships. The first is a bit like reflexivity from the previous chapter: we want to make sure every object is related to itself. Note that we haven't stipulated a particular type of relationship so this can be quite a weak condition. We're not saying "everything is less than itself" or "everyone is a friend of

---

† I'm mostly too lazy to write all those extra characters.

themselves" we're just observing that everything *is the same as* itself and that this most basic relationship should be encapsulated in every category.

So we say that given any object a there must be at least an arrow like this, called an *identity*.

$$a \xrightarrow{\;1_a\;} a$$

We often write it as 1_a, hence the label. We could also draw it going in a little loop, like this.

We often write it as 1_a, hence the label. We could also draw it going in a little loop, like this.

If we draw it the first way, as a straight line, we're not saying that there are two objects called a, we're specifying that the source and target of this particular arrow are both a. Incidentally, when we draw arrows their shape doesn't make any difference to what they are. In a category, the thing that matters is the endpoints of the arrow, not the physical path the arrow takes in between them.

We could draw this arrow in any of these ways and it would be logically the same information, although sometimes it helps our geometrical intuition to draw it different ways up in different situations.

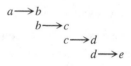

The other thing we want to be able to do with arrows is a bit like transitivity: we want to be able to build up strings of relationships. If we draw this analogously to how we did in the previous chapter it might now come out as shown on the right.

$$a \longrightarrow b$$
$$b \longrightarrow c$$
$$c \longrightarrow d$$
$$d \longrightarrow e$$

However it's a bit tedious to keep writing all those repeated letters, so we could draw it like this.

$$a \longrightarrow b \longrightarrow c \longrightarrow d \longrightarrow e$$

Then we want to get out an arrow like this.

$$a \longrightarrow e$$

But as with transitivity, we only have to specify this for two arrows at a time, the idea being that we can then gradually build up to long strings (subject to some conditions we'll think about in a moment).

So what we actually say is that given two arrows f and g in this "nose-to-tail" configuration

$$a \xrightarrow{\;f\;} b \xrightarrow{\;g\;} c$$

we want to be able to combine them into a single arrow like this:

$$a \longrightarrow c$$

This process of combining arrows is called *composition*. The answer we get out at the end is a new arrow called "f composed with g" and we write it as $g \circ f$ which is a bit annoying because it looks like we've written it backwards. We sort of have, and some people prefer writing it the other way, but the reason I (and many other mathematicians) prefer it this way is that it makes sense for composition of functions. We'll come back to that in the next chapter.

This is similar to transitivity, but different. The relations of the previous chapter were yes/no situations: things just were or weren't related. Now we're in "sense in which" situations, so this is more like saying: if there's a sense in which *a* is related to *b* and a sense in which *b* is related to *c* then we can combine those to get a sense in which *a* is related to *c*. If we think of it as paths then it's like saying "if there's a path from *a* to *b* and a path from *b* to *c* we can do one and then the other to get a path from *a* to *c*".

Example 8.1   We have seen that the relation "is the mother of" is not transitive. But we can *compose* it, and draw the process like this.

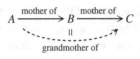

I've drawn the composite as a dotted arrow to remind us that this is the one we added afterwards, deducing it from the others. This is where videos really help with category theory, because the diagrams are often "dynamic" in that you draw part of the diagram first and then you deduce some other part.

Here's another example involving some different family relationships; see if you can practice reading it "dynamically".

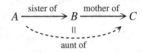

I hope you can now see that the diagrams in Chapter 5 were all secretly diagrams of categorical composition. ("Categorical" means something else in normal life, but in category theory we use it to mean "category theoretical".)

These examples show that composition of arrows is more general than transitivity as it includes examples that transitivity excluded, like "is the mother of". But furthermore, we can apply it to situations where the transitivity question couldn't even be *posed*, like the example producing an aunt.

Composition is also more general than the notion of a binary operation. It does appear to take two inputs (the two arrows we're composing) and produce one output (the composite), but we can't just take *any old* two arrows and produce a composite: they have to be "composable" arrows, that is, where the target of the first one matches the source of the second one. It's a bit like the question of connecting flights, when it appears that you can fly somewhere by changing plane in the middle, but when you click on the details it admits that you also have to switch airport, say from London Heathrow to London Gatwick (which is really quite arduous). We don't want that kind of composite: we want the two paths to *really* meet in the middle.

This makes it a generalization of binary operation. Perhaps it's easier to see that binary operations are a special case of composition, just with nothing really happening at the endpoints. That is, in the special case that all the endpoints are the same, all arrows can be composed so it's just a binary operation.

This is like in Section 5.4 when we saw diagrams like this to think about addition. We were essentially representing numbers as arrows, and composition was the binary operation +.

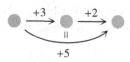

We'll come back to these ideas when we talk about monoids in the Interlude; the main thing to take away now is that composition of arrows is a generalization of the concept of binary operation.

--- **Note on grammar** ---

Note that composition comes with various parts of speech:

- "to compose" (verb): we compose arrows together
- "composition" (noun): the act of composing
- "composite" (noun): the result of composing arrows together is a new arrow called the composite, or composite arrow.

--- **Things To Think About** ---

T 8.1   We introduced the idea of composition with the idea of building up long strings of arrows, but composition only ever refers to two arrows. When we talked about transitivity we saw that we could use two instances of $\square$ to build strings of any length. Can you see why this is harder for composition of arrows? We need another condition, which we'll address in the next section.

8.3 Properties: stipulations on the structure

So far we have our data and our structure. Now we are going to impose some conditions on the structure, the "properties" that will ensure our structure behaves in a most basically sensible way. There is one condition for identities and one for composing more than two arrows. Properties are more formally called *axioms*. They are basic laws governing our structure.

Identities

We said that identities just encapsulate the fact that everything is the same as itself. However so far all we have said is that the source and target are the same. The property we need is a bit like other situations where an element[†]

[†] By "element" I just mean "thing" here.

doesn't *do* anything, and those elements are generally called identities. If I line up these conditions I hope the analogy is clear.

0 is the identity for + on (real) numbers: $\forall a \in \mathbb{R}$ $a + 0 = a$
1 is the identity for × on (real) numbers: $\forall a \in \mathbb{R}$ $a \times 1 = a$

Now instead of addition or multiplication our identity is going to "do nothing" with respect to composition of arrows.

It's a bit like this, where sticking "same as" at the beginning doesn't give us any extra information about the relationship between a and b.

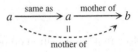

We could stick "same as" at the other end as well, and it still won't make any difference.

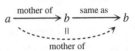

─────────── **Things To Think About** ───────────

T 8.2   See if you can write down the general condition on identity arrows.

For general identity arrows we take the above blueprint but replace "mother of" with a general arrow $a \xrightarrow{f} b$, and replace "same as" with the appropriate identity arrow, giving these two conditions sometimes called *unit laws*.

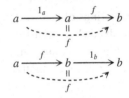

In the first case, the only thing we know in advance is that the composite is $f \circ 1_a$, and in the second case the composite is $1_b \circ f$. So what we are stipulating here is $f \circ 1_a = f = 1_b \circ f$ which shows the analogy with the additive and multiplicative identities from earlier. We have to specify what happens with the identity on either side of f here, because unlike with + and × there isn't any commutativity to rely on.

The expression as a string of symbols makes it look more like the additive and multiplicative identities from before, but drawing arrows helps us see what's going on better, I think. It might be even more visually vivid if I change the geometry a bit, without changing the logical structure.

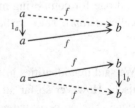

This flexibility of layout is a powerful feature of arrow diagrams, as adjusting the geometry can help us invoke different kinds of intuition.

Composing more than two arrows in a string

Let's go back to family relationships and consider three generations of mothers. There are now
two ways we could compose these
arrows. We could go from A to C
and then include D, or we could go
from B to D and then include A, as
shown in these two diagrams.

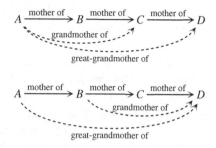

In this situation both ways produce the same relationship between A and D.
This is *associativity* and we will now state it in the general situation.

Suppose we have a string of three composable arrows: $a \xrightarrow{f} b \xrightarrow{g} c \xrightarrow{h} d$.
Then we have the following two ways of composing them:

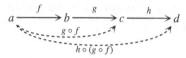   

We want these to be the same, that is, we want
the equation on the right to hold.

$$h \circ (g \circ f) = (h \circ g) \circ f$$

This string of symbols makes it look closely
analogous to associativity of binary operations like addition and multiplication.

$$h + (g + f) = (h + g) + f$$

However it's a little more subtle for arrows because f, g and h have to be
composable arrows, not just any old arrows. The algebraic equation makes it
look familiar, but the arrow diagram gives us a more subtle and also richer
point of view as we'll see shortly.

Note that associativity looks like it's a condition on three composable arrows, and it is, but what we really want to know is that a string of composable
arrows of *any* length has one unambiguous composite.[†] This is a small difference in perspective at this level, but it gets magnified into something quite
serious in higher dimensions. There is an interplay between all the things you
want to be true in the world you're studying, and finding a small tractable set
of conditions you can check, that will ensure all those things are true. There is
then a choice between what you take as the definition and what you take as a
theorem. For associativity the choice is this:

---

† If you know about proof by induction, you could try proving this by induction on the length of
the string; probably the hardest part is writing down precisely what we're trying to prove.

	Approach 1	**Approach 2**
definition:	strings of length 3	strings of length n
theorem:	strings of length n follow	strings of length 3 suffice

In higher levels of math I would say the second type of approach becomes more prevalent. Unfortunately at lower levels the first is more prevalent, which can make things seem rather arbitrary, because the "small tractable set of conditions" that make something broad true can obscure the nature of the broad thing that we really want. It's like the broad thing is the wood (forest) and the tractable set is the trees, and if we focus on the trees we might miss the forest.

8.4 The formal definition

We have talked our way through the definition of category but now we'd better write out the complete definition formally. Drumroll. . .

A *category* $\mathcal{C}$ consists of:

DATA

- Objects: a collection $\mathrm{ob}\,\mathcal{C}$ of objects
- Arrows: for objects $a, b \in \mathrm{ob}\,\mathcal{C}$, a collection $\mathcal{C}(a, b)$ of arrows $a \longrightarrow b$

STRUCTURE

- Identities: for any object $a \in \mathrm{ob}\,\mathcal{C}$ an identity arrow $a \xrightarrow{1_a} a$
- Composition: for any arrows $a \xrightarrow{f} b \xrightarrow{g} c$, a composite arrow $a \xrightarrow{g \circ f} c$

PROPERTIES

- Unit laws: for any arrow $a \xrightarrow{f} b$, $f \circ 1_a = f = 1_b \circ f$
- Associativity: for any arrows $a \xrightarrow{f} b \xrightarrow{g} c \xrightarrow{h} d$, $(h \circ g) \circ f = h \circ (g \circ f)$

Arrows are also called morphisms, and the composite $g \circ f$ is also written gf.

──────────────── **What happened to symmetry?** ────────────────

We talked about how this definition is inspired by equivalence relations, with identities being like reflexivity and composition being like transitivity, so you might be wondering what happened to symmetry. The answer is that we don't *demand* something analogous to symmetry in a category as we want to allow for the possibility of relations that only go one way. Instead of *demanding* this, we will *look for* it. This will be the notion of invertibility, a more subtle notion of sameness in categories, and we will get to that in Chapter 14.

8.5 Size issues

There is a technical issue to do with "size" that we need to be aware of, but we don't need to make a big meal out of it. We just have to make sure we avoid the sort of paradox which arises if you attempt to think about a "set of all sets" or something similarly self-referential. This is the essence of Russell's Paradox, and I've put a few more details in Appendix C in case you want some more explanation. Basically, to avoid getting into a paradox in the foundations of mathematics we have to put a hierarchy of size on collections of things, and in particular the collection of all sets doesn't count as a set: it's a "collection" and is one level up in the hierarchy. This then gives us a concept of different sizes of category, defined as follows:

- If the objects and the morphisms each form a set the category is called *small*.
- If each individual collection $\mathcal{C}(a, b)$ is a set then the category is called *locally small*; in this case the objects might form a collection, not a set.

This becomes more important when one gets into more technical details, but for now if it baffles you I think it's fine to just move on and think about it later if you need to, or have a look in the Appendix if you're interested.

8.6 The geometry of associativity

In the explanation of identities I included a slightly different diagram that brought out some more of the geometry of the situation. I'd like to do the same for associativity now, but it's a bit harder as it's really going to be in three dimensions. The idea to keep in mind all the way through is that with the arrow notation it doesn't matter what physical shape the arrows take, it only matters what the endpoints are. Let's think about the arrow diagrams for associativity again, and this time I've marked in the four regions.

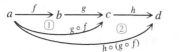   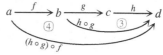

Region 1 has three arrows bordering it, and if we redraw them as straight lines and change the positions a bit, we can see it as a triangle.

Things To Think About

T 8.3 Try and draw out the other three regions as triangles, and label the corners and arrows. If you can, cut them out as actual triangles and then see if you can fit them together as a sort of 3-D jigsaw. Don't forget to use associativity.

The four triangles look like this:

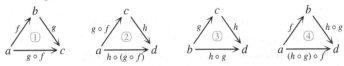

We can then fit them together in a 3-dimensional shape as long as we observe:

- By associativity we know $(h \circ g) \circ f = h \circ (g \circ f)$ so those two edges are allowed to go together.
- Some of the triangles as drawn will face in and some will face out.

This is pretty hard to describe on a 2-dimensional page so I encourage you to cut out some triangles and fiddle around with it a bit. The shape it makes is a tetrahedron, that is, a pyramid with a triangular base, a bit like this.

The geometry of associativity is something that gets studied a lot when we think more about higher dimensions.

─────────────── **Things To Think About** ───────────────

T 8.4   We have a seen progression through dimensions as in this table.

What happens if we try and continue this table up to more dimensions? So at the next level we take 5 objects, with 4 composable arrows, and see what shape we make.

no. of objects	no. of composable arrows	no. of dims	shape
1	0	0	point
2	1	1	line
3	2	2	triangle
4	3	3	tetrahedron

8.7 Drawing helpful diagrams

The diagrams we draw in category theory aren't exactly part of the definition — you can have objects and arrows whether you physically draw them or not. However, drawing diagrams can help by enabling us to invoke our geometric intuition. When we draw the diagrams there are various principles we use to draw helpful, rather than messy, diagrams.

1. The physical positioning of the diagram doesn't affect the logical structure, so we can move things around as long as the arrow connections stay the same. This is key to making geometrically helpful diagrams.

2. We usually don't bother drawing identity arrows because they're always there. Occasionally we draw them if we're trying to emphasize something, for example if two arrows compose to the identity.

3. We omit arrows whose existence we can deduce from composition.

We saw this last point in Chapter 5 where we observed that we don't need to draw relationships that can be deduced, such as in factor diagrams.

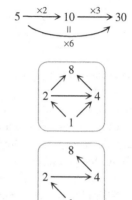

This means, for example, that if we draw a diagram for the factors of 8 we can omit many arrows. It might be tempting to draw something like the diagram on the right.

However, we don't need to draw the arrow $1 \to 4$ because it's the composite of the arrows $1 \to 2 \to 4$; likewise we don't need the arrow $2 \to 8$. This gives us the following diagram.

But now we can improve the geometry: there's no reason for the arrows to zig-zag like that, and we might as well draw them in a straight line as shown. I prefer to draw it going up the page vertically rather than sideways, because in all our lattice diagrams the vertical direction on the page has indicated the increasing number of prime factors at each level.

8.8 The point of composition

There is a philosophy that says composition is the whole point of categories. There are, of course, many philosophies about category theory, and we'll mention various of them along the way. The point of this one is to see what the difference is between a category and just a load of arrows.

The following diagram is a COVID lockdown meme, and is a picture of some points and some arrows, like any kind of flow chart. Ostensibly, it looks a bit like a category because it has objects and arrows. But is there anything illuminating to say about composition here?

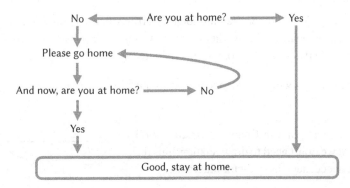

There are several possible paths from "Are you at home?" down to "Good, stay at home", and you could argue that they're not the same as each other, but that's not really the point of the diagram.

Category theory is a lot to do with abstraction and analogies, but the reason composition plays an important role in the definition is that really interesting things happen when composites give rise to interesting information, or give us a way to express interesting information.

When I was growing up we had some family friends who had a situation of family marriages that really intrigued me.

There were two brothers (whom I'll call *A* and *B*) and two sisters (whom I'll call *C* and *D*) and the intriguing thing was that the two brothers married the two sisters. This is perhaps ambiguous linguistically but I hope it's clear what I mean, and will probably be clearer if I draw a diagram.

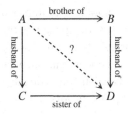

One of the things that really intrigued me about this was the relationship between *A* and *D*. *A* was the brother-in-law of *D*, but in two different ways: he was *D*'s husband's brother, but he was also *D*'s sister's husband. Usually your husband's brother and your sister's husband are two different people, but in this case those two relations had converged on each other.

In the language of category theory the two composite arrows from *A* to *D* are the same.[†]

───────────── **Things To Think About** ─────────────

T 8.5   What other relationships can you think of in this family that would end up being the same, where in other families they're not?

───────────────────────

[†] As this is normal life not abstract mathematics there is a slight ambiguity here that I haven't ironed out. Depending on how we define these arrows we could say that there are two different arrows from *A* to *D* and that *A* is related to *D* in two different ways. Or not.

Diagrams like this where two composites are (perhaps unexpectedly) the same are called *commutative diagrams* in category theory and are very typical of how we deduce things.

Definition 8.2 A *diagram* in a category is a collection of objects and arrows from the category, possibly not all of them. A diagram is said to *commute* if any two composable strings with the same endpoints produce the same composite.

Note that we refer to diagrams in category theory as shapes just based on how many arrows go around the outside.

So if there are four arrows around the outside we call it a square, regardless of the lengths of the sides and the angles — because those don't affect the logic of the diagram anyway. Commutative squares are the classic building block of category theory.

$$
\begin{array}{ccc}
a & \xrightarrow{f} & b \\
{\scriptstyle s}\downarrow & & \downarrow{\scriptstyle g} \\
c & \xrightarrow{t} & d
\end{array}
$$

In that square the only composable strings with the same endpoints are these two paths from a to d.

$$a \xrightarrow{f} b \xrightarrow{g} d$$
$$a \xrightarrow{s} c \xrightarrow{t} d$$

For the diagram to commute those two composites must be equal, that is

$$g \circ f = t \circ s.$$

If you're very used to writing out equations in algebra like that then you might feel more comfortable with it, but in category theory we don't tend to write those equations, we just draw the diagrams and say they commute. This is because as the diagrams get more and more complex it is often a case of fitting parts together like a jigsaw, and the equations as strings of symbols then become pretty incomprehensible.

When I was thinking about our family friends when I was little I did not think I was doing category theory, but sometimes you're already doing category theory even if you don't realize it, just like you're sometimes doing math even if you don't realize it.

These sorts of real life category theory situations are admittedly not exactly rigorous, and they're not really profound category theory either. However, they are a way in which I can clarify my thinking about the world around me by drawing on the techniques of category theory and the way of thinking that it provides. In the next part of the book we'll take an interlude and delve into some more rigorously mathematical examples.

INTERLUDE

A TOUR OF MATH

9

Examples we've already seen, secretly

Some of the real life situations we've already looked at, re-focused through the lens of categories.

"Examples first" is a mantra urging us to introduce examples before theory when teaching math. However, I personally prefer seeing theory before examples, so I've always been a bit ambivalent about that mantra. Some people prefer it one way round and some people prefer it another.

Another issue with "examples first" is that it often makes some guesses or assumptions about what sort of examples will be motivating to the audience, and the reality can be wildly variable. The examples you pick might not be interesting or even familiar to your audience. If the examples are unfamiliar then it's a bit pointless doing them before the theory; rather, if the theory is interesting then it helps us understand unfamiliar examples.

As I've said, the usual approach to category theory says that you "need" to understand a whole load of examples from mathematics before you can understand this theory that relates them all to each other. However, I am deliberately not assuming that you know any of those examples. Personally, I did know some of those examples before learning category theory, but often didn't feel fluent or familiar with them, and some of them I didn't know at all. Then when I met category theory I saw those other examples in the light of this abstract theory I really liked, and finally felt able to become familiar with them.

In this Interlude I'm going to introduce some mathematical structures as examples of categories. You might have encountered them before but I'm going to act like you haven't. A helicopter tour of a city you know is fun, as you can see familiar things fitting together at a broad scale. But it can be fun even if you don't already know the city, and it might inspire you to go back to earth and explore some things you caught a glimpse of from on high.

I'll start by going back over some of the examples we've already seen, which were "secretly" all examples of categories; later in this interlude we'll move on to more advanced mathematical examples. You could arguably skip this whole interlude and go straight into Part Two if you're not an examples sort of person; you can always come back here later if something comes up.

9.1 Symmetry

In Section 5.2 we looked at symmetry, and informally drew diagrams of arrows like this one depicting how rotations combine.

Formally we can take any shape and make a category of its symmetries. It has only one object, which we could regard as being the shape itself. The arrows are then the symmetries, and composition is given by combining symmetries in the way described in that section.

The identity arrow is the "trivial symmetry" which is technically "rotation by 0°". This is one of the reasons that mathematicians like considering that to be a symmetry although you might not usually do that in normal life. In normal life you'd say "this thing has no symmetry" rather than "this thing has one symmetry, the trivial symmetry".

9.2 Equivalence relations

We briefly mentioned in the previous chapter that categories can be seen to be "inspired" by equivalence relations. I now want to stress the rigorous way of putting this: equivalence relations are a special case of categories. Suppose we are thinking about a set S with an equivalence relation $\bigcirc$ on it. Then we can make it into a category with

- objects: the elements of S
- arrows: there's an arrow $a \longrightarrow b$ whenever $a \bigcirc b$.

Things To Think About

T 9.1   Can you check that this is a category? What is the identity on each object and what is composition? How do we know it satisfies the axioms? Conversely, can you put some conditions on a category to ensure that it "is" an equivalence relation?

In the category coming from an equivalence relation the identity on each object is the relation $a \bigcirc a$, which we know is true from reflexivity. Composition comes from transitivity. The axioms are all true trivially, because in this category there just either *is* or *isn't* an arrow $a \longrightarrow b$ for any objects a and b, as an arrow is the assertion $a \bigcirc b$. The axioms for a category all declare that some arrow must equal some other arrow with the same endpoints, but there is only one arrow between any two endpoints so there is nothing to check.

That thought process gives us some clues for how to tell if some random category is in fact giving an equivalence relation. At the very least, to be able to regard arrows as assertions $a \, \boxed{\;} \, b$, there must be at most one arrow $a \longrightarrow b$ for any objects a and b. This is to make sure we have a "yes/no" situation, not a "sense in which" situation. So we have no arrow, or one arrow, but not more.

The relation expressed by such a category would automatically be reflexive and transitive, from the identities and composition of the category, so it remains to ensure that it is symmetric. This means that if there is an arrow $a \longrightarrow b$ then there must be an arrow $b \longrightarrow a$. We are later going to see that this is the beginning of the definition of an *inverse* of an arrow.

Things To Think About

T 9.2   Consider such a category, with at most one arrow from any a to b.

Consider a pair of objects with arrows going both ways.
$$x \underset{g}{\overset{f}{\rightleftarrows}} y$$

What can we say about these two composites that result?
$$x \xrightarrow{\ f\ } y \xrightarrow{\ g\ } x$$
$$y \xrightarrow{\ g\ } x \xrightarrow{\ f\ } y$$

First note that as there can be at most one arrow from any a to b, there can in particular be at most one arrow $x \longrightarrow x$ (as a and b didn't have to be distinct). And we know there is an identity $x \xrightarrow{1_x} x$, so any other arrow $x \longrightarrow x$ must be the identity. This means that the first composite above has to be the identity 1_x. Similarly the second composite has to be the identity 1_y.

We will see that this is how we define an inverse for an arrow — we ask for an arrow going "backwards" such that both of those composites are the identity. In a category coming from equivalence relations the composites are *automatically* the identity so we don't have to specify it.

Finally here's a table summing up how we regard equivalence relations as categories. These are deep ideas and might be difficult at first. Note the "mismatch" between what counts as data, structure and properties on each side.

	equivalence relation		category	
DATA	objects	$- - - \blacktriangleright$	objects	DATA
STRUCTURE	relations	$- - - \blacktriangleright$	arrows	
PROPERTIES	reflexivity	$- - - \blacktriangleright$	identities	STRUCTURE
	symmetry	$- - - \blacktriangleright$	inverses	
	transitivity	$- - - \blacktriangleright$	composition	
			unit laws	PROPERTIES
			associativity	

This is a sort of dimension shift that happens quite often in category theory. As we add more nuance to situations, things that were previously properties (often with a true/false value) become structure ("sense in which") and then new properties are needed at a higher level to govern the structure.

9.3 Factors

In Section 5.6 we drew some diagrams of factors which were in fact diagrams of categories. We start by picking a number, say 30 as we did in that section. Then we make a category with

- objects: all the factors of 30
- arrows: an arrow $a \longrightarrow b$ whenever a is a factor of b.

We could also do this the other way round and say there's an arrow $a \longrightarrow b$ whenever a is a *multiple* of b, and the information in the diagram will be the same; it will just be depicted with the arrows pointing the other way. Turning the arrows around is called taking the *dual* in category theory, and we'll look into that more in Chapter 17.

In this category there is again a maximum of one arrow from a to b. This is because the arrows are actually coming from a relation "is a factor of" and a is either related to b or it isn't. There aren't different senses in which it might be related. It can't be related in two different ways. It's just yes or no. This means that all the diagrams commute, just like in the categories coming from equivalence relations.

Remember, a diagram is said to "commute" if all the paths through it between the same endpoints commute. In Chapter 5 we saw that the square faces in the diagrams of factors all had the form in this diagram.

If we put in the results of doing those multiplications, starting at the bottom from some number y, we can follow the two different paths around the square and see that *a priori*[†] we get the two different outcomes shown on the right.

The fact that this actually meets up in a square is then down to the fact that $yab = yba$, which is commutativity of multiplication.[‡] So in this case the "commutativity" of the diagram comes from another form of commutativity that we're more familiar with.

[†] When we say "*a priori*" we are referring to what we know in advance, from the outset, without doing any logical deduction. Usually we then follow up by deducing more things.

[‡] There's also some associativity of multiplication at work here.

We can generalize this to factors of other numbers. Given $n \in \mathbb{N}$ we can form a category with

- objects: all the factors of n
- arrows: $a \longrightarrow b$ whenever a is a factor of b.

───────────────── Things To Think About ─────────────────

T 9.3   What would happen if we took our objects to be the whole of $\mathbb{N}$ instead?

When we do this for just the factors of n we get a shape whose number of dimensions corresponds to the number of distinct prime factors of n. If we take as objects all of $\mathbb{N}$ instead of just the factors of some fixed n we will need a dimension for every prime number, so we will have infinite dimensions.[†]

9.4 Number systems

Taking all the natural numbers as objects in the previous section resulted in something rather complicated involving infinite dimensions. That's because we were basing the structure on prime factorization. The fact that this yielded a complicated infinite-dimensional structure is somehow related to the fact that the study of prime numbers is so complicated; at least, that's how my categorical brain thinks of it.

That is not the only way to put categorical structure on numbers. It's an important part of category theory that you can take the same objects and regard them as a category in different ways depending on what you take as your arrows. The "natural numbers" are called that because they're supposed to be rather natural. But their naturalness really comes from thinking in terms of addition, not multiplication. We will look at this more in Chapter 20.

For now I want to construct much simpler categories of numbers where we base the relationships just on size rather than on factorization. Here's a way to do it for the natural numbers:

- objects: the natural numbers
- arrows: $a \longrightarrow b$ whenever $a \le b$.

───────────────────────

[†] This sounds like we might get an infinite-dimensional coordinate system with points corresponding to natural numbers. If you know about countable and uncountable infinities you might worry that we appear to get uncountably many points, despite the fact that $\mathbb{N}$ is countable. However, although there exist infinitely many primes, any given natural number only has a finite string as its prime factorization. So the natural numbers only correspond to points in this space where at most finitely many coordinates are non-zero.

T 9.4   Try drawing a picture of this category. Remember to omit arrows that can be deduced; that is, we don't need to draw identity arrows or composites. Does it make a difference whether we include 0 or not? Could we use < instead of ≤? What if we used ≥ instead?

If we think about all the arrows in this category we know that we need arrows like these.

$$0 \longrightarrow 1 \longrightarrow 2 \longrightarrow \cdots$$

But we also know $0 \leq n$ for all natural numbers, so we need all the arrows shown here; observe the "loop" depicting $0 \leq 0$, which is the identity arrow.

But also $1 \leq n$ for all natural numbers except 0, so we need all these arrows.

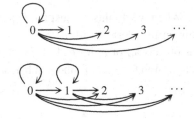

Evidently this approach is going to make a huge mess, so that turned out to be not a very helpful way of thinking about it. The thing is to remember to omit arrows we don't need as they can be deduced: identities (the "loops" above) and composites. This "tidies up" the diagram to look like this instead:

$$0 \longrightarrow 1 \longrightarrow 2 \longrightarrow 3 \longrightarrow 4 \longrightarrow 5 \longrightarrow \cdots$$

This diagram doesn't just look tidier, it also better matches our usual intuition about the natural numbers and how they sit on a line.

Note that it doesn't really matter whether or not we include 0 — the general shape of the category will be the same, with a starting point on the left, and then everything pointing to the right and going on forever. There's no ending point as there is not a largest natural number. We'll come back to that feature when we're talking about universal properties in Chapter 16.

We couldn't use < instead of ≤ to make a category of natural numbers because then there wouldn't be any identities. That's because for any number a, it is *not* true that $a < a$. However, we could use ≥ instead of ≤. This just means all the arrows would point the other way:

$$0 \longleftarrow 1 \longleftarrow 2 \longleftarrow 3 \longleftarrow 4 \longleftarrow 5 \longleftarrow \cdots$$

Later we will talk more about the idea of turning all the arrows around in a category. If we turn all the arrows around but leave everything else the same it's called the *dual* category. It hasn't really changed the structure of the category, it's just presented it differently. The first version and the dual version above encapsulate the same information. This small and apparently irrelevant flip in perspective actually turns out to be rather profound and often useful; we'll come back to it in Chapter 17.

10

Ordered sets

Categories with particularly tidy arrow structures.

In the category of natural numbers we just saw, everything sits neatly in a straight line. This is another special kind of category, corresponding to a mathematical structure that is studied by other mathematicians in its own right, called a totally ordered set.

10.1 Totally ordered sets

An ordered set is a set equipped with an ordering. An ordering is a way of organizing all the objects in an order. A *totally* ordered set is one that is "totally organized". However, in order to study these things mathematically we need to say it in a more rigorous way than that. This definition takes the *visual* idea of a category being "all in a straight line" and identifies an *algebraic* property that corresponds to it.

Definition 10.1 A *totally ordered set* is a category in which for all objects a, b there is exactly one arrow between them.

That is, given any objects a and b there is an arrow $a \longrightarrow b$ or there is an arrow $b \longrightarrow a$ but not both, unless $a = b$ in which case the identity arrow is both of these things.

─────────────── **Things To Think About** ───────────────

T 10.1   Try checking that this is true of the category of natural numbers with arrows $a \longrightarrow b$ given by $a \leq b$.

Given natural numbers a, b, it is definitely true that $a \leq b$ or $b \leq a$, and they can't both be true unless $a = b$. So this category is a totally ordered set. In fact, as $\leq$ is a motivating or quintessential example of this kind of ordering, we often write an ordering as $\leq$ regardless of what kind of relation it really was.

In the normal progression of learning mathematics, you encounter ordered sets before you encounter categories. In that case the definition has to be more direct and does not go via a category. The definition looks a bit like this.

Non-categorical definition of ordered set

An *ordered set* is a set S equipped with a binary relation $\leq$ satisfying

- reflexivity: $a \leq a$ for all $a \in S$
- antisymmetry: if $a \leq b$ and $b \leq a$ then $a = b$
- transitivity: if $a \leq b$ and $b \leq c$ then $a \leq c$
- trichotomy[†]: for all $a, b \in S$, $a \leq b$ or $b \leq a$.

Note that in math we typically use the "inclusive or" which really means "and/or".
So "$a \leq b$ or $b \leq a$" means "$a \leq b$ or $b \leq a$ or both". If we mean one of them but
not both, that's called the "exclusive or". There are various reasons for preferring
the inclusive or, but one is that it comes up more often and so it's more efficient to
take it as the default.

───────────────── **Things To Think About** ─────────────────

T 10.2   1. In fact, the reflexivity condition above is redundant. Can you see
how to deduce it from the other conditions?

2. In the categorical definition we appear only to have specified one condition.
What happened to the other conditions?

Reflexivity is a special case of trichotomy: if we put $a = b$ in the definition
of trichotomy we get "$a \leq a$ or $a \leq a$", which is reflexivity.[‡] It's useful to state
it as a condition though, because when we relax the conditions later we *will*
need to specify it.

In the categorical definition we don't need the other conditions because:

- Reflexivity comes from the category having identities.
- Transitivity comes from composition in the category.
- Antisymmetry and trichotomy both come from there being exactly one arrow between a and b: antisymmetry comes from there being *at most* one,
and trichotomy from there being *at least* one.

In normal life we don't often think of "exactly one" to mean "at least one
and at most one" but in math this is often how it pans out, as in the above
example. Sometimes "at least one" is stated as "existence", and "at most one"
is "uniqueness", but sometimes we prove that there is at most one of something
before proving that there is at least one, and it's a bit odd to think of proving
that something is unique before proving it exists.

[†] "Trichotomy" ought to mean there are three cases where this looks like there are only two;
however deep down it is coming from the three cases $a < b$ or $a = b$ or $a > b$.

[‡] If we say "X is true or X is true" then it's logically equivalent to saying "X is true", although it
is linguistically convoluted.

Things To Think About

T 10.3   How many ways can you think of for a category to *fail* to be an ordered set? You could try doing this by algebra or by pictures. By algebra you'd say: what ways are there for a category to fail to fulfill the definition? By pictures you'd say: what pictures can I draw of a category that isn't just a bunch of arrows in a straight line? You don't need to say what the objects and arrows represent — it can just be a picture of dots and arrows.

Here's one to get you going. Can you see in what way this is failing to be an ordered set? I haven't labeled the objects or arrows but I'm taking them all to be different.

It can sometimes seem that math is all about getting right and wrong answers, but often in higher math if something doesn't work it's really interesting to try and classify all the possible ways in which it could fail. This often gives us some deeper understanding and points us in the direction of a helpful generalization in which we relax one criterion slightly in order to count some of those failures as an interesting structure in their own right.

Here are some pictures of categories that are not arrows all in a line (even if we could draw them all in a physical line they wouldn't point the same way, so wouldn't be abstractly "arrows all in a line"). I haven't labeled the objects or the arrows but I'm taking them all to be different.

 This one fails because it has two arrows branching out from the same place, so the arrows aren't going along in order. Formally, the definition fails as the two objects on the right have no arrow between them.

 This one fails for "the same" reason as the one above, just with the branching going in rather than out. In fact we've just turned around the arrows, which is an example of what is called the dual in category theory.

 Here we have parallel arrows rather than a single straight line. This fails the definition as these two objects have two arrows between them instead of just one.

 In this case we have multiple arrows even though there's only one object. They're still technically "parallel" arrows, although they're really loops.

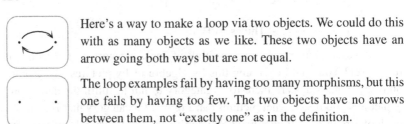

Here's a way to make a loop via two objects. We could do this with as many objects as we like. These two objects have an arrow going both ways but are not equal.

The loop examples fail by having too many morphisms, but this one fails by having too few. The two objects have no arrows between them, not "exactly one" as in the definition.

Any disconnected category will fail, like this one. "Disconnected" means there are separate parts with a complete gap between them (but is stated more rigorously).

10.2 Partially ordered sets

We just saw that a totally ordered set has "exactly one" arrow between any pair of objects, and thus there are two ways to fail:

- a pair of objects with *more* than one arrow between them, or
- a pair of objects with *fewer* than one arrow between them (i.e. none).

We could now try relaxing this definition by maintaining the first condition but dropping the second, so we are thinking about categories in which there is *at most* one arrow between any two objects, but possibly none.

We have seen some of these already, such as any of the categories of factors from Chapter 5. Here's the one for 30 again:

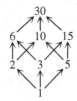

Things To Think About

T 10.4 It might look like some pairs of objects have more than one arrow between them. For example the two composites on the right. Can you see why there is actually only one arrow there?

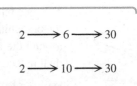

These two composites are equal, because an arrow $a \rightarrow b$ is *the assertion* that a is a factor of b. Put another way, as this is a true/false relationship, there is at most one arrow $a \rightarrow b$, so those two composites must be the same.

This is part of a more general picture: whenever there is at most one arrow between any two objects in a category then all diagrams in the category must commute. This is because if there can only be one arrow between two objects, all composites with those endpoints must be equal.

In the example of factors there are some objects that have no arrows between them, such as 6, 10 and 15. As a result the picture is no longer a straight line but is still quite organized. It consists of several straight line paths that join up and separate in places, and no loops. These structures are interesting enough to have a name.

Definition 10.2   A *partially ordered set* is a category in which for any objects *a*, *b* there is at most one arrow between them.

For emphasis, the previous, more rigid version is called a *totally* ordered set. And these are sometimes abbreviated to "toset" and "poset" because mathematicians are so lazy.

As with tosets, you might usually meet posets before categories, via a definition that doesn't mention categories. Just as the categorical definition of poset came from relaxing the categorical definition of toset, we could do that to the non-categorical definition as well.

─────────────── **Things To Think About** ───────────────

T 10.5   Look back at the non-categorical definition of toset and see what condition(s) we need to drop to make a non-categorical definition of poset.

The condition we need to drop is trichotomy, as we now want to allow pairs of objects that are not related either way round. Note that unlike with tosets we now can't drop the reflexivity condition, because that was only redundant in the presence of trichotomy. This is one reason to prefer keeping it in the toset definition, so that we can then get to the poset definition by simply dropping one condition, which is more mathematically satisfying when you're making a generalization. (Of course, I would say that this is a reason to prefer the categorical definition.)

So here's the non-categorical definition.

Non-categorical definition of partially ordered set

A *partially ordered set* or *poset* is a set S equipped with a relation $\leq$ satisfying

- reflexivity: $a \leq a$ for all $a \in S$
- antisymmetry: if $a \leq b$ and $b \leq a$ then $a = b$
- transitivity: if $a \leq b$ and $b \leq c$ then $a \leq c$.

Things To Think About

T 10.6   Look back at our pictures of categories that were not tosets. Which ones are actually posets?

The ones that are posets will be the ones that failed to be tosets only by having too *few* arrows somewhere, not too many. So not the ones with parallel arrows or loops.

This way of generalizing a definition by relaxing a condition is typical in math. We took a definition with several conditions on a structure, say conditions A, B, C, D. Then we relaxed the definition by dropping condition D and looking at structures that only satisfy A, B, C. Logically, anything that satisfies A, B, C, D definitely satisfies A, B, C and so anything that is the first type of structure will definitely be the second type, which is why the second type is called a generalization; conversely the first type is a special case.

Then it's good to ask whether there actually are any structures of the second type that aren't the first type. We have just seen some small examples of posets that are not tosets, so we know that this is a genuine generalization.

We could also encapsulate this in a Venn* diagram, as shown, or we could use logical terms and say "toset $\Rightarrow$ poset" although this isn't quite grammatical because "toset" and "poset" aren't statements. However, I'm not pedantic enough to make them into full statements at the expense of concision and clarity.[†]

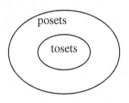

Things To Think About

T 10.7   In Chapter 5 we moved from lattices of factors to lattices of subsets. Can you make a formal definition of a category of subsets of S (for some fixed set S, based on Section 5.6? Can you then show that it is a poset?

We start by fixing a set S. We can then construct a category with

- objects: all the subsets of S
- arrows: given subsets A, B, we have an arrow $A \longrightarrow B$ whenever A is a subset of B, which we write as $A \subseteq B$.

[†] In *The Art of Logic* I characterized pedantry as precision that does not increase clarity.

Everything else follows more or less as for factors. The set of all subsets of S is called the *power set* of S, and is often written $\mathcal{P}(S)$.

> Does it seem confusing that the objects of this category are now sets? It's worth just thinking about that for a second. We are going to see that one of the powerful aspects of category theory is that we can zoom in and out and use it at different levels. The objects of a category can be points, sets, sets of sets, whole categories, entire mathematical worlds... And I do like the fact that many types of historically pre-existing mathematical structure can be encapsulated as a special type of category.

────────────── **Things To Think About** ──────────────

T 10.8   Look back at Chapter 5. Can you see any other posets that we looked at in that chapter?

The diagram of quadrilaterals and their generalizations was another poset. In the next chapter we're going to look at some other examples of mathematical structures that arise naturally in their own right, like posets, but aren't as tidily organized.

11

Small mathematical structures

Examples of small structures expressed as categories, before we zoom out and move on to look at large structures.

11.1 Small drawable examples

In the previous chapter we drew some examples of "very small" categories like this one. This is a very abstract way to define a category — it's not based on a particular example.

The objects and arrows have no meaning here; all we're specifying is the structure of the category. This is different from situations where we take the objects to be "all the factors of 30" or "all the natural numbers" in which case those objects come with properties and behaviors already.

Small drawable examples of categories are useful as we can use them to think about structural features that we might find inside a specific category. We are going to see that we can then map these small abstract categories into large meaningful ones in order to study features inside them.

When we specify a category by drawing it we often adopt the convention that we don't bother drawing the identity arrows, because they're always there (and would make the diagram more of a mess).

So there are categories that we might draw as shown here, although they really have an identity arrow on each object.

You might think these look just like sets, as there are no arrows between objects. This is true, and is a way in which any set can be realized as a special case of a category in which there are no arrows except identity arrows.

Definition 11.1 A category is called *discrete* if it has no arrows except identity arrows.

In this way categories are generalizations of sets. Incidentally, we can have an empty category, which could also be thought of as the discrete category

124

on the empty set. It has no objects and no arrows. This is the only way to have a category with no arrows at all, because as soon as you have any objects you must have identity arrows. However we sometimes call identity arrows "trivial", so we can think of discrete categories as having no *non-trivial* arrows.

Here is the "quintessential" category consisting of one arrow. As in the previous section, the convention when we draw things like this is that all the objects and arrows are distinct, even though we haven't labeled them with anything.

If we wanted to make it so that those objects were the same we could draw a loop, like this, but then this diagram would be ambiguous and we'd have to specify more things.

First we'd need to specify that that loop is not the identity. Then we'd need to specify what happens when we compose it, as it is now composable with itself. If we go round the loop twice does that make a new arrow? Does it just keep making a new arrow every time we go round it? If so, we're not really in a small *drawable* category any more, but in a small mathematical structure.

11.2 Monoids

We have already seen how some mathematical structures can be expressed as special cases of categories. We've seen this with equivalence relations, and also tosets and posets.[†] The loop example at the end of the previous section leads us into some more such structures.

Suppose every time we compose that loop with itself we get a new arrow. In that case this category has only one object but infinitely many arrows. If we call the basic arrow f then:

- There is also $f \circ f$ which we might write as f^2.
- Then there is $f \circ f \circ f$ which we might write as f^3.
- In fact there's f^n for every $n \in \mathbb{N}$.

Note it makes sense to include 0 in $\mathbb{N}$ here, with $f^0 = 1$, the identity.

This structure basically encapsulates the structure of arithmetic on the natural numbers, in the way that we thought about it in Sections 5.3 and 5.4 when we drew diagrams like the one on the right.

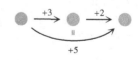

We could now take those gray blobs to be really the same object, and draw every arrow as a loop. Then we have one arrow for every natural number,

---

[†] Totally ordered sets and partially ordered sets.

and composition is given by addition. This is the same structure we described with the f notation above, except that then all the arrows were written as f^n rather than n. Composition was like multiplication, but really it was addition of natural numbers in the exponent: $\boxed{f^n \circ f^m = f^{n+m}.}$

It is yet another way to regard the natural numbers as a category, with emphasis on a different aspect of the structure.

We have now seen these three ways of putting a category structure on the natural numbers, and each one gives us a very different shape of category.

1. Using factors.
2. Using size.
3. Using addition.

In the first two cases, the objects of the category were the natural numbers, but they used different arrows. The third case is a bit different as it now takes the *arrows* to be natural numbers, and there was just one object which was a sort of dummy object. This is a form of "dimension shift" that can feel a little destabilizing but which is very powerful. We'll come back to it. Categories with only one object are quite special so they deserve a name.

Definition 11.2   A category with only one object is called a *monoid*.

Possibly the most striking consequence of having only one object is that all arrows are composable. We mentioned this in Chapter 8, when we were talking about the fact that composition is a generalization of a binary operation. When there is only one object, composition actually *is* a binary operation on the arrows.

As with tosets and posets, one might typically encounter monoids before categories, and so there is a direct non-categorical definition. It uses binary operations as follows.

Non-categorical definition of a monoid

A *monoid* is a set M equipped with an identity 1 and a unital and associative binary operation $\circ$. The monoid is sometimes written fully as $(M, \circ, 1)$.

As usual:
- *unital* means that for all $m \in M$, $1 \circ m = m = m \circ 1$
- *associativity* means that for all $a, b, c \in M$, $(a \circ b) \circ c = a \circ (b \circ c)$.

—————————— **Things To Think About** ——————————

T 11.1   Can you "translate" between this definition and the categorical one?

The key here is that the *objects* (also called elements) of the monoid in the non-categorical definition become the *arrows* of the monoid in the categorical definition. I call this a "dimension shift" because objects are 0-dimensional and arrows are 1-dimensional.

	non-categorical	$- - - - \rightarrow$	categorical
DATA	objects	$- - - - \rightarrow$	*"dummy" object* arrows
STRUCTURE	identity object binary operation	$- - - - \rightarrow$ $- - - - \rightarrow$	identity arrow composition
PROPERTIES	unitality associativity	$- - - - \rightarrow$ $- - - - \rightarrow$	unitality associativity

I hope you can see from the table that once you've done the dimension shift at the level of data, everything else just sort of automatically follows. In the data, the dummy object comes from nowhere and has no meaning. In any category with a single object it doesn't matter what that object is; it mostly just serves to ensure that all arrows are composable. So we often ignore it.

──────────────── **Note on the dimension shift** ────────────────

This is a mental shift in perspective and abstraction. It's important so we'll be coming back to it several times. If you find it confusing then I think you may well be on the right track. I still remember feeling completely confused by this shift the first time I saw it, and feeling slightly dizzy and vertiginous. I think that leaps of abstraction can often feel like that, and that the vertigo is a bit like looking over the edge of an abstract cliff. One possible response is to run away, and that's what I do with real cliffs. But some people love looking over the edge and enjoy the de-stabilizing feeling. I like doing it for abstract cliffs. And if you sit there for long enough the vertiginous feeling may go away and you'll have made it to another level of abstraction.

The important thing, I think, is to know that if it feels weird you're not wrong, and you're not doing badly. If it feels completely fine to someone else it might be because they're brilliant at abstraction but it might also be because they haven't got the point.

──────────────────────────────

The natural numbers are the essential example of a monoid (under addition) and later we will give this idea a formal meaning by saying they are the "free monoid on one generator". This encapsulates the fact that we arrived at the natural numbers right at the start of this section by taking one object and one arrow (in a loop) and composing it repeatedly making a new arrow each time.

If instead we imposed an equation on some of those composites then it would no longer be a "free" monoid (because we've imposed some constraints). For example we could declare that f^2 is a new arrow and f^3 is a new arrow but that $f^4 = 1$.

Then by associativity we find various other relationships, as shown, and the only four arrows in this category are now $1, f, f^2, f^3$.

$$f^5 = f^4 \circ f = 1 \circ f = f$$
$$f^6 = f^4 \circ f^2 = 1 \circ f^2 = f^2$$
$$\vdots$$

Things To Think About

T 11.2   Can you fill in the following composition table for the arrows of this category? Does the resulting pattern remind you of anything we've seen in earlier chapters?

$\circ$	1	f	f^2	f^3
1				
f				
f^2				
f^3				

The table has "the same" diagonal pattern we saw for the addition in modular arithmetic. In fact this table looks just like the table for the integers mod 4, as shown at the far right.

$\circ$	1	f	f^2	f^3
1	1	f	f^2	f^3
f	f	f^2	f^3	1
f^2	f^2	f^3	1	f
f^3	f^3	1	f	f^2

	0	1	2	3
0	0	1	2	3
1	1	2	3	0
2	2	3	0	1
3	3	0	1	2

Formally we can "translate" from the second table to the first by raising f to the power of each entry. Our monoid with the constraint $f^4 = 1$ "is" the integers mod 4 in the same way that our original one "is" the natural numbers.

Things To Think About

T 11.3   1. How would we impose a constraint to make $\mathbb{Z}_n$?
2. Why was the original version $\mathbb{N}$ and not $\mathbb{Z}$? That is, why did the constraint produce the *integers* mod n and not "natural numbers mod n"?

We could generalize this to $\mathbb{Z}_n$ with the constraint $f^n = 1$. For the second point we need to think about inverses, which takes us to the topic of groups.

11.3 Groups

The integers are crucially different from the natural numbers in having negative numbers, which are inverses with respect to addition, also called "additive inverses". We can in general ask whether inverses exist for any particular element in a monoid. Inverses "undo" the binary operation with respect to that element, taking us back to the identity, so the definition refers to both the binary operation and the identity for it.

Definition 11.3 An *inverse* for an element[†] a of a monoid $(M, \circ, 1)$ is an element b such that $a \circ b = 1$ and $b \circ a = 1$.

Monoids do not necessarily have inverses for every element; those that do are special. The monoid of natural numbers (with binary operation +) does not have inverses for any element except 0. However in the monoid of *integers* under + every element does have an inverse: the inverse of a is $-a$. In fact that's the definition of negative numbers.

We'll now see that $\mathbb{Z}_n$ has additive inverses for everything even though it doesn't seem to have any negative numbers.

Things To Think About

T 11.4 It's illuminating to think about this in both presentations of $\mathbb{Z}_n$: as numbers on the n-hour clock, and as arrows f^k where $f^n = 1$.

1. Can you find an inverse for each element of $\mathbb{Z}_4$ under addition? So for each $a \in \mathbb{Z}_4$, find $b \in \mathbb{Z}_4$ such that $a + b \equiv 0 \pmod 4$. (The condition for $b + a$ follows by commutativity.) The elements here are $0, 1, 2, 3$. What about $\mathbb{Z}_n$ in general?

2. In the monoid generated by the arrow f with constraint $f^n = 1$ can you find an inverse for each element f^k? Here the identity is 1.

This table shows the inverses in $\mathbb{Z}_4$. Remember, in $\mathbb{Z}_4$ any multiple of 4 is "the same" as 0. So if two numbers add to 4 in the normal integers, then they will add to 0 in $\mathbb{Z}_4$. As you can see from this table we have inverses for everything even though there are not exactly any "negative numbers"; another way of thinking of it is that 3 in some sense "is" the negative of 1, that is $3 \equiv -1 \pmod 4$.

a	inverse
0	0
1	3
2	2
3	1

We can check this formally from the definition $a \equiv b \pmod 4 \iff 4 \mid b - a$.

This gives us a hint for how to generalize to n. If two numbers add to n in the normal integers then they will add to 0 in $\mathbb{Z}_n$. So given any $a \in \mathbb{Z}_n$, we can take its inverse to be $n - a$ (with a few caveats[‡]).

[†] I often use "element" when I'm not emphasizing anything dimensional about a monoid, and "object" when I am.

[‡] There are other ways to present $\mathbb{Z}_n$.

If instead we think of the presentation using arrows f^k, then this all comes to the same thing, just at the level of exponents instead. The arrow f^k has inverse f^{n-k} since $\boxed{f^k \circ f^{n-k} = f^n = 1.}$

Putting the constraint $n \equiv 0$ on the natural numbers has incidentally given everything an inverse. A monoid in which every element has an inverse is an important structure so gets a name.

Definition 11.4   A *group* is a monoid in which every element has an inverse.

As with everything else in this chapter, groups are usually something you would meet before categories, and you would meet them via a direct definition starting with a set of objects/elements.

───────── **Things To Think About** ─────────

T 11.5   Can you produce the non-categorical definition of a group?

Non-categorical definition of a group

A *group* consists of a set G equipped with a binary operation $\circ$ and an identity element 1 satisfying unit laws and associativity, and such that every element has an inverse with respect to $\circ$.

───────── **Things To Think About** ─────────

T 11.6   Can you organize this into data, structure and properties, and then see how the *structure* corresponds to the *properties* for an equivalence relation?

Comparing equivalence relations with groups directly can be confusing, and I find it easier to do by expressing them both as special cases of categories; the category framework then unifies them. The following table extends the "dimension shift" table for equivalence relations.

	equivalence relation		category		group	
	objects	$---\rightarrow$	objects			
	relations	$---\rightarrow$	arrows	$\leftarrow---$	objects	
PROPERTIES	reflexivity	$---\rightarrow$	identities	$\leftarrow---$	identity	STRUCTURE
	symmetry	$---\rightarrow$	inverses	$\leftarrow---$	inverses	
	transitivity	$---\rightarrow$	composition	$\leftarrow---$	binary operation	
			unitality	$\leftarrow---$	unitality	
			associativity	$\leftarrow---$	associativity	

Note that in a category coming from an equivalence relation the arrows are restricted — there can be at most one from any object a to an object b. By contrast in a category coming from a group the *objects* are restricted: there is just one, the "dummy" object.

——————————————— Note on binary operations ———————————————

In other texts on introductory group theory you might see an extra condition for groups called "closure". This demands that when you combine two elements by the binary operation, the result is still inside the group. However, in abstract mathematics a binary operation is defined to be a function $S \times S \to S$, so closure is built in. Here $S \times S$ denotes the set of pairs of elements of S, which we'll come back to in Chapter 18.

Small finite groups are often expressed by means of a table, as we did in Chapter 5 with these tables for rotations of a square and addition in the integers modulo 4.

	0	90	180	270
0	0	90	180	270
90	90	180	270	0
180	180	270	0	90
270	270	0	90	180

	0	1	2	3
0	0	1	2	3
1	1	2	3	0
2	2	3	0	1
3	3	0	1	2

Tables arising from the binary operation of a group are called *Cayley tables* and have this important feature: each element appears exactly once in each row and each column. Squares with this property are called *Latin squares.*[†]

The pattern in the table gives us an indication of the abstract structure of the group. In the above two diagrams, the pattern is the same even though the actual entries in the tables are different: in each case it's a pattern of entries repeating in shifting diagonals. We are going to see that this means the two groups are in some sense "the same" even though they arise from different situations. This is the sort of thing that category theory will look for: when things look different on the surface but are "the same" deep down.

———————————————— Things To Think About ————————————————

T 11.7   Try filling in these multiplication tables, in the context of the integers mod 8 and the integers mod 10:

1. the numbers $1, 3, 5, 7$ in $\mathbb{Z}_8$ 2. the numbers $1, 3, 7, 9$ in $\mathbb{Z}_{10}$

One of these has "the same" pattern as the above two tables, but only if you list the elements in a different order. The other one does not have the same pattern; how can we be sure? We will come back to this in Chapter 14.

———

[†] Note that not every Latin square is a group though, as the conditions of being a group impose more constraints on the possible patterns in the square.

Some broader context

I have introduced groups as a special case of categories for the purposes of exposition, but I want to stress that they are an important field of research that has been around for much longer than category theory.

Groups are often introduced directly via symmetry. The idea is just like what we did with the symmetries of a square in Chapter 5, but now with any shape we choose: any two symmetries of a given shape can be combined to produce another symmetry of that shape, and that structure forms a group.

This also gives us a way to generalize the idea of symmetry to more abstract situations that don't involve the obvious types of visible symmetry on shapes. For example, equations can have symmetry: if we switch the x and y in the equation $x^2 + 2xy + y^2 = 0$ then the equation stays the same (assuming that addition and multiplication are commutative). This is the beginning of the idea behind Galois theory, which connects equations with groups via some "abstract symmetry" involving their solutions.

Group theory is a classic research field of pure mathematics with a vast body of knowledge and techniques. For this reason many advances in other areas of math have come from finding a way to make a link with group theory. This is often indicated by putting the word "algebraic" before a field of research, because group theory is a form of algebra. (Algebra in high school means something completely different and rather mundane, usually to do with manipulating equations and solving for x. I'm sorry that the word "algebra" has been taken over in this way in high school.)

For example algebraic topology starts from the idea that you can take a topological space and encapsulate some of its structure as a group (we'll talk about what topological spaces are in a moment). Algebraic geometry comes from studying geometry by encapsulating some of its structure as a group. Algebraic number theory studies numbers via group structures.

There is one further generalization of group I want to mention which is mind-boggling if you think of the non-categorical definition of group, but quite natural if you think of the categorical definition. It starts with thinking of how we would put conditions on a category to ensure that it is a group.

─────────── **Things To Think About** ───────────

T 11.8   Can you make a definition of a group directly as a category satisfying some conditions?

Categorical definition of a group

A group is a category with only one object, such that every arrow has an inverse.

The first condition ensures that we have a monoid, and the second says it is a monoid with inverses. Now, we can relax this definition by dropping the second condition, and we'll get back to monoids. But we could instead relax this definition by dropping the *first* condition, as follows.

Definition 11.5   A *groupoid* is a category in which every arrow has an inverse.

Trying to understand the concept of a groupoid directly from a group can be confusing, but going via categories makes it in some ways more natural. Groupoids have been studied for less long than groups and there is a smaller body of knowledge about them, but in many ways they have been coming into their own recently as they generalize better to higher dimensions. I mentioned briefly the idea behind "algebraic topology" as the study of topological spaces via algebra, classically groups. However, more recently this has been via groupoids and their higher-dimensional generalizations. At this point I'd better say something more about what topological spaces are.

11.4 Points and paths

One of the things mathematicians study along with numbers, equations and patterns is shapes and the shape of space. We can try and study the shape of a space with obstacles in, or holes, or other places that we can't go. This makes it more complicated than just a wide open space.

I grew up in a house that had a fireplace and chimney in the middle, and all the rooms sort of went round it in a circle — not an actual curved circle, but a loop. The floor plan was notionally something like this, although the scale of this is definitely very off. The fireplace and chimney could then heat all the rooms from the center (although by the time I lived there we did have central heating).

The best thing was that as the rooms all linked together in a loop, my sister and I could chase each other round and round in circles.[†] The shape of that space is quite different from a house in which you can't run around in circles, like everywhere I've lived since then. As I write this we're all in lockdown for the COVID-19 pandemic, and it would be quite nice to be able to walk in

---

[†]   I wrote about this in *Beyond Infinity* because the loop meant it was like having an infinite house in which to run around.

non-trivial circles around my apartment, but I am making do with going back and forth from the kitchen to the front door.

This is the sort of thing that is studied in topology: what is it that makes the shape of one space really different from the shape of another? If you just moved a wall inwards slightly then the size of the space would change but in some sense not really the *shape*. Likewise if you shrank the entire building.

Mathematicians seek ways to bring rigor to a situation that seems to make intuitive sense, and one way of doing that in topology is to think about the different paths you can take between two points in a space. In my childhood home, there were two really different paths you could take from the front door to the kitchen — the more direct route through the hall and past the staircase, or the more circuitous route through the sitting room and dining room. But in my current apartment there is only a direct route, and the different paths just consist of veering a bit more to the left or right, which is not *interestingly* different. We can study this via categories.

Given any space X we can try and make a category with:

- objects: the points in X
- arrows: an arrow $a \longrightarrow b$ is a path from a to b in the space.

However there is a technical question of exactly how we define a "path" in order to be able to get an associative composition, and an ideological question of whether we want to count two paths as genuinely different if they're only a slight wobble away from each other.

———————— Things To Think About ————————

T 11.9   Here is an abstract diagram of my childhood home, with the chimney obstruction in the middle, the approximate position of the front door marked as a and the kitchen marked as b. What paths can you think of from a to b that are not really that different from one another? What paths can you think of that are genuinely different?

On the left are some paths that I would say are not really that different, whereas on the right are two that I think are really quite different.

And let's not forget the possibility of actually going all the way round the chimney, possibly several times.

These ideas are all from the study of *topological spaces*. I think the technical definition of a topological space is too much of a digression for me to go into here (see Appendix D for a little more detail), but the idea is that there is a notion of "continuous deformation" where we think about whether or not we can continuously morph one thing into another without *really* changing it or breaking it at any stage. The three paths from a to b in the left-hand diagram above would formally count as "the same" in the sense that we could define each one as a function and then define something called a "homotopy"[†] that continuously deforms one function into the other. This also helps us deal with the technical question of associativity.

With the two paths in the right-hand diagram we wouldn't be able to do that because our attempt at a continuous deformation would be doomed to crash into the chimney in the middle. Understanding the formality of this requires the formality of

- sets and functions, which we will get into later,
- calculus, or some formal way to understand what "continuous" means.

As with posets, monoids and groups, there is a direct non-categorical definition based on sets, and it starts with a notion of "closeness" so that we can think about which points in the space count as close together and which do not. There are also various categorical definitions. All of them are too hard for right now, but here are the ideas I hope you'll take away.

1. Points and paths in a space can be encapsulated as a category.
2. There is a more nuanced notion of when paths count as "the same" according to when they can be gently morphed into each other.
3. Thinking about what genuinely different paths there are in a space tells us about the overall shape of the space.

With these ideas I can then say that my sketch of the floor plan of my house was *topologically* correct even though it was geometrically quite wrong.

[†] I pronounce this with the stress on the first syllable, but some people put the stress on the second, although I think that is less common.

12

Sets and functions

A first example of a large category of mathematical structures. This is one of the fundamental motivating examples of category theory.

So far the examples we have seen have been either tiny drawable examples or examples of individual mathematical structures expressed as a category. However, one of the most powerful aspects of category theory is its ability to zoom out and look at huge worlds of structures rather than just individual structures. Put another way, one of the great insights of category theory is that a totality of mathematical structures of a given type can itself be regarded as a mathematical structure.

That is a lot of words, so here's an example. We saw that a poset is a set of objects equipped with some sort of ordering on it. We expressed that as a category, where the objects are the elements of the given poset. But we could now look at the collection of *all possible posets* and regard that as a category. The objects of this category are now *the posets themselves* and the arrows will be some form of map between posets. Here's a schematic diagram.

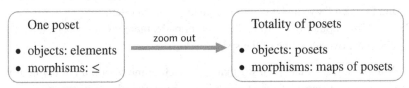

We could do this for monoids as well:

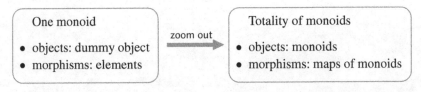

We could also look for a category of groups and maps between groups, or spaces and maps between spaces.

The starting point for all of these structures is an underlying set. We then say that the set is special in some way, by having an ordering, or by having a binary operation, or by having a notion of closeness. The starting point for the *maps* between structures is then the concept of a function, the basic notion of a map between sets. We then find a good notion of map between special structures by looking at maps that "preserve" the special properties in question. We'll look at that, but first we need to think about functions.

12.1 Functions

The category of sets and functions is one of the key motivating examples in category theory. Sets and functions are an essential starting point of the whole of mathematics, and the techniques mathematicians have built up for dealing with them are things we can then try and do in more complicated and nuanced mathematical worlds.

In a way, numbers are simplified versions of sets. You might think that numbers are the fundamental concepts of mathematics, and in a sense that is true both historically and in terms of the order in which we tend to learn and understand math as children. However, when mathematicians tried to make the idea of numbers very precise, they found they had to deal with sets first, and then build numbers out of particular sets. When mathematicians became really serious about developing mathematics completely rigorously, they started with sets. Set theory was developed as a *foundation* of mathematics.

Category theory is sometimes considered to be a competing foundation. I don't really like competitions and don't see why everything has to be made into a competition, so I don't want to start a set theory vs category theory flame war (there are plenty of those on the internet). But I'll just say this. Set theory starts with the concept of *membership* as a basic concept, and builds up from the idea of elements being members of sets. By contrast, category theory builds up from the notion of *relationships* which is why our basic concepts in category theory are elements (objects) and relationships (arrows) between them, rather than membership.

Ideologically I like to think of this as the difference between looking at people's intrinsic characteristics and looking at how they behave in different contexts; I wrote about that at length in my book $X + Y$. But a more mathematical ideology to consider is that category theory takes *sets and functions* as the basic starting point, rather than sets and membership.

The two approaches are in some sense equivalent, in that you can build the concept of a function from the concept of set membership, and conversely you can build the concept of set membership from the concept of a function. It's really a matter of preference which concept you take to be more fundamental.

Many people, including me, find that the categorical approach more closely matches my actual mathematical practice. Gathering together mathematical structures and thinking about maps between them seems to happen all over pure math, and so taking functions as a starting point seems more illuminating to me. I think this is part of why the language and formalism of category theory have turned out to be so fruitful across many branches of math.

Anyway, this means we'd better develop an understanding of functions. Functions are quite misunderstood. The trouble is that, like with so many parts of math, the way they're typically treated in school math is not really like the way they're treated in research math. For that reason this chapter might actually be easier if you've never encountered functions before; if you've encountered them before in high school math classes then you might have a bit of un-remembering to do (unmembering?) in order to get your head around functions in the way that category theorists think of them.

In school you might have met functions as things like $f(x) = x^2$ or sin, cos, tan or e^x. Those are all examples of functions but they are not really expressed categorically, as this doesn't tell us what sets we are relating to each other. In a categorical approach, a function is a relationship between sets. It goes from one set to another. We can think of those sets as set of inputs and a set of potential outputs. It's like a machine, where you can feed in any of the inputs and one of the outputs will pop out. The inputs and outputs could be any sets: they don't have to be numbers, although functions that deal with numbers are a big part of math and those tend to be the kinds of functions you meet in school. Moroever, functions don't have to be given by any kind of formula although, again, those are the kind you tend to meet in school.

A function could be something like a vending machine.

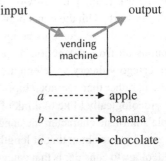

Perhaps the machine has three buttons, labeled a, b, c, and you can buy an apple, a banana, or chocolate. This chart tells us what each button produces.

This is the most "sensible" way to set up this particular machine, but it doesn't have to be like that. It could instead be set up in any of the following ways, for example. The third might happen if there are no bananas today.

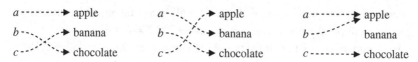

It's important to note that in the last case the banana still counts as an output even though there's no way of achieving it, so it might be better to think of the outputs as *potential outputs*.

It's also important to note that all the inputs have to produce something. If we ran out of bananas and just left the "*b*" slot empty then it would not count as a function any more.

a -------→ apple

b

c -------→ chocolate

Also we can't put two different items in the same compartment: when we press a button we have to know exactly what we are going to get out.

a -------→ apple

b ----→ banana

c -------→ chocolate

Those other situations do exist, they're just not counted as functions. So functions have an asymmetry between the role of the inputs and the outputs:

- Every input must produce an output. However, there could be outputs that are not produced by inputs.
- We can't have one input producing two different outputs. However, we can have one output being produced by two different inputs.

Functions with less asymmetry in the role of input and output are interestingly well-behaved; we'll look at them later when we think about special properties of functions.

If there is a different number of inputs from outputs we are destined to have either some outputs left out, or some outputs hit by more than one input.

For example, if chocolate is completely removed from the vending machine options, we'll only have two possible outputs. Any function will need to have at least two inputs producing the same output, like here.

--- Things To Think About ---

T 12.1 How many different functions are possible between these three inputs and two outputs above? You could try drawing them all or you could think through the problem logically. What if there were p inputs and k outputs?

To see how many functions there are we could take each input in turn. The first input has two possible choices for the output. The second input also does, and so does the third. The choice of output for a does not affect any of the other possible outputs, so these choices are all independent, which means we can just multiply the number of choices to get $2 \times 2 \times 2 = 8$.

We could make a list of all the possible functions in a table as follows. Here, each column represents one function: the top entry is the output for a, the middle entry is the output for b, and the bottom entry is the output for c.

a	apple	apple	apple	apple	banana	banana	banana	banana
b	apple	apple	banana	banana	apple	apple	banana	banana
c	apple	banana	apple	banana	apple	banana	apple	banana

If we generalize to p inputs and k outputs we see that each input has a choice of k outputs, so the total number of possible functions is: $\underbrace{k \times \cdots \times k}_{p \text{ times}} = k^p$.

Things To Think About

T 12.2   What does this formula say if the set of inputs is the empty set? What if the set of outputs is the empty set? Is this weird? Can you interpret it?

The empty set (which we write as $\varnothing$) is a set containing 0 elements. So if we have 0 inputs then $p = 0$ and the formula tells us that the total number of possible functions is k^0 which is 1 no matter what k is. This seems to be telling us that no matter what set of outputs B we pick there is one function $\varnothing \longrightarrow B$.

How can that be? If there are no inputs then how can there be a function? This is an example where it may be necessary to take the definition formally rather than informally. The formal definition says something like "f produces, for every object in A, an object of B". The key is in the phrase "for every object in A". If A has no objects then we are already done: whatever comes after that clause, we have already done it for every object in A, because there aren't any. This is called having a condition that is "vacuously satisfied". For example all the elephants in my house are purple. This is true because there are no elephants in my house. The "vacuous" function from the empty set to any set is something we'll come back to when we talk about initial objects. It's not exactly interesting in its own right, but it is an interesting and indeed important function to have around in the context (category) of sets and functions.

On the other hand if we have 0 outputs then we put $k = 0$ in the formula and the total number of possible functions is 0^p, which is 0 unless there are also zero inputs, in which case the answer is 1. That is not surprising — a function has to produce an output for every input, so if there are no outputs we'll be really stuck unless there are also no inputs.

The above discussion has, I hope, made it clear that we really should define functions a bit more formally. Here is a more formal definition. You might find the wording strange, especially the grammar involving the word "assigns", which is not really the way we use that verb in normal English. I think we tend to word it this way round in math so that the inputs come before the outputs in the sentence.[†]

Definition 12.1   Let A and B be any sets. A *function* $A \xrightarrow{f} B$ assigns to every element $a \in A$ an element $f(a) \in B$.

In very formal set theory the definition happens by means of "ordered pairs": we construct a new set from A and B consisting of pairs of elements, one from A and one from B, representing the input–output pairs that define f. We will not need to go to that level of formality here.

──────────────── **Notation and terminology** ────────────────

In category theory thinking, a function *always* comes with the information of what its sets of inputs and outputs are. Functions are going to be the arrows of a category, as you have probably guessed. So we might call A and B the source and target of the function f, or the domain and codomain. They are sometimes called the domain and range, but I don't like "range" because it is used ambiguously in school math.

We might write this function as $f : A \longrightarrow B$ or as $A \xrightarrow{f} B$. That is at the level of sets. If we start thinking about the action of f on elements inside the sets we write things like $a \in A, f(a) \in B$. If $f(a) = b$ we might say b is the *image* of a under f, and we might write $f : a \longmapsto b$. The vertical bar on the arrow is to indicate that we are now talking about the action of the function on elements, not about its source and target sets.

We might sum up the whole situation as shown on the right; note the two different types of arrow meaning different things with respect to the same function f.

$$f : A \longrightarrow B$$
$$a \longmapsto b$$

In school math, functions are often defined without specifying what their source and target sets are, and then you are somehow supposed to infer what their domain and "range" are based on where it is possible to define them. That makes me very uncomfortable categorically (as well as linguistically).

─────────────

Sometimes functions can be defined by a formula, but not always, and sometimes their inputs and outputs aren't actually numbers.

───────────

[†] Perhaps a more natural way to phrase this in normal English would be something like "A function assigns an element of B to each element of A."

Examples Here are some examples from life and then some examples from math. We will think more about these examples in Chapter 15 when we think about particular properties of functions. Each one is a function $A \xrightarrow{f} B$.

1. A = people in Chicago
    B = countries in the world
    $f(a)$ = where a was born.

2. A = people in New York City
    B = the set of pairs {name, date}
    $f(a)$ = the name and date of birth of the person.

3. $A = \{1, 2, 3\}$
    $B = \{1, 2, 3, 4, 5\}$
    $f(a) = a + 1$

4. $A = \mathbb{Z}$
    $B = \mathbb{Z}$
    $f(a) = a + 1$

We are of course going to assemble sets and functions into a category. The objects will be all possible sets and the arrows $A \rightarrow B$ will be all the possible functions $A \rightarrow B$. An aside is probably called for here on what a set is. I don't intend to get into the foundations of set theory and it's not *really* necessary in order to think about the category of sets. I think we just need to agree that we have a notion of set such that a set is entirely determined by its elements, and such that Russell's paradox doesn't happen, as we mentioned in Chapter 8. (Again, see Appendix C for more details.)

Note that a set is completely determined by its elements and a function is completely determined by what it does to elements. So it doesn't matter how it produces the outputs — it could be by a formula, by a machine, by looking at a list — if two functions take the same inputs and produce the same outputs for each one then they count as the same function.[†] (Note that in category theory they must also have the same source and target sets before we can even ask if they are the same function.)

So for example we could define two functions $\mathbb{R}_+ \rightarrow \mathbb{R}$ like this:

- $f(x) = \frac{x}{x}$ - $g(x) = 1.$

Note that $\mathbb{R}_+$ means all the *positive* real numbers, so we don't have to worry about what $\frac{0}{0}$ is. Anyway for all $x \in R_+$, we have $f(x) = g(x) = 1$ and this means that $f = g$ as functions, even though they have different formulae.

Definition 12.2 Let f, g be functions $A \rightarrow B$. We say $f = g$ as functions whenever $\boxed{\forall a \in A \ \ f(a) = g(a).}$

We now have the data for our category of sets and functions, that is the objects (sets) and the morphisms (functions); next we need to sort out the structure and the properties.

[†] This is the essence of the definition of a function in set theory by input–output pairs.

12.2 Structure: identities and composition

Given any set A there is an identity function $A \longrightarrow A$ which just sends every element of A to itself. We can call it 1_A and define it by $\boxed{\forall a \in A \ \ 1_A(a) = a.}$

For composition we need to think about functions $A \xrightarrow{f} B \xrightarrow{g} C$. We can think about defining the composite function as if we are joining two machines together.

Here's a picture showing that we can put an input into f, and then whatever the output is, we immediately feed that into g and see what the output is.

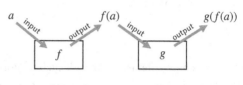

Here's the more mathematically rigorous presentation.

$$A \xrightarrow{\ f\ } B \xrightarrow{\ g\ } C$$
$$a \longmapsto f(a) \longmapsto g(f(a))$$

So we define the composite function $g \circ f$ by: $\boxed{(g \circ f)(a) = g(f(a)).}$

─────────────── **Notation for composites** ───────────────

This finally explains why the composition notation for arrows in a category looks "backwards": when we annotate functions, we put the variable (input) on the right so that the function can act on it from the left. But that means that a subsequent function will need to act on the left of that, which means that composites build up to the left even if the arrows are heading towards the right. Some people try and make this consistent by drawing arrows pointing to the left, or by writing functions acting on the right. I prefer to just give up on the idea of consistency. Incidentally I think that is the most widespread practice among mathematicians. Drawing arrows pointing left is quite unusual; writing functions acting on the right is less unusual but still not mainstream (although it may be more common in some countries or in some subjects).

─────────────── **Things To Think About** ───────────────

T 12.3   To show that sets and functions form a category we need to check that the identity really behaves like an identity, both on the left and on the right, and that composition is associative. We'll do that next, but you might want to see if you can check for yourself first.

12.3 Properties: unit and associativity laws

For properties we have to check unit laws and associativity.

For the unit laws we need to check that the identity function behaves like the identity on each side, that is, given a function $A \xrightarrow{f} B$ we need to check these two equations hold.

$$A \xrightarrow{1_A} A \xrightarrow{f} B$$
$$\parallel$$
$$f$$

$$A \xrightarrow{f} B \xrightarrow{1_B} B$$
$$\parallel$$
$$f$$

We can "chase" elements through each diagram as shown here. This means we consider an element in the set at the start of the diagram and follow it through arrow by arrow to the end to see where it arrives.

$$A \xrightarrow{1_A} A \xrightarrow{f} B$$
$$a \longmapsto a \longmapsto f(a)$$

$$A \xrightarrow{f} B \xrightarrow{1_B} B$$
$$a \longmapsto f(a) \longmapsto f(a)$$

In both cases the element a arrives at $f(a)$, showing that in each case the composite is indeed the same as f. In more traditional algebra we might write it like this: for any $a \in A$

$$(f \circ 1_A)(a) \;=\; f(1_A(a)) \quad \text{by definition of } \circ$$
$$=\; f(a) \qquad \text{by definition of } 1_A$$

as required. (And similarly on the other side.)

To check that composition of functions is associative we need to consider three composable functions $A \xrightarrow{f} B \xrightarrow{g} C \xrightarrow{h} D$ and show that

$$(h \circ g) \circ f \;=\; h \circ (g \circ f)$$

that is, $\forall a \in A$: $((h \circ g) \circ f)(a) \;=\; (h \circ (g \circ f))(a)$.

This is really just a case of carefully writing out the definitions and getting the parentheses in the right places. I recommend you try it yourself to see if you really understand; in any case you might find that less painful than trying to read through this printed version, but here goes anyway. For any $a \in A$:

The left-hand side gives us   The right-hand side gives us

$$((h \circ g) \circ f)(a) \;=\; (h \circ g)(f(a))$$
$$=\; h(g(f(a)))$$

$$(h \circ (g \circ f))(a) \;=\; h((g \circ f)(a))$$
$$=\; h(g(f(a)))$$

So the two sides are equal as required.

12.4 The category of sets and functions

We now have all we need to construct the very important category of sets and functions. This category is typically written as **Set** and is given by:

- objects: all sets
- morphisms: a morphism $A \longrightarrow B$ is a function $A \longrightarrow B$
- identities: given a set A, the identity is the identity function 1_A
- composition: composition of morphisms is composition of functions.

This category is not only the starting point for building many other categories, it is also the prototype for a "well-behaved" category and the motivation for many of the structures we seek to express in a general category. The idea of **Set** as a starting point is that any mathematical structure that starts with an underlying set together with some other stuff can then be built from the category **Set** by looking at "sets with stuff" and "functions interacting well with that stuff". We'll look at this next.

The idea behind **Set** as a prototype is that this category is the basic world in which much of mathematics is done. So we know it is a good world in which we can do many things we like (if we like doing math at all). Then if we move to work in a different world we might want to be able to do some of those same things, so we work out exactly what aspect of the category **Set** enabled us to do those things, and then look for those features in the new category we want to work in. In order to do that we have to express those features just using the objects and arrows of the category **Set**, not any particular details about sets themselves. This is the idea behind expressing things categorically, rather than at the level of elements (of sets), and it will come up a lot when we start working with category theory.

13

Large worlds of mathematical structures

A further exploration of large categories, based on sets and functions.

We will now develop some categories of sets-with-structure, starting with monoids. The morphisms will then use the idea of "structure-preserving maps".

13.1 Monoids

As we said at the beginning of Chapter 12, we previously zoomed in and looked at an *individual* monoid as a (small) one-object category, but we can also zoom out and look at the *totality* of monoids. In this large category, the objects are monoids themselves, and the morphisms are functions between them.

Our aim now is to come up with a definition of "function between monoids" that is a sensible form of relationship for monoids. Recall[†] that a monoid has

- data: a set of objects
- structure: an identity and a binary operation
- properties: unitality and associativity.

So it can be broadly summed up as a "set equipped with extra structure (satisfying some properties)". Whenever we have "sets-with-structure" we can look for "sensible" functions between them, that is, functions that respect that structure. An arbitrary function might not respect the structure, in which case it wouldn't be such a useful way of relating two different examples to each other.

For example, let's start with the monoid of natural numbers $\mathbb{N}$ under addition, and the monoid of integers $\mathbb{Z}$ under addition.[‡] First let's consider the function $f \colon \mathbb{N} \longrightarrow \mathbb{Z}$ with $f(n) = n$, so every natural number is mapped to itself as an integer. This "respects" addition in the following sense: we can add two numbers together before mapping them across, or we can add them together after mapping them across, and we get the same answer.

[†] I know, it's tricky when mathematicians say "recall" and you feel like it's something you've never heard of before. But what I mean is "I said this earlier and I'm hoping you have some recollection of it but if you don't it's in an earlier chapter for you to refer to".

[‡] We're taking the natural numbers to be the non-negative whole numbers; the integers are all of these and also the negative numbers.

Adding them together before mapping gives this: $\quad f(n + m) \; = \; n + m$
Mapping them first before adding them gives this: $\quad f(n) + f(m) \; = \; n + m$

So it doesn't matter which way we do it, and this is one of the key conditions for being a "sensible" function between monoids. The condition is expressed abstractly like this: $\boxed{f(n + m) = f(n) + f(m).}$

<hr>

Things To Think About

T 13.1   Another possible function $f : \mathbb{N} \longrightarrow \mathbb{Z}$ is given by $f(n) = 2n$. See if you can check that this also satisfies the above condition.

<hr>

For this function $f(n + m) = 2(n + m)$, and $f(n) + f(m) = 2n + 2m$. These are equal, by the distributivity of multiplication over addition.

There are functions that are sensible and useful in other senses, just not in the sense of respecting addition. For example $f(n) = n + 1$ defines a perfectly sensible function $\mathbb{N} \longrightarrow \mathbb{Z}$, it just doesn't respect addition.[†] However, the function is sensible in the sense of being order-preserving (see Section 13.3).

Next we'll look at a function $\mathbb{N} \longrightarrow \mathbb{Z}$ that is useful and illuminating as it shows that we can "pair up" the natural numbers with the integers perfectly, even though on the face of it the natural numbers seem to be only half of the integers (the non-negative ones).

This function pairs the natural numbers with the integers by alternating between positive and negative numbers: the odd numbers are paired with positive numbers and the even numbers are paired with negative numbers (other than 0, which is paired with itself).[‡]

$$
\begin{array}{ccc}
0 & \dashrightarrow & 0 \\
1 & \dashrightarrow & 1 \\
2 & \dashrightarrow & -1 \\
3 & \dashrightarrow & 2 \\
4 & \dashrightarrow & -2 \\
5 & \dashrightarrow & 3 \\
6 & \dashrightarrow & -3 \\
& \vdots &
\end{array}
$$

<hr>

Things To Think About

T 13.2   See if you can show that this function does not respect addition.

<hr>

The idea is that this is an unnatural process, switching between positives and negatives like that, so you might already feel like this is not respectful towards the structure of numbers. Of course, "respect" here doesn't mean politeness, it means something rigorous, but these visceral feelings can often be a decent starting point for the math. In fact, developing a visceral sense for things in a mathematical world is an important part of learning math. It's very hard to get anywhere without that, a bit like driving round a city relying *entirely* on GPS.

<hr>

[†] $\; f(n + m) = n + m + 1$ but $f(n) + f(m) = n + 1 + m + 1 = n + m + 2$
[‡] This is not a rigorous definition of the function but I hope the idea is clear. If you're interested in formality it's good practice to try expressing this function as a formula but I don't think the formula is particularly illuminating.

Anyway, we only have to find one place where this goes wrong in order to show that something has gone wrong, so let's take $n = 1$ and $m = 2$. Then $f(n + m) = f(3) = 2$ but $f(n) + f(m) = 1 + (-1) = 0$. So this function does *not* respect addition. However, it is still illuminating in the sense that it shows us that the natural numbers can be perfectly paired up with the integers. This is called a *bijection* and we'll come back to it in Chapter 14 as it gives us a pertinent notion of "sameness" for sets.

The above two non-examples were to illustrate that different functions can count as "sensible" or "illuminating" depending on which aspects of structure we are trying to illuminate. The idea of category theory is that while we are thinking about monoid structure, the "sensible" notion of morphism is the one that respects monoid structure.

There is one more aspect to a monoid structure other than addition (or more generally, the binary operation) — the identity. So if we're going to respect the entire monoid structure we need to respect that as well, which means that the identity should map to the identity. In our examples above, that means we need $f(0) = 0$, which is indeed the case.

―――――――― **Things To Think About** ――――――――

T 13.3   Mapping into the integers is a bit of a special case because it turns out that respecting addition forces a function to respect the identity. Can you prove that? We'll come back to it when we're talking about groups.

We have now basically seen the entire definition of a *monoid homomorphism* or just *monoid morphism*. "Morphism" is another word for "arrow" in a category, and "homomorphism" is a generic mathematical word for "morphism that leaves things the same", that is, a structure-respecting morphism, also sometimes referred to as structure-preserving.

Definition 13.1   Let A and B be monoids, and write the identity as 1 and the binary operation as ∘ in each case.[†] Then a *monoid homomorphism* from A to B is a function $f : A \longrightarrow B$ on the underlying sets, such that

- f respects ∘: for all $x, y \in A$ we have $f(x \circ y) = f(x) \circ f(y)$, and
- f respects the identity: $f(1) = 1$.

In general not every function on the underlying sets will respect the monoid structure, so this is a restriction to some special functions.

――――――――――――――

[†] Some authors would studiously always call the monoid $(A, 1, \circ)$ but I find that a little pedantic although it is arguably more correct. I think as long as it's unambiguous, we can just call the monoid A with impunity. Sometimes we say "by abuse of notation", like a confession.

─────────────────── **Note on "obvious axioms"** ───────────────────

Category theorists sometimes have a habit of referring to "the obvious axioms" rather than actually listing them, which can frustrate other people. However, if you get really into the spirit of category theory you'll start understanding what counts as "obvious axioms". When it comes to maps between sets-with-structure, the axioms ensuring that structure is preserved might be thought of as the "obvious axioms" for a map.

───────────────

Our aim here was to form a category for the totality of monoids and their morphisms, and there are now a couple of things to check.

─────────────────── **Things To Think About** ───────────────────

T 13.4   What do we need to check to show that monoids and their homomorphisms form a category? Can you check it?

We need to check that there are identities and composition, and that they are unital and associative. Now, monoid morphisms are special functions between monoids. So if we can show that the usual identity functions and composition work, then the axioms will automatically hold, as they hold for all functions.

For the identities, we need to check that the identity function on any monoid is a monoid morphism. The identity function leaves everything the same, so it is indeed structure-preserving; that part is usually pretty easy when checking a category with structure-preserving maps.

Intuitively, composition should be immediate too: we start with composable functions f and g, and have to show that if they're each structure-preserving, then the composite is structure-preserving. It would be quite odd to do one structure-preserving thing followed by another and find that structure had not been preserved overall. But we can (and should) still check. The most confusing thing in what follows is probably notation, as there is the potential for using ∘ too many times. So I'm going to use ∘ for composition of the functions and ∗ for the binary operation in the monoids. First we show $g \circ f$ respects ∗:

$$
\begin{aligned}
(g \circ f)(x \ast y) &= g(f(x \ast y)) && \text{by definition of } \circ \\
&= g(f(x) \ast f(y)) && \text{as } f \text{ respects } \ast \\
&= g(f(x)) \ast g(f(y)) && \text{as } g \text{ respects } \ast \\
&= (g \circ f)(x) \ast (g \circ f)(y) && \text{by definition of } \circ
\end{aligned}
$$

Next we show that $g \circ f$ respects 1:

$$
\begin{aligned}
(g \circ f)(1) &= g(f(1)) && \text{by definition of } \circ \\
&= g(1) && \text{as } f \text{ respects } 1 \\
&= 1 && \text{as } g \text{ respects } 1
\end{aligned}
$$

So $g \circ f$ is a monoid morphism as required. Thus we have a category **Mnd** whose objects are all the monoids, and whose morphisms are all the monoid morphisms between them.

We quite often name categories like this: naming them after their objects, in abbreviation, in boldface. This is when it is sort of "obvious" what morphisms we are using. When it's not, then we might name the category after their morphisms; we'll see some examples of that later.

13.2 Groups

We have looked at homomorphisms of monoids as "structure-preserving functions", and we have looked at groups as monoids with an extra condition. So let's now try and combine those ideas to think about homomorphisms of groups. The idea is that we have been gradually building up structure on sets.

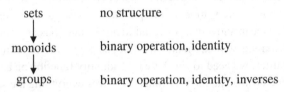

Note that at the moment those arrows I've drawn are just schematic arrows but, as you might have guessed, later we are going to see that they have a formal meaning and they are indeed arrows in some very large category, and that this process of building up structure is an intuitive idea that we make very precise.

―――――――――― **Things To Think About** ――――――――――

T 13.5   Can you invent the definition of "group homomorphism" for yourself, as a function respecting the group structure?

When we're dealing with groups it turns out to be redundant to ask for the identity to be preserved. It's also redundant to ask for the inverses to be preserved. Can you prove this?

A group is a monoid with some extra structure, so a group homomorphism should *at least* be a monoid homomorphism, but it should also preserve the extra structure. So we might think that a group homomorphism $f: A \longrightarrow B$ should be a function on the underlying sets such that

- the binary operation $\circ$ is respected: $\forall x, y \in a,\ f(x \circ y) = f(x) \circ f(y)$,
- the identity 1 is respected: $f(1) = 1$, and
- inverses are respected: $\forall x \in A,\ f(x^{-1}) = (f(x))^{-1}$.

However it turns out that the last two conditions are redundant, that is, we can deduce them from the first one.

For identities, first note that $1 = 1 \circ 1$, so we have

$$
\begin{aligned}
f(1) &= f(1 \circ 1) && \text{applying } f \text{ to both sides} \\
&= f(1) \circ f(1) && \text{by the first condition} \\
\text{thus} \quad 1 &= f(1) && \text{multiplying both sides by } (f(1))^{-1}
\end{aligned}
$$

> Remember that multiplying anything by its inverse takes us back to the identity. I have sort of jumped over a step involving identities here. Also I've referred to $\circ$ as "multiplication" generically.

For the inverses, given any $x \in A$ we know $x \circ x^{-1} = 1$. So we have the following chain of logic. (I admit that this is one of those calculations that might not seem very illuminating overall, but I hope you can at least follow each step. It's more like a jigsaw puzzle where you put together the pieces in the only way that they fit, and then you're done.)

$$
\begin{aligned}
\text{First} \quad f(x \circ x^{-1}) &= f(1) && \text{applying } f \text{ to both sides} \\
&= 1 && \text{as } f \text{ preserves identities} \\
\text{But also} \quad f(x \circ x^{-1}) &= f(x) \circ f(x^{-1}) && \text{as } f \text{ respects } \circ \\
\text{thus} \quad f(x) \circ f(x^{-1}) &= 1 && \text{combining the two parts} \\
\text{so} \quad f(x^{-1}) &= (f(x))^{-1}
\end{aligned}
$$

Note that the last step is by left-multiplying both sides of the equation by $(f(x))^{-1}$. (I specified *left*-multiplying as $\circ$ might not be commutative.)

──────── **Technicality on preserving identities** ────────

Proving that the condition on identities was redundant depended on the existence of inverses, which is why it doesn't happen for general monoids. When we mapped into the integers we were regarding it as a monoid but it secretly did have inverses, which is why identities were "automatically" preserved. There do exist monoid maps that respect the binary operation without respecting the identity. Here's a tiny example, in case you're curious.

- Let A be a monoid with two elements 1 and a, with $a^2 = 1$.
- Let B be a monoid with two elements 1 and b, with $b^2 = b$.

Now define a function $f \colon A \longrightarrow B$ by $f(1) = f(a) = b$. This does not preserve the identity, but it does respect the binary operation.

These monoids may seem a little contrived, but the first one involves an element that acts like a reflection, because if you do it twice you get back to the identity. The second one involves an element that acts like a projection,[†] because doing it twice is the same as doing it once. This example is not exactly important, it's just here to show that the condition on preserving identities really is necessary for monoids, and thus also for categories.

——— Things To Think About ———

T 13.6   Do we have to check anything else to show that groups and group homomorphisms form a category?

There isn't *really* anything else to check because this is really just the same as the definition of monoid morphism, and identities and composition will work just the same way. So we can declare we have a category **Grp** of groups and group homomorphisms.

13.3 Posets

Posets (partially ordered sets) are another example we've seen of "sets with structure". In this case the structure was an ordering. We'll think about these in two ways: directly, and through the concept of structure-preserving maps.

——— Things To Think About ———

T 13.7   Consider the following totally ordered sets, where the ordering is $\leq$.

$$A = \{1, 2\}, \qquad B = \{1, 2, 3\}$$

The diagrams below depict some functions from A to B, showing the action on elements. Which ones do you think deserve to be called "order-preserving", and why? What visual features do you notice? What other functions can you think of that deserve to be called order-preserving? Can you draw all the possible functions (first working out how many there are) and decide which ones are order-preserving? What about for other examples of A and B?

Can you make up a definition of "order-preserving function" for posets in general just by analogy with the structure-preserving maps we've seen already?

[†] A projection is like a shadow. If you're on a cartesian plane in XY-coordinates, you could project onto the X-axis by just taking the X-coordinate. If you then do it again you just get the same thing as you're already on the X-axis.

One way of thinking of "order-preserving" for a totally ordered set is to think about whether the elements stay in the same order after they've been mapped over to B. You could imagine two people in a queue in positions 1 and 2 in queue A, and then they're asked to move into queue B and the function shows the new positions they end up in. Are they in the same order or not? You can see this in the pictures from whether the dotted arrows cross over or not. In the first and second examples above they don't, and the two people do stay in the same order in the notional queue. In the third example they cross over, so the order has changed. The last example is a bit ambiguous — they've ended up in the same place. They haven't changed order but they haven't exactly stayed in order either.

We know from what we saw before that with 2 inputs and 3 outputs there should be 9 possible functions. Here are the other five:

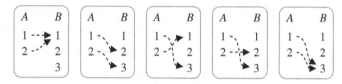

The second one definitely keeps everything in the same order, the third and fourth definitely don't, and the first and last are ambiguous.

Sometimes when there's an ambiguous case, looking at the formalism can help us sort out a good way of thinking. Here we have posets with an ordering $\leq$ which is the "structure". A structure-preserving function $A \xrightarrow{f} B$ might then be reasonably thought of as one where for all $x, y \in A$ we have

$$x \leq y \implies f(x) \leq f(y).$$

─────────────── **Things To Think About** ───────────────

T 13.8   Does this match the definition we came up with from the pictures? Does it allow the ambiguous cases to count or not?

This formal definition includes the ambiguous cases as order-preserving, because the ordering includes equality as a possibility. If we didn't want to allow it we'd have to do something "stricter" and perhaps use $<$ instead. The trouble is that it's not such a categorical notion. (Recall that when we tried to express $<$ as arrows in a category we found we would not have any identities.)

Another way we could eliminate the ambiguous cases would be to put in a condition saying that different elements of A are not allowed to land on the same place in B, which is what we'd say if we really were asking people to move queue. In fact this is a sensible, interesting and illuminating condition to think about for any functions, not just order-preserving ones. It's called "injectivity" and we'll come back to it in Chapter 15.

─────────────── **Things To Think About** ───────────────

T 13.9   Do we have to check anything to show that we have a category of posets and order-preserving functions?

As with monoids and groups, we need to check that the identity function is order-preserving and that the composite of two order-preserving functions is order-preserving. And likewise nothing much happens here.

Let's quickly write out the part about composites: suppose we have maps $A \xrightarrow{f} B \xrightarrow{g} C$ which are order-preserving, and consider $x, y \in A$. Then

$$
\begin{aligned}
x \leq y &\implies f(x) \leq f(y) && \text{since } f \text{ is order-preserving} \\
&\implies g(f(x)) \leq g(f(y)) && \text{since } g \text{ is order-preserving} \\
&\implies (g \circ f)(x) \leq (g \circ f)(y) && \text{by definition of composition}
\end{aligned}
$$

as required.

So we have a category **Poset** of posets and order-preserving functions.

─────────────── **Things To Think About** ───────────────

T 13.10   Think again about the poset of factors of 42 from Chapter 5. At that time we observed that 6 < 7 but 6 is higher than 7 in this diagram. What two orderings are we considering on the same set of numbers here, and what map is it that is not order-preserving?

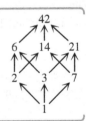

There are actually three orderings floating around in this diagram.

1. The one that we've actually drawn in arrows, where $a \rightarrow b$ whenever a is a factor of b; remember this is written $a \mid b$.
2. The usual ordering according to size, $a \leq b$.
3. The physical levels on the page.

This last ordering really goes according to how many terms there are in the prime factorization of each number.

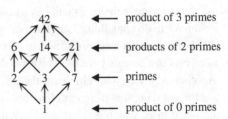

It's a bit long-winded to explain without some formal notation, so let's use this: given a whole number a, write
$N(a)$ for the number of terms in its prime factorization.[†]
Here is a table showing the values of N at each level of
the cube. We will write $a \,\square\, b$ whenever $N(a) \leq N(b)$.

a	$N(a)$
42	3
6, 14, 21	2
2, 3, 7	1
1	0

Now things are slightly tricky as we have one set but three
different orderings on it. This is where it helps to declare what
ordering we're using, so if we write A for the set of factors of
42 then we have the three posets shown on the right.

1. $(A, \,|\,)$
2. $(A, \leq)$
3. $(A, \,\square\,)$

We can now think about the identity function on A and ask whether or not it
is order-preserving between these different posets. The issue of 6 being above
7 is formally this: $7 \,\square\, 6$ although $6 \leq 7$. Formally this is saying that the
identity function on A is not an order-preserving map between these posets:

$$(A, \,\square\,) \to (A, \leq).$$

In the case of the analogous cube of privilege it becomes a source of deep
social conflict. Here are the analogous structures side by side. I have flipped
the diagram of factors to emphasize which parts of the structure I'm seeing as
analogous to 6 and 7.

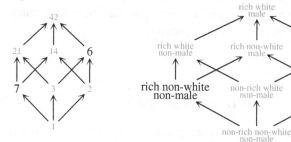

The point here is that, just like with the factors of 42, there are two separate
hierachies in the diagram of privilege: one by height on the page, which now
represents the number of types of privilege someone has, and the other by
absolute amounts of privilege or comfort that someone has in the world.

When ordered by number of types of privilege, poor white men have two
types and so are "higher" on the page than rich black women; poor white men
are in the position of 6 and rich black women are in the position of 7 with
only one type of privilege. However, when it comes to absolute privilege I am

---

[†] For numbers with repeated prime factors we have to be careful to count the terms here because
something like $2 \times 2 \times 3$ needs to count as having three terms. However here we can just think
about the number of prime factors each number has, as they're all distinct.

sure that a sufficient amount of wealth can greatly mitigate for the very real disadvantages of being black and female, where no amount of maleness and whiteness can counteract being very poor, perhaps unemployed or homeless. Thus as with the factors of 42 we have an identity function that does not preserve order. This has helped me understand why some poor white men are particularly angry about the theory of privilege, as they are told they are very privileged, but they see people with technically less privilege doing better than them in society. So they think the theory of privilege is nonsense, because they don't feel the manifestation of any privilege.

I think it is much more productive to understand the source of this anger rather than simply get angry in return. This poset and the map that doesn't preserve order are how I found clarity around this issue, and I have found that the abstraction helps to keep emotional reactions out of the way and enable a less divisive discussion. I currently consider this to be the most important "application" of category theory I have found, even though it's really an application of categorical thinking rather than of a piece of category theory itself.

Incidentally the analogy between these structures is not just an informal similarity, it is a rigorous form of sameness between categories that we will come back to in Part Two of the book.

It is worth looking out for any non-order-preserving functions in life. Whenever you have two hierarchies on the same set of people and those hierarchies don't agree it is a potential cause of antagonism.

13.4 Topological spaces

We only briefly looked into topological spaces so we'll only briefly look into their structure-preserving maps here. Again, I think it is worth sketching some of the ideas here even though a rigorous treatment is beyond the scope of this book; see Appendix D for more details.

Topological spaces are sets with "a notion of closeness" (which is actually called *open sets*). So a sensible notion of function between them is going to be "functions that preserve closeness", which is the concept of a *continuous function*. The idea is that a function deserves to be called continuous if it doesn't break things apart. So for a function $A \longrightarrow B$ if two points were close together in A then they should not end up far apart in B. But note that "closeness" in topology isn't really to do with distance, it's more to do with connectedness.

For example let's think about functions $\mathbb{R} \longrightarrow \mathbb{R}$ so that they're a bit more familiar, and we can draw them as a graph and use visual intuition.

In this exponential graph $f(1)$ and $f(1.1)$ are much further apart than 0.1, because the graph of an exponential is so steep (and gets steeper all the time). However the points are connected so it sort of doesn't matter.

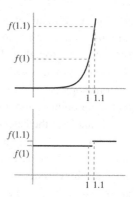

Whereas in this one $f(1)$ and $f(1.1)$ are closer together in terms of distance, but there's a gap in the graph in between meaning that in topological terms they're infinitely far apart as there's no way to get from one to the other.

Continuous functions (also called continuous maps) are typically defined in calculus via a tortured-looking formula that you might have seen at some point in your life, like this:

$$\forall a \in \mathbb{R} \ \forall \varepsilon > 0 \ \exists \delta > 0 \ \text{s.t.} \ |x - a| < \delta \implies |f(x) - f(a)| < \varepsilon$$

I think this definition is brilliant, but unfortunately its symbols, the machinations involved in using it to prove anything, and the trauma associated with calculus classes can serve to obscure the elegance of its ideas and the point of continuous maps, which is the preservation of the closeness structure. That last point is all you really need to absorb for our purposes.

In calculus one goes to great lengths to prove that continuous functions compose to make continuous functions. That is very important, and together with the (easier) fact that the identity function is continuous, it gives us a category of topological spaces, written **Top**, with

- objects: all the topological spaces
- morphisms: a morphism $A \longrightarrow B$ is a continuous function $A \longrightarrow B$.

In practice topologists (and thus category theorists) often want to eliminate some really "pathological" weirdly behaving spaces from their study, and so we might restrict the category **Top** to include only better-behaved topological spaces, but keep all the morphisms between them.

In fact continuous functions are arguably not quite the point of topology. The definition of topological space is there to support the definition of continuous function, which is in turn there to support the definition of "continuous deformation". When we talked earlier about continuously nudging paths around to see if they were genuinely different or not, that was an example of a continuous deformation. We can do this thing to entire spaces to see whether they count as genuinely different or not. This is the idea behind the infamous topological result that "a coffee cup is the same as a doughnut". Note that this coffee cup has to have a handle and the doughnut has to have a hole. The idea is that if these were made of playdough you could turn one into the other by continuously squidging it around, without ever breaking or piercing it or sticking separate parts together. This is a more subtle notion of sameness appropriate for topology, and is technically called homotopy equivalence.

Homotopy is the notion of continuous deformation which is in some sense really the point of topology, but it requires more dimensions of thought than are available in the category of spaces and continuous maps. One way to deal with those higher dimensions is to go into higher-dimensional category theory, which we will briefly discuss at the end of the book. Another way is to try and work out how to detect traces of the higher dimensions in lower-dimensional algebra so that we can avoid creating new types of algebra to deal with it.

Algebraic topology largely takes the approach of detecting each dimension as a group structure, via *homotopy groups*, *homology* or *cohomology*.[†] This involves building some sort of relationship or relationships between the category **Top** of spaces and continuous maps, and the category **Grp** of groups and group homomorphisms. This means we need morphisms between categories.

13.5 Categories

We have been talking about large categories formed by totalities of mathematical structures. Our examples have included monoids, groups, posets and topological spaces. All of those are things we already had as small examples of mathematical structures.

That is, an individual monoid is an example of a category, as is any individual group, poset or topological space. We have a schematic diagram like this:

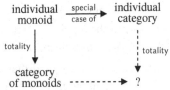

———————— **Things To Think About** ————————
T 13.11   What might you expect to find at the bottom right corner? What might the bottom dotted arrow represent?

This diagram is just schematic, that is, it's a diagram of ideas and thought processes, but it does hint to us that if the totality of monoids can be assembled into a category then we should be able to assemble a totality of categories themselves.[‡] In that case monoid homomorphisms should be a special

---

[†] I've just included those terms in case you're interested in looking them up.
[‡] We will use size restrictions to avoid Russell's paradox; see Appendix C for more details.

case of the arrows between categories. So should group homomorphisms and order-preserving functions. The totality of monoids should sit inside the totality of categories as something we might call a "sub-category".

We need a concept of "morphism between categories", and we can arrive at the definition by our usual method of thinking about structure-preserving maps. However, the situation is now more complicated because our underlying data isn't just a set: even if we restrict to small categories, we have a set of objects but also for every pair of objects a set of arrows from one to the other. As our data has two levels, our underlying function needs two levels as well: we need a function on objects and a function on arrows.

─────────────── **Things To Think About** ───────────────

T 13.12   Can you think of a sensible starting point for a definition of a morphism of categories $F: \mathbb{C} \rightarrow \mathbb{D}$? It should map objects to objects and arrows to arrows, but if we start with an arrow $x \xrightarrow{f} y$ in $\mathbb{C}$ what should the source and target of $F(f)$ be in $\mathbb{D}$?

- We definitely want for every object $x \in \mathbb{C}$ an object $F(x) \in \mathbb{D}$.
- Given an arrow $x \xrightarrow{f} y$ in $\mathbb{C}$ we want an arrow $F(f)$ in $\mathbb{D}$ and it should live here

$$F(x) \xrightarrow{F(f)} F(y).$$

Sometimes we get a little bored of all these parentheses so we might omit them and just write Fx, Fy and Ff trusting that using uppercase and lowercase letters makes the point that F is applied to the lowercase letters.

We now need to think about what "structure-preserving" means.

─────────────── **Things To Think About** ───────────────

T 13.13   Can you come up with the definition of a morphism between categories, starting with the action of F on objects and arrows above, and then proceeding by making sure it is "structure-preserving"? To do this, you need to be clear what the structure of a category is: identities and composition. Once you've done this, can you check that this gives us a category of small categories and all morphisms between them?

These morphisms between categories are so important that they have an actual name rather than just "morphisms of categories": they are called *functors*. The property of preserving identities and composition is called *functoriality*. We will give the full definitions in Chapters 20 and 21. For now it's enough to keep in mind that functors are a good type of relationship for categories, via structure-preserving maps.

13.6 Matrices

Most of the examples we've seen so far have involved morphisms/arrows that are some sort of function or map, preserving structure. I want to stress that this need not be the case. We've seen a few small examples of that where the arrows were relations, or "assertions". In the case of factors an arrow $a \longrightarrow b$ was the *assertion* that a is a factor of b. In the case of ordered sets an arrow $a \longrightarrow b$ is the assertion $a \leq b$.

Another type of example was the category of natural numbers expressed as a monoid, where we had just one (dummy) object and the morphisms were the numbers themselves. One of my favorite examples is the category of matrices. In this category again it's not the objects that are really interesting but the morphisms: the morphisms are matrices. That is, we regard matrices as a sort of map.

A matrix is a grid of numbers like these examples. It doesn't have to be square — it could have a different number of rows and columns, like the second one.
$$\begin{pmatrix} 1 & 2 \\ 2 & 5 \end{pmatrix} \quad \begin{pmatrix} 1 & 2 & 3 \\ 2 & 5 & 1 \end{pmatrix}$$

The second one is called a 2×3 matrix as it has 2 rows and 3 columns; in general an $r \times c$ matrix has r rows and c columns. In basic matrices the entries are all numbers, but they could come from other suitable mathematical worlds; we won't go into that here.

One of the baffling things about matrices when you encounter them in school is how they get multiplied. Addition isn't too bad because you just add them entry by entry. But multiplication involves some slightly contorted-looking thing involving twisting the columns around onto the rows.

It is not my aim to explain matrix multiplication here, but if you do remember it you might remember that we could multiply the matrices shown, by sort of matching up the row and column shown, and continuing.
$$\begin{pmatrix} 1 & 2 & 3 \\ 2 & 5 & 1 \end{pmatrix} \times \begin{pmatrix} 1 & 2 & 3 & 4 \\ 3 & 7 & 1 & 0 \\ 0 & 2 & 4 & 2 \end{pmatrix}$$

The important point for current purposes is that the *width* of the first matrix has to match the *height* of the second. In general we can multiply an $a \times b$ matrix with a $b \times c$ matrix to produce an $a \times c$ matrix.

─────────────── **Things To Think About** ───────────────

T 13.14   Does this remind you of composition in a category? How?

In a bit of a leap of imagination we can regard an $a \times b$ matrix as a morphism $a \longrightarrow b$, in which case a $b \times c$ matrix is a morphism $b \longrightarrow c$ and the condition of composability matches the condition of when we can multiply matrices. We get a category **Mat** with

- objects: the natural numbers
- arrows: an arrow $a \longrightarrow b$ is an $a \times b$ matrix
- composition: matrix multiplication.

─────────────── **Things To Think About** ───────────────

T 13.15 If you've ever seen matrices before you might like to think about what the identity is and how we check the axioms.

This might seem like a weird category until we consider that one of the points of matrices is actually to encode linear maps between vector spaces.[†] In that way, matrices really do come from some kind of map, it's just that the encoding has taken them quite far away from looking like a map. That typically makes it very handy for computation (especially if you're telling a computer how to do it) but can be quite baffling for students if they're just told about matrices for no particular reason — like so much of math.

Incidentally, you might wonder if there's a category of vector spaces and linear maps, and whether there's a relationship (perhaps via a functor) between this category and the category of matrices. The answer is yes, and if you spontaneously wondered those things then that's fantastic: you're thinking like a category theorist.

Having a framework that works at all scales

This tour of math has gone on quite long enough now and I just want to end by remarking that we have seen examples of categories at many scales: small drawable ones, individual mathematical structures, and large totalities of mathematical structures. This way in which category theory can work at many different scales is one of its powerful aspects.

In the next part of the book we are going to investigate things that we actually do in categories and in category theory. In some sense it doesn't matter whether or not you understand or remember any of the examples we saw in this tour. If you're interested in abstract structures, the things we see in the next part can be of interest entirely in their own right.

───────────────

[†] I mention all this in case you've heard of it; don't worry if you haven't.

PART TWO

DOING CATEGORY THEORY

In Part One we built up to the idea of categories, warmed up into mathematical formalism, and then met the definition of a category. We then took an Interlude to meet some examples of categories and see how various branches of mathematics could be seen in a categorical light. In Part Two we are going to "do" category theory. This means we are going to build on the basic definition of a category and think about particular types of structure we might be interested in, inside any given category. The point of the definition of a category is to give ourselves a framework for studying structure, so I would say that "doing category theory" means studying interesting structure, using the framework. The framework involves a certain amount of formality in order to achieve rigor, and so we are going to use that formality in what comes next. Part Two might therefore seem significantly harder than what came before, and I hope you will congratulate yourself for reading some quite advanced mathematics.

14

Isomorphisms

The idea of sameness and how category theory provides a more nuanced way of dealing with it.

14.1 Sameness

The idea of sameness is fundamental to all of mathematics. This is and isn't like the widely-held view of math as "all about numbers and equations". The objects we study in math *start* with numbers, and the idea of sameness *starts* with equality and equations, but as things develop the objects become much more complicated and subtle, and so does the notion of sameness.

Numbers aren't very subtle, which is a good and a bad thing. The simplification of real world situations into abstract numbers means that results can be stated clearly, but with a necessary loss of nuance. That is fine as long as we're aware we're doing it. The simplification means that numbers really don't have many ways of being the same as each other: they are equal or they aren't. That said, if we're dealing with decimal fractions we do talk about things being the same up to a certain number of decimal places, but that's usually for practical rather than theoretical purposes.

At the other extreme, *people* are about as complicated and nuanced as it is possible for anything to be, but we still try to talk about equality between people, and we get terribly confused in the process. On the one hand equality is a basic principle of decent society, but on the other hand no two people are the same. How can men and women be equal when they're different? We need a more subtle notion of sameness that doesn't demand that objects are *exactly* the same but finds some pertinent *sense in which* they're the same, and gives us a framework for using that sense as a pivot in similar ways to how we use equations to pivot between two different points of view.

This is an important way of getting out of black-and-white thinking that only has yes and no answers. That sort of thinking can get us into divisive arguments and push us further apart to extremes, when really we're all in a gray area somewhere. Sometimes it actually endangers humans, as in the COVID-19 pandemic when some people argued that there's no point wearing masks as they're not 100% effective — as if anything less than 100% counts as 0.

Things To Think About

T 14.1   For each of the following pairs of situations, can you think of a sense in which they're the same and a sense in which they're not the same? It's good to be able to think of both, in as nuanced a way as possible. That is, try and find things to say that encapsulate the real crux of the matter, not just "these are both numbers".

1. a) $6 + 3$
    b) $8 + 1$

2. a) $1 + 4$
    b) $4 + 1$

3. a) $-(-4)$
    b) $\dfrac{1}{\frac{1}{4}}$

4. a) $12 = 2 \times 2 \times 3$
    b) $12 = 3 \times 2 \times 2$

5. a) Christian bakers being forced to bake a cake for a same sex wedding
    b) Jewish bakers being forced to bake a cake for a Nazi wedding

6. a) COVID-19 spreading as a global pandemic
    b) A grumpy cat meme going "viral" online

This isn't about right and wrong answers, this is about what we can say about these situations to highlight what is interesting or subtle about them.

The first three examples are all pairs of ways of producing the same result, but something increasingly nuanced is going on. In (1) it's just two different ways of adding numbers to get the same answer. In (2) it's two different ways of adding *the same* two numbers to get the same answer. In (3) it's two different ways of doing an inverse twice to get back to where you started: in part (3a) it's the additive inverse of the additive inverse, and in part (3b) it's multiplicative inverses. Note that if we write things like $6 + 3 = 8 + 1$ as an equation, we may tend to focus on the sense in which the two sides are the same (they both come to 9) when really the point of the equation is that there's also a sense in which the two sides are different. Sometimes I like to say "all equations are lies" but really I mean that all equations involve some things that are in some sense not the same, apart from the equation $x = x$, which is useless. The point of an equation is to use the sense in which two sides are the same to pivot between the senses in which they aren't. Category theory says that the "sense in which they're the same" can be more relaxed than equality as long as it still gives us a way to pivot between two sides.

Example 4 is a case of this which you might have seen. These are two expressions of the prime factorization of 12. The Fundamental Theorem of Arithmetic says that every whole number can be expressed as a product of prime numbers *in a unique way*, but in that case these two examples have to count as "the same": changing the order of the factors doesn't count as different. Later we're going to see that once we've relaxed the notion of sameness, the notion of "uniqueness" gets a bit more relaxed as well.

We dealt with the vile argument of example 5 in Section 2.6. For (6), the two scenarios are obviously very different in content and importance, but they can both be modeled by the same math which is why viral memes are called viral. The idea is that each affected person goes on to affect R people on average, who then also each go on to affect R people on average, and we model that by an equation taking into account how many people are affected so far, and how many more potential victims remain. This equation is called the logistic equation and is often studied in calculus classes (but apparently not enough for any significant proportion of the population to understand it when we're faced with a real pandemic).

Equivalence relations are a version of sameness that is more subtle than equality but still somewhat crude; in the end they just amount to putting all objects in compartments and declaring that everything in the same compartment counts as the same.

Category theory gives us a framework in which we can say something more subtle. As we'll later see, when we go into higher dimensions, each extra dimension gives us another layer of subtlety. Categories have one extra dimension over sets. That extra dimension is where we express relationships between objects, and we are now going to use that dimension to express this particularly strong form of relationship, that is, categorical sameness.

14.2 Invertibility

We are going to consider two objects in a category to be "sort of the same" if they have a particularly strong relationship between them: an *invertible* arrow between them. This is like the idea of symmetry in an equivalence relation, except that instead of demanding that it is true everywhere, we just look for it and then sort of go "Oh look!".

Inverting a process is about undoing that process. An invertible process is one that we can undo, in such a way that we're essentially back to where we started, as if we hadn't done anything. For example if you write something in pencil, you can erase it, and you'll be pretty much where you started unless you're very picky about clean pieces of paper. If you freeze water and then thaw it you basically have normal water again. However, freezing can't be inverted so easily in all circumstances. If you freeze milk and thaw it then it might separate and look very strange like it's gone bad.

One thing that definitely can't be inverted is cracking an egg. Once you've cracked it you can't put it back together again (see Humpty Dumpty).

Things To Think About

T 14.2 1. When I was little I thought that pepper was the inverse of salt. Was I right or heading for lifelong disappointment?
2. In what sense is a pardon the inverse of a criminal conviction, and in what sense is it not? In what sense is a divorce the inverse of a marriage?
3. Can a wrong ever be inverted by a wrong?

In higher math we get to make choices about what to count as "the same", and this is a crucial point. In some categorical situations we make a category and then look for the things that count as the same inside it, but in other situations we start with some things that we want to count as the same, and then we construct a category in which that will be true.

We already met inverses when we were talking about groups in Section 11.3. But we also met them earlier without explicitly saying so, in Section 2.6 when we were thinking about negative numbers and fractions.

Those may seem to be just numbers, but if we think of each number as a process we have relationships like those shown on the right.

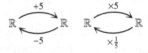

Here the arrows going in opposite directions depict processes that are inverses of each other. That is, whatever number you start with on the left, if you go over to the right and come back again, you get back to the number you started with. Also, if you start on the right, come over to the left and then go back right, you also get back to where you started; more precisely, that composite process is the same as doing the identity.

Things To Think About

T 14.3 Why do these following processes on   1. Squaring.
the real numbers not have an inverse?   2. Multiplication by 0

For squaring, we are trying to produce a process as shown here by the dotted arrow.

The first problem is that the way to undo squaring is to take a square root, but this is not a function $\mathbb{R} \longrightarrow \mathbb{R}$; we either have to include complex numbers (which I don't want to get into) or restrict to non-negative numbers and make a function $\mathbb{R}_{\geq 0} \longrightarrow \mathbb{R}$. But we still face the fact that there are positive and negative numbers that square to the same thing, so that if we try and undo the process we don't know where we're supposed to go.

We could just pick the positive one, which is the usual way of defining a square root function $\sqrt{\ }$ and we'll get the promising-looking relationship shown here.

However, the composites only produce the identity *sometimes*. Starting on the right is fine, but starting on the left goes wrong if we start with a negative number. For example if we start with -3 on the left we'll go right to 9 and then left back to 3, which is not where we started.

Multiplication by 0 has a similar problem but even worse: it sends everything to 0, so it's not just that pairs of numbers go to the same place on the right, it's that *all* numbers do.

This means it is impossible for us to define an inverse arrow going back again because on the right we always land at 0, so how can we be sure of coming back to where we started when the arrow going back can only pick one? It's hopeless. So 0 has no multiplicative inverse, which is in fact the content of the idea that you "can't divide by 0".

14.3 Isomorphism in a category

We are now ready to make the formal definition of "nuanced sameness" in a category, called isomorphism. It goes via the idea of inverses. In this definition, remember that 1_x denotes the identity morphism on an object x.

Definition 14.1 Let $a \xrightarrow{f} b$ be an arrow in a category $\mathcal{C}$. An *inverse* for f is an arrow $b \xrightarrow{g} a$ such that $g \circ f = 1_a$ and $f \circ g = 1_b$.

I think this is much more vivid with pictures. The situation we have is this pair of arrows.

We need the composites both ways round to equal the identity, as shown.

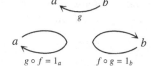

Note that the definition is symmetric in f and g, so that if g is an inverse of f then f is also an inverse of g and we can say they are inverses of each other.

Definition 14.2 Let $a \xrightarrow{f} b$ be an arrow in a category $\mathcal{C}$. If f has an inverse we say that f is *invertible* and call it an *isomorphism*, and we say that a and b are *isomorphic*. We use the notation $\cong$ as in $a \cong b$, or we write $a \xrightarrow{\sim} b$.

In the next section we're going to see senses in which isomorphic objects count as "the same" in a category even when they're not actually the same object. However the first thing we should do is check that this generalization of sameness hasn't thrown out the old notion of sameness, equality.

─────────────────── **Things To Think About** ───────────────────

T 14.4   Can you show that any object in a category is isomorphic to itself?

Given any object a in a category, the identity 1_a is an isomorphism and is its own inverse, as $1_a \circ 1_a = 1_a$. This deals with the composite both ways round as they're the same.

─────────────────── **Things To Think About** ───────────────────

T 14.5   Is it possible for an arrow to have two different inverses? It may help to think about the analogous question for additive inverses of numbers.

First we'll show that a number can only have one additive inverse. By definition, an additive inverse for a is a number b such that $a + b = 0$ (assuming addition is commutative so we don't have to insist on $b + a = 0$ separately). A typical way to show that something is unique in math is to assume that there are two and prove that they must be equal. So we can assume we have two inverses, b_1 and b_2. So this means $\boxed{a + b_1 = 0 = a + b_2.}$

We can then subtract a from both sides, and conclude that $b_1 = b_2$. More abstractly this consists of "canceling out" a using an additive inverse. It takes a little effort to make that rigorous, depending on how far towards first principles you want to go, but I mainly wanted to give a flavor to inspire us, as the proof for categorical inverses is analogous.

Note that for composition in a category we do have to be careful about the order, but the sides can be confusing so instead of talking about composing on the "left" or "right" of f I will say

- *post-composing* by g if we compose g *after* f, giving this: $a \xrightarrow{f} b \dashrightarrow^{g} c \ = \ a \xrightarrow{g \circ f} c$

- *pre-composing* by g if we compose g *before* f, giving: $x \dashrightarrow^{g} a \xrightarrow{f} b \ = \ x \xrightarrow{f \circ g} b.$

Just like when doing something to both sides of an equation in numbers, if we pre- or post-compose by g on both sides of an equation, those sides will still be equal. For example: if we know $s = t$ then we can deduce $g \circ s = g \circ t$.

Proposition 14.3 ("Inverses are unique.") *Let $a \xrightarrow{f} b$ be an arrow in a category $\mathcal{C}$. Suppose g_1 and g_2 are both inverses for f. Then $g_1 = g_2$.*

Proof Since g_1 and g_2 are both inverses for f we know

$$f \circ g_1 = 1_b = f \circ g_2 \qquad (1)$$
$$\text{and} \quad g_1 \circ f = 1_a = g_2 \circ f \qquad (2)$$

We now "cancel" f from both sides of (1), using (2), as follows.

Post-composing both sides of (1) by g_1 gives:

$$g_1 \circ f \circ g_1 = g_1 \circ f \circ g_2 \qquad (*)$$
$$\text{thus} \quad 1_a \circ g_1 = 1_a \circ g_2 \qquad \text{by (2)}$$
$$\text{so} \quad g_1 = g_2 \qquad \text{by definition of identities.} \qquad \square$$

────────── **Things To Think About** ──────────

T 14.6   Can you see how this proof is analogous to the one for additive inverses? Also see if drawing it out using arrows makes it more illuminating.

I think the point of the proof is clearer with a diagram. Here's the configuration we're using. $b \underset{g_2}{\overset{g_1}{\rightrightarrows}} a \xrightarrow{f} b \xrightarrow{g_1} a$

In Chapter 15 we will see diagrams like this again, and study the property of f we used here when we "canceled" it out.

As we go along we'll see quite a few uniqueness proofs that work like this, where we assume there are two things and then show they're the same. Once we have shown that something is unique we can use some notation for it, and for the inverse of f we typically write f^{-1}.

Finally, before we move on, note that inverses are unique but isomorphisms aren't — you can have more than one isomorphism between two objects. So an isomorphism doesn't just tell us that two objects are "the same" it tells us a way in which they are the same. In this way, saying two objects "are isomorphic" is different from actually producing an isomorphism between them. This is an important general principle that we will come back to periodically.

14.4 Treating isomorphic objects as the same

Isomorphism is the "more nuanced version of sameness" of category theory. Let's now see how category theory treats isomorphic objects as the same.

The point of the framework of categories is to study objects in a category via their relationships with other objects, not via their intrinsic characteristics. So it doesn't matter what the object is called or what it looks like or what it "is", we just look at what morphisms it has to other objects and how those morphisms interact with each other.

Isomorphic objects are treated as the same by the rest of the category because they have the same relationships with other objects in the category.

Suppose a and b are isomorphic objects via f and g like this.

Now consider some other object x in the category. We're going to show that x "can't tell the difference" between a and b because whatever relationships x has with a, it has the same system of relationships with b.

This is quite a deep idea, and is a bit like how I tell people apart, if I'm going to be honest. A lot of people look the same as each other to me in terms of physical appearance (especially white men) and I can only tell them apart via personal interaction with them.

─────────── **Things To Think About** ───────────

T 14.7   Can you think of how to show that x has the same relationships with a that it does with b? Consider a morphism $x \longrightarrow a$ and use the isomorphism to produce a morphism $x \longrightarrow b$. Then go back the other way.

We could draw the situation as shown on the right, with x sort of "looking at" a and b. We'll show how to switch back and forth between looking at a and looking at b.

Now, if we have a morphism $x \xrightarrow{s} a$ we can use it to produce a morphism $x \longrightarrow b$ by post-composing it with f. Conversely, given a morphism $x \xrightarrow{t} b$ we can produce a morphism $x \longrightarrow a$ by post-composing with g as shown.

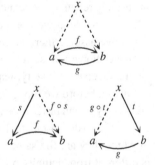

So we have a correspondence between morphisms $x \longrightarrow a$ and morphisms $x \longrightarrow b$. Moreover, the correspondence makes a perfect matching between the morphisms in each case.

That is, if we start with a morphism to a, turn it into a morphism to b and then turn it back into a morphism to a we really do get back the one we started with, as shown.

$$s \longmapsto f \circ s \longmapsto g \circ (f \circ s) = s$$

Note that $g \circ (f \circ s) = s$ as g and f are inverses so compose to the identity.

Things To Think About

T 14.8   Can you do the analogous construction for morphisms *to* *x* rather than morphisms *from x*? The object *x* cares about all its relationships with *a* and *b*, in both directions. What will be different in this situation?

> Note that what follows is pretty much one of those "jigsaw puzzle" type proofs where you fit pieces together in the only way that they'll fit. Try not to get too stuck on the words "pre-compose" and "post-compose", but follow the arrows round the diagrams instead. I use the words just to give you a chance to get used to them.

If we look at morphisms to *x* instead, we need to pre-compose with the isomorphisms rather than post-compose, to get the correspondence going. The diagrams would be like this.

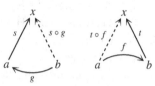

If writing out this version or looking at the diagrams feels not very substantially different to you that's good intuition. They are essentially the same diagrams, just with the arrows turned around to point the other way. This is the categorical notion of "duality" which we'll discuss further in Chapter 17. If we turn all the arrows around then we haven't really changed anything abstractly, so anything we just proved should still be true in the new version. However, the new version does give us new information, so sometimes we get a sort of BOGOF[†] on proofs, because we can just say "and dually" for the second one.

This is actually a step towards the "Yoneda embedding" and the "Yoneda Lemma". We'll come back to that in our grand finale, Chapter 23, but I'm pointing out this connection now in case you've heard of Yoneda elsewhere.

14.5 Isomorphisms of sets

When we have constructed a category we often start wondering things like: what are the isomorphisms in this category? We will do this any time we have defined a new categorical property. In the coming chapters, for example, we will define products, and then look to see what the products are in various other categories. We will define terminal objects and then wonder what the terminal objects are in various other categories. And so on. We have just defined isomorphisms, so we can look in various categories and wonder what the isomorphisms are. Let's start with the category **Set** of sets and functions.

[†] Buy One Get One Free, possibly "BOGO" in the US. Maybe this should be POGOF for Prove One Get One Free.

Consider sets A and B and inverses f and g as shown.

Here f takes inputs from A and produces outputs in B, and g does it the other way round. Remember that the composite both ways round needs to be the identity. This says if you take an element in A, apply f, and then apply g, you should get back your original element in A. And also if you take an element in B, apply g and then apply f, you should get back your original element in B.

Things To Think About

T 14.9   1. Take $A = \{a, b, c\}$ and $B = \{1, 2, 3\}$. Can you construct an isomorphism between them? How many different ones are there?
2. Now take $A = \{a, b, c\}$ and $B = \{1, 2\}$. Why is it not possible to construct an isomorphism?
3. Can you come up with a theory of when it is and isn't possible to construct an isomorphism between sets?

To construct an isomorphism we need to give one function going in each direction and show that they compose to the identity both ways round. Here is a pair of such functions for the first example.

$$A \xrightarrow{f} B \qquad B \xrightarrow{g} A$$
$$
\begin{array}{ll}
a \dashrightarrow 1 & 1 \dashrightarrow a \\
b \dashrightarrow 2 & 2 \dashrightarrow b \\
c \dashrightarrow 3 & 3 \dashrightarrow c
\end{array}
$$

Now if we start in A and go along f and then g, each element a, b, c will end up back at itself. The same is true if we start in B and go along g and then f. So this is indeed an isomorphism.

However with $A = \{a, b, c\}$ and $B = \{1, 2\}$ we're somewhat stuck for how to construct f. There are 3 inputs and only 2 outputs, so some of the inputs are doomed to have to land on the same outputs. That means that when we try to construct an inverse we're going to be in trouble — if an output was arrived at from two inputs, where should it go back to?

Suppose we tried to make f as shown here. How could we construct g going back again? 1 can just go back to a. However 2 "wants" to go back to both b and c, but a function can't send an input to two different outputs.

This means that there cannot be an isomorphism between these two sets. You might have worked out by now that there can be an isomorphism whenever the two sets have the same number of elements, and not if they don't.

Note on infinite sets

There is an important caveat here: this is only true of *finite* sets. Infinite sets work differently. In fact, we turn things around and use isomorphism of sets to define what it means for an infinite set to have the "same number" of elements as another. This gives us the notion of cardinality of an infinite set, and the idea that there are different orders of infinity: this is when infinite sets have no isomorphism between them, so have different "sizes" of infinity.

Note that not every function between isomorphic sets is an isomorphism. Consider the one shown on the right — this has no inverse. The sets are still isomorphic; they just need a different function to exhibit an isomorphism. Declaring two sets to be *isomorphic* and exhibiting *an isomorphism* thus involve different amounts of information.

And note further that there are several possible isomorphisms. You can try drawing them all, and here they are:

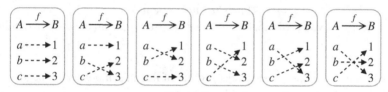

Perhaps in the course of drawing them all out you noticed your thought processes as you went along. One way you might do it is to decide where a is going to go (out of 3 options), and then decide where b is going to go, but there are now only 2 remaining options, and then there's no choice about where c goes as there's only one possible output left that hasn't already been taken. That means that the number of possible isomorphisms is: $3 \times 2 \times 1 = 6$.

There is a possible question here about why we multiply the number of options together rather than add them. The answer is that each option gives rise to several options afterwards. We could draw it as a decision tree as shown on the right. So first we make a decision about the output for a, and then *whichever* choice we made, there are 2 remaining options for the output for b. So the choices proliferate, and we multiply rather than add.

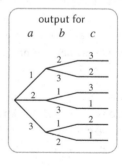

The functions that are isomorphisms between sets are called *bijections*. They might remind you of permutations — in fact permutations can be seen as bijections between a set and itself. As usual, bijections are typically something you meet long before you meet category theory, and so bijections have a direct characterization, as follows.

Non-categorical definition of bijection

A function $A \xrightarrow{f} B$ is a bijection if for every element $b \in B$ there is a *unique* element $a \in A$ such that $f(a) = b$.

The intuitive idea here is that if we were trying to construct an inverse for f we'd need to be able to go backwards from B back to A, but this only works if the elements on each side match up perfectly. This characterization is non-categorical because it refers to the elements of the sets A and B rather than just the morphisms between them. The way we expressed isomorphism in a category only referred to the objects and morphisms in the category. Indeed in an arbitrary category the objects might not be like sets so might not have elements. One theme of category theory is that we take definitions that refer to elements of sets, express them using only morphisms in a category, and can then immmediately apply them in any category we like, not just **Set**.

────────────── **Things To Think About** ──────────────

T 14.10   Can you prove that a function is an isomorphism of sets if and only if it is a bijection?

I suspect that reading someone else's proof of this in symbols is not nearly as enlightening as doing it yourself, but here goes.

Suppose first that we have a pair of inverse functions as shown. We aim to show f is a bijection. Let $b \in B$.

> We need to show that there is a unique $a \in A$ such that $f(a) = b$, which has two parts: existence and uniqueness. There is one way to get an element of A from b here, which is by applying g. So we do that, and check it "works".

- Existence: Let $a = g(b)$. Then $f(a) = f(g(b)) = b$ as f and g are inverses.
- Uniqueness: Suppose $f(a) = f(a') = b$. Thus $g(f(a)) = g(f(a'))$. But f and g are inverses, so this equation gives $a = a'$ as required.

Conversely suppose that f is a bijection. We aim to construct an inverse for it. Consider $b \in B$. We know there is a unique element $a \in A$ such that $f(a) = b$, so take this element to be $g(b)$ (so $f(g(b)) = b$ by definition). Again we need to show two things: that the composites both ways round are the identity.

- Given $b \in B$ we know $(f \circ g)(b) = f(g(b)) = b$ by definition, so $f \circ g = 1_B$.
- Given $a \in A$, $(g \circ f)(a) = g(f(a))$. What is $g(f(a))$? By definition $g(b)$ is the unique element in A such that $f(a) = b$. In this case we're doing $b = f(a)$ so we're looking for the unique element a such that $f(a) = f(a)$, which must be a itself. So $g(f(a)) = a$ and $g \circ f = 1_A$ as required.

14.6 Isomorphisms of large structures

We have looked at isomorphisms in the category **Set**, and we have also seen various large categories of structures based on sets, such as monoids, groups, posets and topological spaces. So we could now look at what isomorphisms are in all of those categories, which will give a good notion of sameness for the structures in question. First note that in each of these cases a morphism is a special function (one that respects structure) and so an inverse for that is going to be a structure-preserving function going back again, such that the composites are the identity both ways round. This means that there must at least be an underlying isomorphism of sets (which moreover must preserve structure) so we can at least start by knowing that our underlying function must be a bijection.

─────────────── **Things To Think About** ───────────────

T 14.11   Can you show that a morphism $A \xrightarrow{f} B$ of monoids is an isomorphism in the category of monoids if and only if it is a bijective homomorphism? The content of this is that we don't have to insist on the inverse being structure-preserving because it will automatically follow.

Can you show the analogous result is true for posets? If you've studied topology can you show that the analogous result for topological spaces is *not* true?

Isomorphisms of monoids

Let's start with monoids. Intuively two monoids should count as "the same" if there's a way of perfectly matching up their elements in a way that also makes the structure match up, so that all we've really done is re-label the elements. We'll now see that this is indeed what the formal definitions give us.

Suppose we have a monoid morphism $A \xrightarrow{f} B$ which is also a bijection. We know we have an inverse function $B \xrightarrow{g} A$ but the question is whether it preserves structure. We'll write the identity as 1 and the binary operation as ∘.

> Idea: f is a bijection so in particular if we apply it to two things and get the same answer then those two things must already have been equal.

- Identities: We need to show $g(1) = 1$. Applying f to the left gives 1, as f and g are inverses. Applying f to the right gives 1, as f preserves identities. So the left and right must be equal.[†]

[†] Note that identities are unique.

- Binary operation: We need to show that $g(b_1 \circ b_2) = g(b_1) \circ g(b_2)$. Applying f to the left gives $b_1 \circ b_2$ as f and g are inverses. Applying f to the right gives

$$
\begin{aligned}
f(g(b_1) \circ g(b_2)) &= f(g(b_1)) \circ f(g(b_2)) \quad \text{as } f \text{ respects } \circ \\
&= b_1 \circ b_2 \quad\quad\quad\quad\quad \text{as } f \text{ and } g \text{ are inverses}
\end{aligned}
$$

So the left and right must be equal.

We have shown that if f is a bijective homomorphism then it is an isomorphism of monoids. For the converse we have to show that if f is an isomorphism of monoids then it is a bijective homomorphism. This is immediate as f is a homomorphism, and it has an inverse, so it is certainly an isomorphism at the level of sets and thus must be a bijection.

That is the technical description of a monoid isomorphism, but the *point* of it is that if two monoids are isomorphic then they have "the same" structure, just with different labels. Imagine having one copy of the natural numbers painted red and another copy painted blue. They're still just the natural numbers, they just happen to be painted different colors. Their behavior as numbers would still be the same. You could translate them into a different language and their behavior as numbers would still be the same. In fact you could invent entirely different words instead of "one, two, three, …" and as long as you didn't change the interaction between the numbers, you could call them anything you wanted and the only problem would be communication with other people. This is the idea of isomorphisms of structure — that if we only change the labels, not the actual way in which the structure behaves, then it shouldn't really count as different. We'll talk about this more in the context of groups.

Conversely, monoids could have the same elements but different structures, in which case they are not isomorphic as monoids, only as sets.

――――――――――――― **Things To Think About** ―――――――――――――

T 14.12   Here are two monoids with the same set of elements $A = \{1, a\}$ but different structure. See if you can show that they are not isomorphic as monoids.

1. $(A, \circ, 1)$ has $a \circ a = 1$.
2. $(A, \odot, 1)$ has $a \odot a = a$.

We show that no bijective homomorphism is possible here. We'll try to construct a bijective homomorphism $f : (A, \circ, 1) \longrightarrow (A, \odot, 1)$. We know that the identity will have to be preserved and so $f(1) = 1$, so in order to be a bijection we must have $f(a) = a$. But now we can show that the binary operation is not preserved:

$$
\begin{aligned}
f(a \circ a) &= f(1) = 1 \\
\text{but} \quad f(a) \odot f(a) &= a \odot a = a
\end{aligned}
$$

In summary, the *names* of the objects are not important at all, it's the *interaction* between the objects which determines the structure. I like to think this is why I am bad at remembering people's names; because that's just a superficial labeling. I do remember people's characters, and sometimes genuinely recognize someone by their character rather than their face.

Isomorphisms of groups

Something similar is true for groups, as we know that a homomorphism of groups is just a homomorphism of the underlying monoids. Thus a group isomorphism must just be a bijective homomorphism. But again, the real point is that groups are isomorphic if they have the same structure. This same structure is indicated by the Cayley table patterns that we saw in Section 11.3.

We saw two groups with the tables shown on the right, and said that they "have the same pattern". We can now be more precise using the concept of a group isomorphism.

	0	90	180	270
0	0	90	180	270
90	90	180	270	0
180	180	270	0	90
270	270	0	90	180

	0	1	2	3
0	0	1	2	3
1	1	2	3	0
2	2	3	0	1
3	3	0	1	2

──────── **Things To Think About** ────────

T 14.13   Can you construct a group isomorphism between these groups? Can you construct two different ones? What is the *meaning* of this?

The group on the left is the group of rotations of a square, so I'll call it **Rot**. The one on the right is the integers mod 4 (under addition) which is written $\mathbb{Z}_4$.

We can construct an isomorphism by looking at the pattern and matching things up. We can see that things in the pattern correspond as shown. We can use this scheme to define a bijection and show it's a homomorphism. The fact that it makes the patterns correspond shows that it is respecting the binary operation of the group (though that's not quite a rigorous argument).

Rot	$\mathbb{Z}_4$
0 ◄---► 0
90 ◄---► 1
180 ◄---► 2
270 ◄---► 3

Another approach instead of sheer pattern-spotting is "thinking inside our head" as I like to call it, that is, thinking about deep reasons and meaning. In abstract math it often helps to take both approaches and see if they match up. (Sometimes when I'm trying to prove something I'll keep oscillating between the two.) Here we might notice that we can move from $\mathbb{Z}_4$ to the rotations by multiplying by 90.

This is related to the "deep meaning" of the situation which
is that if we count rotations in quarter-turns instead of degrees
then we really get the integers mod 4. Or, to look at it the other
way round, if we put the integers mod 4 on a 4-hour clock they
are really indicating the four possible rotations of a square.

To me this indicates "why" those two groups are isomorphic, in the sense of a
deep structural reason rather than a superficial "look we can switch the labels
around" reason. This is the sort of deep reason that I look for in abstract math,
and the sort of structural explanation that I seek in category theory.

In this particular case it also helps us
see that there is another possible isomorphism: I could have put the numbers on the clock going the other way
round, and that would produce a different correspondence with the angles.

Rot		$\mathbb{Z}_4$
0	◄----►	0
90	◄----►	3
180	◄----►	2
270	◄----►	1

Another way of thinking of this is that we could
re-order either table and get "the same" pattern,
as before, as shown here for the table of rotations. This is really just a sign that the choice
of clockwise or anti-clockwise is arbitrary and
doesn't affect the group structure.

	0	270	180	90
0	0	270	180	90
270	270	180	90	0
180	180	90	0	270
90	90	0	270	180

I like to think this means that, from an abstract mathematical point of view, it is
somehow correct that I get so confused between clockwise and anti-clockwise.

The other situations I
suggested thinking about
in Section 11.3 were mul-
tiplication in a couple
of other modular arith-
metic settings. Here are
the multiplication tables.

$\mathbb{Z}_8$	1	3	5	7
1	1	3	5	7
3	3	1	7	5
5	5	7	1	3
7	7	5	3	1

$\mathbb{Z}_{10}$	1	3	7	9
1	1	3	7	9
3	3	9	1	7
7	7	1	9	3
9	9	7	3	1

In general, $\mathbb{Z}_n$ is only a group under addition, not multiplication, as it might not have
all multiplicative inverses. Even if we omit 0 we might not have a multiplicative
inverse for everything: only for each number that has no common factor with n
except 1. So the above selections do in fact form groups.

It looks like neither of these has the same pattern as rotations of a square, but if we re-order the elements of the second one as shown on the right, the pattern appears. Re-ordering the rows and columns doesn't change the structure of the group, it just changes our presentation of it.

$\mathbb{Z}_{10}$	1	3	9	7
1	1	3	9	7
3	3	9	7	1
9	9	7	1	3
7	7	1	3	9

Note that there is no way to re-order the $\mathbb{Z}_8$ example to make the same pattern. We can tell this because it has identities all the way down the leading diagonal (top left to bottom right) — every element squares to the identity.

Things To Think About

T 14.14   Can you prove that if one group has every element squaring to the identity and another group does not, they can't be isomorphic?

We could prove formally that this prevents the groups from being isomorphic, but the structural reason is that isomorphic groups have the same deep structure, so any structural feature that one of them has must be there in the other as well — the only thing that's changed is the labels. When thinking categorically I would say it is the *reason* that really convinces us the thing is true; the formal proof is there for rigor.[†]

Groups can be referred to by their deep structure rather than their superficial labels. For example, the group of rotations of a square or the integers mod 4 under addition or anything else with the same structure is the *cyclic group of order* 4. When I was first studying groups I got very confused about whether there is one cyclic group of order 4 or many different ones coming from different places. There are theorems saying things like "The only possible group with two elements is the cyclic group of order 2" and yet there are many different versions of this group, so how is it unique? In retrospect I believe I was thinking like a category theorist: not wanting to consider those as genuinely different as they're only superficially different. We are going to see that "unique" in category theory means that there might be isomorphic ones but not *really* different ones, a bit like factorizations of a number into primes not counting as different if you re-ordered the factors.

Isomorphisms of posets

For isomorphisms of posets we do something very similar. We will try to show that a morphism of posets is an isomorphism if it is both order-preserving and

---
[†] If you're interested in formality it's still worth trying to write this out. I suggest showing that any group isomorphism must preserve the property of every element squaring to the identity.

a bijection, the point being that the inverse is automatically order-preserving. So we consider posets A and B, and an order-preserving bijection $A \xrightarrow{f} B$. We know there is an inverse function $B \xrightarrow{g} A$ and we need to show that it is order-preserving. In fact we're going to show the contrapositive[†] of the statement of order-preserving-ness.

original statement: $b_1 \leq b_2 \implies g(b_1) \leq g(b_2)$

contrapositive: $g(b_1) > g(b_2) \implies b_1 > b_2$.

Now, f is order-preserving so, using the contrapositive of the definition, we know: $\boxed{g(b_1) > g(b_2) \implies f(g(b_1)) > f(g(b_2)).}$

Since g is inverse to f the right-hand side gives $b_1 > b_2$ as required.

Isomorphisms of topological spaces

The situation for topological spaces is different: a function that is continuous can have an inverse function that is not continuous, so being bijective is not enough to ensure that a continuous map is an isomorphism in the category of spaces and continuous maps. Here is an example, which might give the idea even if you don't know the formal definition of continuity.

We are going to take A to be a half-open unit interval: a portion of the real number line from 0 to 1, including 0 but not including 1. We often write this with a square bracket for the "closed" end and a round bracket for the "open" end, as shown on the right. Formally we're taking all numbers $x \in \mathbb{R}$ such that $0 \leq x < 1$.

For B we will take a circle.

The function $f : A \longrightarrow B$ is going to wrap the interval around the circle so that the ends meet up. They won't overlap because one end contains the endpoint and the other doesn't, a bit like a jigsaw fitting perfectly together with one sticking-out part and one sticking-in part. We could write down a formula for that but I'm not concerned with the formula or the proof here, just the intuition.

Now, f is continuous, essentially because it doesn't break the interval apart. It is bijective because it wraps the interval around the circle without any overlap, and without any gaps. However, its inverse as a function is *not* continuous because it "breaks" the circle apart to go back to being an interval. This is not in the slightest bit rigorous, but captures the intuition; see Appendix D for a little more detail.

[†] The contrapositive of "P implies Q" is "not Q implies not P" and it is logically equivalent. It's like saying: if P is true, must Q be true? Well if Q *isn't* true then P can't be true, so if P is true that shows that Q must have been true all along.

Incidentally, isomorphism is sort of the "wrong" notion of sameness for topological space. When it's defined directly in topology (rather than via categories) it gets the name *homeomorphism* (not to be confused with homomorphism[†]) but often the type of sameness we're interested in is the kind we talked about earlier where you continuously deform things, and this is called homotopy equivalence. This doesn't mean our notion of isomorphism is wrong, it means that the basic category structure we've put on topological spaces is not sensitive enough to detect the more nuanced type of sameness. But it's still a decent starting point. (And homeomorphisms aren't completely useless.)

Sets with structure

These examples were all about some categories that are based on **Set** but where the sets have some extra structure and conditions, making them a poset, or monoid, or group, and so on. There's a further level of abstraction that unifies all these examples. If we generically call such a category **Extra**, then what we really have is a functor (morphism of categories) $\boxed{\textbf{Extra} \xrightarrow{F} \textbf{Set}}$ which "forgets" the extra structure and gives us back the underlying set in each case. We can then ask what the relationship is between isomorphisms in **Extra** and isomorphisms in **Set**. We have seen the following:

- An isomorphism in **Extra** is definitely at least an isomorphism in **Set**, that is, when it is mapped across by F it is still an isomorphism.
- However, if a morphism is mapped to an isomorphism in **Set**, that does not necessarily mean it was already an isomorphism in **Extra**. It was the case for monoids, groups and posets, but not for topological spaces.

These are the concepts of *preserving* and *reflecting* isomorphisms, and we will come back to this in Chapter 20. For now the point I want to make is that not only do we look for structure in different categories, but we then look to see how that structure carries across to other categories via functors.

Isomorphisms of categories

We will come back to talking about isomorphisms of categories in Chapter 21 as we need to define functors properly first. For now I just want to stress that, as with topological spaces, isomorphisms of categories are sort of the "wrong" notion of sameness for categories. As with spaces, this is because the basic category structure we put on (small) categories is not sensitive enough to detect the "better" more subtle form of sameness we're interested in. For that, we need another dimension, and we'll eventually get there.

[†] I know, terminology in math can be a bit of a mess. Sorry.

An isomorphism of categories provides a *perfect* matching of objects and morphisms so that the object and arrow structure in each case is exactly the same. This is too strict for categories because it involves invoking equalities between objects. Categories don't just happen to see isomorphic objects as being "the same" — that is the *correct* notion of sameness in a category, because it is genuinely the scenario in which a category can't tell objects apart.

Thus any time we invoke an equality between objects in a category we have done something not very categorical, or not really in the spirit of category theory.[†] When you get used to categorical thinking I hope you will feel a sort of shudder of distaste any time you see an equality between objects, and at least a slight ringing of alarm bells any time you see an equality at all, just in case. Incidentally this distaste for equalities makes me feel particularly put out by the assumption that mathematics is all about numbers and equations. Not only do I not do equations in my research, but I am actively horrified by them.

14.7 Further topics on isomorphisms

There are many more things to say about isomorphisms that are beyond the scope of this chapter, so I'll just hint at some of the further topics.

Groupoids

We have seen that isomorphisms generalize the notion of symmetry for relations, except that we don't demand them everywhere, we just look for them. However, if they *are* everywhere then we have a "groupoid" — a category in which every morphism is an isomorphism. Groupoids are more expressive than equivalence relations as things can be related in more than one way.

They are related to topological spaces if we think about the zoomed-in version of spaces, where each space produces a category of points and paths between them. Every path in a topological space has a sort of "inverse" which consists of going backwards along the exact same path. This means that the category theory associated with traditional topological spaces is really all about groupoids, and there's a whole branch of higher-dimensional category theory that focuses on that rather than the more general case where some things are not invertible. There are newer theories of "directed space" in which not all paths can be reversed, somewhat like having a one-way street in a city. These sorts of spaces then do need general categories rather than groupoids.

[†] Some authors call this "evil" but I don't really believe in evil, even in jest.

Categorical uniqueness

In math we often try to characterize things by a property and then ask whether something is the *unique* object fulfilling that property. For example the number 0 is the unique number with the property that adding it to any other number doesn't do anything. We have seen that a typical way to prove uniqueness is to assume that there are two such things and then show that they must be equal.

Now alarm bells should ring because I used the word "equal". Indeed, if we're talking about objects in a category we have just done something uncategorical. The categorical thing to do would be to assume there are two such things and then see if they must be isomorphic. This is in fact the categorical version of "uniqueness", with one further subtlety — we would like the two objects not just to be isomorphic but to be *uniquely* isomorphic, that is, that there is a unique isomorphism between them. We might prove that by assuming there are two isomorphisms and showing they must be equal. Thus the "equals" has moved up to the level of morphisms, and this is fine for now as this is the top dimension of our category.

We will come back to categorical uniqueness when we talk about universal properties.

Incidentally, I previously said we use "categorical" to mean "category theoretical", but perhaps more subtly we don't just mean "in the manner of category theory", but "in the manner of *good* category theory", or perhaps "in the true spirit of category theory, not just the technicalities".

Categorification

That process we just went through of replacing an equality with an isomorphism is a typical part of a process known as "categorification".[†] This usually refers to a process of putting an extra dimension into something to give it more nuance via some morphisms. But we don't just give it some morphisms — we then take every part of the old definition, find all the equalities between objects, and turn those into isomorphisms instead. Then, just as with categorical uniqueness, we may need the isomorphisms to satisfy some conditions of their own. This is one of the key processes in higher-dimensional category theory, which we'll come back to in the final chapter.

[†] The word was introduced in L. Crane, "Clock and category: is quantum gravity algebraic?", *Journal of Mathematical Physics* 36:6180–6193, 1995; it was further popularized by J. Baez and J. Dolan, "Categorification", *Higher Category Theory* (Contemporary Mathematics, no. 230), 1–36, American Mathematical Society, 1998.

15

Monics and epics

Another example of doing things categorically, that is, moving away from elements and expressing everything in terms of morphisms.

There is a general process of "doing" category theory like this:

Doing category theory, type 1
1. Become curious about a structure somewhere, often in sets and functions.
2. Express it categorically, that is, using the objects and arrows of the category **Set**, never referring to elements of sets.
3. Take the definition to other categories to see what it corresponds to there.

The idea is that as basic math happens in **Set**, it could be fruitful to take any feature that we often use in there and look for it in other categories, so that we can transfer our techniques from **Set** to somewhere else.

When studying category theory in its own right, without necessarily going through the usual preliminary stages of math first, sometimes this process gets turned around because you might not have seen the supposedly "motivating" phenomenon in **Set** ever before in your life. It's like the classic research seminar where the speaker tells you that something you've never heard of before is just an example of this other thing you've never heard of before.[†]

If you haven't met the "motivating" examples before, the categorical structure may instead have its own intrinsic motivation. In that case we're bypassing step 1 in the above scheme, and doing something more like this:

Doing category theory, type 2
1. Fit some logical pieces together to create an interesting-looking structure in a general category.
2. Take the definition to some examples of categories to see what it gives.

I think of the first type of approach as "externally motivated" whereas the second type is "internally motivated" as we're motivated entirely from within category theory by something that fits together in an abstractly interesting way,

---

[†] I believe I heard this succinct description of seminars from John Baez.

and we then see where we can use it, rather than starting with a specific use as a goal.

I have written before[†] about how this is like two different ways of being a tourist in a new city. The external way is to decide on the places you want to visit, and then work out how to get to them. The internal way is to just plonk yourself in the middle of the city and start following your nose. Of course one often does a mix of the two: perhaps you decide on one place to visit to start the day and then follow your nose from there. The "following your nose" approach works better in some cities than others.

We are going to continue from the previous chapter where we were looking at isomorphisms of sets. When we characterized these as bijections the definition implicitly had two parts, giving two distinct ways in which a function might fail to be a bijection; this is in turn related to the asymmetry in the definition of a function. In this chapter we're going to focus on those two aspects of functions, and see what abstract structure they translate to in category theory.

15.1 The asymmetry of functions

When we first introduced functions we drew some pictures showing inputs going to outputs and observed that some features aren't allowed. Those features were not symmetrical. Here they are in pictures.

Every input must produce an output. So inputs with no arrow attached are not allowed.

not allowed

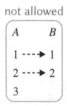

There can be outputs that are not achieved. So outputs with no arrow attached are allowed.

allowed

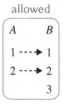

One input cannot produce multiple outputs. So arrows meeting on the left are not allowed.

not allowed

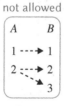

One output can be achieved by multiple inputs. So arrows meeting on the right are allowed.

allowed

While this asymmetry is fine for functions in general, it is what prevents general functions from being isomorphisms. This makes sense as isomorphism is a generalization of symmetry of relations.

[†] *How to Bake Pi*, Profile Books (Basic Books in the US), 2015.

An isomorphism of sets is when the inputs and outputs match perfectly, with arrows neither meeting up nor omitting objects. It gives pictures such as the one on the right.

$$
\begin{array}{cc}
A & B \\
1 \dashrightarrow 1 \\
2 \dashrightarrow 2 \\
3 \dashrightarrow 3
\end{array}
$$

Mathematicians take some interest in the features that *prevent* this from happening, and look more closely at how to iron them out.

--- **Things To Think About** ---

T 15.1   Think about the following functions, which are not isomorphisms. Exactly which of the two pertinent features is getting in the way (or is it both)? Note that although these sets are infinite you could still draw a picture showing the general features, which might help you.

1. $f: \mathbb{N} \longrightarrow \mathbb{N}$ where $f(x) = x + 1$.

2. $f: \mathbb{Z} \longrightarrow \mathbb{N}$ where $f(x) = |x|$. This function takes the *absolute value* of x, that is, it ignores any negative signs.

3. $f: \mathbb{Z} \longrightarrow \mathbb{Z}$ where $f(x) = 0$.

Here are pictures of the general pattern of each function.

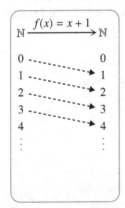 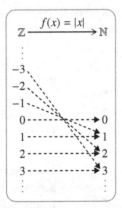 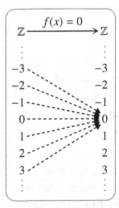

For the first function we see that the output 0 is not hit by any input, so the function is not invertible. The function "+1" is invertible on the integers, but the problem here is that we have no negative numbers. In fact the problem is really that we have a *smallest* number, and this output will never be hit by the "+1" function as long as you have the same set of numbers on both sides.

The second function fails to be invertible because we have multiple inputs going to the same output, that is, arrows meet up on the right. The third function is quite extreme and exhibits both of these problems: there are outputs with no arrow landing on them, and also multiple arrows meeting at the same output. In fact *all* the arrows meet at the same output. This is what happens when we multiply by 0, and shows why we can't invert that process or "divide" by 0.

We will now look at each of those problems separately.

15.2 Injective and surjective functions

Isomorphisms are the best-behaved type of function, where both types of problematic behavior are ruled out. But there are "in between" type functions where just one type has been ruled out. If you already know about injective and surjective functions you could go through this section quickly, just to pick up the spirit of how I like to think of these things, before moving on to the categorical approach.

Injective functions

A function is called *injective* (or an injection) if it does not have the problem of "arrows meeting on the right". There are various equivalent ways of thinking about this and it's good to be able to think via any of them.

- Arrows do not meet on the right.
- Every output is hit at most once.
- If an output is achieved, there is a unique input that went to it.
- Given any output there is *at most* one input that goes to it.
- If two inputs go to the same output they must have been the same already.[†]

The last way of thinking takes us to the formal definition.

Definition 15.1 A function $f : A \longrightarrow B$ is called *injective* if

$$\forall x, y \in A, \quad f(x) = f(y) \implies x = y.$$

Injective functions are sometimes called "one-to-one", perhaps on the grounds that it is a more intuitive term to try and remember. However I don't like that terminology because I think (to my intuition, anyway) that makes it sound like a perfect correspondence, that is, a bijection.

The formal definition is arguably not intuitive at all, but is in a convenient form for doing rigorous proofs. It comes back to that principle of proving something is unique by assuming there are two and showing they are the same.

[†] Remember that "two inputs" in math can be referring to the same input twice; we didn't say "two distinct inputs".

For example, if we take $f\colon \mathbb{N} \longrightarrow \mathbb{N}$ given by $f(x) = x + 1$, here's how we could prove that the function is injective:

$$f(x) = f(y) \quad \text{means} \quad x + 1 = y + 1, \quad \text{which implies} \quad x = y.$$

To prove that a function is *not* injective we have to prove the negation[†] of the definition of injective, that is: $\boxed{\exists\, x, y \in A \text{ such that } f(x) = f(y) \text{ but } x \neq y.}$

For example if we define $f\colon \mathbb{Z} \longrightarrow \mathbb{N}$ by $f(x) = |x|$, then we can show this is not injective by observing that $f(-1) = f(1)$ although $-1 \neq 1$. Translating between intuition and formality in math is one of the challenges, but if you can get your head around it this opens up huge worlds of complex and nuanced logical arguments that are too complicated to follow by intuition alone.

Surjective functions

A function is called *surjective* (or a surjection) if it does not have the problem of "outputs with no arrow to them". Here are some equivalent ways of thinking about it.

- There are no outputs without an arrow to them.
- Every output is hit at least once.
- Given any output there is at least one input that goes to it.

The last one takes us to the formal definition.

Definition 15.2 A function $f\colon A \longrightarrow B$ is called *surjective* if

$$\forall b \in B \ \exists\, a \in A \text{ such that } f(a) = b.$$

─────────────────── Note on terminology ───────────────────

Surjective functions are sometimes called "onto" to encapsulate the idea that they take the inputs onto everything, but I always think it sounds strange to use a preposition as an adjective. The phrase "this function is onto" makes me want to go "onto what?". I think this other terminology is there because of the idea that "injective" and "surjective" sound scary, or perhaps it's difficult to remember which way round they are. I think of medical injections, and you definitely don't have two needles going into the same spot. For "surjection" I think of the root "sur" as in "on".

To prove that something is surjective we can consider an arbitrary element of B and then exhibit an element of A that lands on it. For example, if we consider $f\colon \mathbb{Z} \longrightarrow \mathbb{Z}$ defined by $f(x) = x + 1$, we can show it is surjective like this:

$$\text{Given any } n \in \mathbb{Z}, \text{ we have } f(n - 1) = n.$$

[†] See Appendix B if you need help with negating statements like this.

That is, we took $b = n$ and $a = n - 1$ in the general form of the definition.

To show that something is not surjective we need to negate the definition, which gives us this: $\boxed{\exists\, b \in B \text{ such that } \forall a \in A \ f(a) \neq b.}$

For example if we take $f : \mathbb{N} \longrightarrow \mathbb{N}$ and $f(x) = x + 1$ we can show this is not surjective as follows:

Given any $n \in \mathbb{N}$, we know $f(n) > 0$. Therefore $\forall n \in \mathbb{N}, \ f(n) \neq 0$.

Bijective functions

Finally note that if a function is both injective and surjective then it is called *bijective* (or a bijection). Here are some ways of bringing together the separate ideas of injective and surjective functions to make bijective functions.

- Arrows neither *meet* on the right nor *leave anything out* on the right.
- Every output is hit *exactly* once.
- Given any output there is *exactly one* input that goes to it.

Combining the formal definitions we get this.

Definition 15.3 A function $f : A \longrightarrow B$ is called *bijective* if it is injective and surjective, that is: $\boxed{\forall b \in B \ \exists! a \in A \ \text{ such that } \ f(a) = b.}$

Here the exclamation mark ! is mathematical notation for "unique".[†] Note that the uniqueness applies to everything after it: it's not a unique element in A, it's a unique "element in A such that $f(a) = b$".

Here again we have "exactly one" broken down into "at least one" and "at most one", or existence and uniqueness, and encapsulated by the $\exists!$ notation.

Examples

Injectivity and surjectivity are logically independent, that is, in general, functions can be injective or surjective or both or neither.[‡]

─────────── **Things To Think About** ───────────

T 15.2 Try and interpret the following situations using the concept of injectivity and surjectivity. (So start by constructing a function for the situation.)

1. Someone steps on your foot in the train.
2. Some people have multiple homes but others are experiencing homelessness.

[†] The symbol ! after a natural number means factorial, but those rarely arise in category theory.

[‡] In some specific situations injectivity might force surjectivity, or the other way round, say if the source and target set are both finite with the same number of elements.

3. In New York City, court records appear to be kept according to name and
date of birth pairs. Lisa S Davis wrote about repeatedly being summoned to
court for "unpaid tickets" that were not hers, and she eventually figured out
that they were for another Lisa S Davis with the same birthday.[†]

Example 1 If someone steps on your foot then their foot has landed on yours.
That's two feet in the same place, which feels like injectivity has been violated.
We can realize this by a function from the set of feet on the train to the set of
positions on the train, with the function giving the position of each foot.

Example 2 It might be tempting to realize this one as people being mapped
to homes, but that would require some inputs to have multiple outputs in order
to capture the fact that some people have multiple homes; also people with no
home would then have no output. Instead we can take the inputs to be homes
and the outputs to be people. We need to be a bit subtle (or vague) about the fact
that multiple people can live in a home, but glossing over that fact we now get
that the failure of injectivity comes from some people having multiple homes,
and the failure of surjectivity is from some people experiencing homelessness.
(We might worry about abandoned houses that have no people living in them,
but I would just not call it a "home" if nobody lives there.)

Example 3 Here we can take the set of inputs to be the people in New York
City, and the outputs are pairs of information {*name, date*}. The function is
not surjective: there are definitely dates on which nobody currently in NYC
was born (for example, January 1, 1500). The story of Lisa S Davis (×2) is
about the fact that this function is not injective, but unfortunately the DMV
and NYPD[‡] seem to act on the assumption that the function *is* injective, and
take this combination of information to be an identifier of a unique person.

The part of the story with a moral is that the tickets the author received were
all for trivial infractions, and the other Lisa S Davis was Black. The Caucasian
Lisa had never had a ticket for anything, but not because she'd never done
anything. When she went to court to try and sort it out she was surrounded by
black people who had been summoned for trivial offences. The judge didn't
believe that she wasn't the correct Lisa S Davis, and she realized "For me, this
was an inconvenience and an aberration. But I was beginning to understand
that, for most of the people there, injustice was a given."

[†] Lisa Selin Davis, "For 18 years, I thought she was stealing my identity. Until I found her." *The Guardian*, April 3, 2017.
https://www.theguardian.com/us-news/2017/apr/03/identity-theft-racial-justice
[‡] Department of Motor Vehicles and New York Police Department.

─────────────── **Things To Think About** ───────────────

T 15.3   Now try these mathematical examples. See if you can get the intuitive idea, and then see if you can prove it rigorously if you like. But I think getting the intuitive idea is the important part. Again each one is a function $f : A \longrightarrow B$.

4. $A = \mathbb{Z}, \quad B = \mathbb{Z}$ 6. $A = \mathbb{Z}, \quad B =$ the even numbers
   $f(a) = a + 1$ $f(a) = 2a$

5. $A = \mathbb{Z}, \quad B = \mathbb{Z}$ 7. $A = \varnothing, \quad B = \mathbb{Z}$
   $f(a) = 2a$ f is the empty function.

Example 4 This function is bijective. Adding 1 to numbers is always injective, and now that we have no smallest number, every number n can be achieved starting from $n - 1$, so the function is also surjective.

Example 5 This function is injective: if $2a = 2a'$ then certainly $a = a'$. However it is not surjective; for example there is no integer a such that $2a = 1$.

Example 6 This function is injective by the same proof as the previous example, but it is now surjective as well, so it is a bijection. For the surjective part observe that every even number is of the form $2k$ for some $k \in \mathbb{Z}$, and $f(k) = 2k$. This gives us a bijection between all the integers and the even numbers. This is a little counterintuitive as you might think that the even numbers are just "half" of the integers. But this is one of the weird and wonderful things about infinite sets — you can take half of the set and still have the "same number of things". We try not to call it a "number" at this point, because our intuition about numbers is very different from how infinities behave. This is really telling us that the even numbers and the integers give the same level or "size" of infinity. By contrast, it is a profound fact that there is no bijection between the integers and the real numbers, and this leads to the idea that there is a hierarchy of bigger and bigger infinities.

Example 7 The empty function is the one that is vacuous, and it might be hard to use intuition to decide if it is injective or surjective. Here the formal definition is what we need. For injective, the definition starts "for all $x, y \in A$". We can now stop because A is empty, so whatever happens this condition is vacuously satisfied. Thus the function is injective. For surjective there is a part of the definition that says "$\exists\, a \in A$ such that..." and then we're in trouble because there is no element in A. The only way we could survive this is if the condition is vacuously satisfied, that is: the condition starts with "$\forall b \in B$" so if B is empty then the condition is vacuously satisfied. Here B is not empty so it is not surjective, but the empty function $\varnothing \longrightarrow \varnothing$ is both injective and surjective. As you might expect it is a bijection between the empty set and itself.

As we have seen, functions ostensibly given by the same formula can take on different characteristics depending on what we take their source and target sets to be. This is one of the reasons I prefer the categorical approach where functions always come with a source and target as part of their information. We are now going to give a categorical approach to injectivity and surjectivity.

15.3 Monics: categorical injectivity

The definitions of injectivity and surjectivity had elements of sets all over them. To express them categorically we need to express them using only sets and *functions*, not sets and *elements*. So we need to avoid saying anything like "for any element in A" or "there exists an element in A".

We consider a morphism $a \xrightarrow{f} b$ in an arbitrary category $\mathcal{C}$.

──────────────── **Note on notation** ────────────────

I prefer lower case letters for objects of a category. I might use upper case if I know they're sets and therefore have elements inside them which I may then denote with lower case letters. I know this sort of thing can be confusing when you're learning new math, but it is essentially impossible to keep consistent notation across all of math. We can try and stick to some general principles, while also remaining flexible about what individual letters can denote.

Instead of considering pairs of elements being mapped to the same place, we can look at pairs of morphisms going in, as shown in this key diagram called a "fork".

$$m \underset{t}{\overset{s}{\rightrightarrows}} a \xrightarrow{f} b$$

Definition 15.4 A morphism $a \xrightarrow{f} b$ in a category $\mathcal{C}$ is called a *monomorphism* or *monic* if, given any fork diagram as above in $\mathcal{C}$,

$$f \circ s = f \circ t \implies s = t.$$

Note that the $\circ$ notation for composition gets a bit tedious and redundant so we might just write this condition as: $fs = ft \implies s = t.$

Now, this definition is supposed to be

1. a categorical version of injectivity. and
2. part of being an isomorphism.

So it would be sensible to check both of those things.

──────────────── **Things To Think About** ────────────────

T 15.4   Can you work out what it means to check both of those things, and then check them?

Checking that monics are a categorical version of injectivity amounts to checking that in the case of **Set** monics correspond to injective functions.

Proposition 15.5   *A function is injective if and only if it is monic in* **Set**.

Proof   Consider an injective function $f: a \longrightarrow b$. We aim to show f is monic.

So we consider functions s and t as in this fork diagram such that $fs = ft$, and want to show $s = t$.

$$m \underset{t}{\overset{s}{\rightrightarrows}} a \overset{f}{\longrightarrow} b$$

> Note that I am going to be a bit lazy about brackets here. I will write $fs(x)$ to mean both $(fs)(x)$ and $f(s(x))$. The point now is that $s(x)$ and $t(x)$ are two elements mapped to the same place by the injective function f, so they must be the same.

Now $fs = ft$ means: $\quad \forall x \in a \quad fs(x) = ft(x)$

but f is injective, so this implies: $\quad \forall x \in a \quad s(x) = t(x)$

thus $\quad s = t$

so f is monic as required.

Conversely suppose $f: a \longrightarrow b$ is monic in **Set**. We aim to show that f is injective. So consider $x, y \in a$ such that $f(x) = f(y)$; we want to show $x = y$.

> We need to re-express this situation using functions s and t into a as in the fork diagram so that we can invoke the definition of monic. We can do this by making m a one-element set: functions out of that will then just pick out one element of a.

Let 1 denote a set with one element $*$. We set up a fork diagram with s, t defined by $s(*) = x$ and $t(*) = y$.

$$1 \underset{t}{\overset{s}{\rightrightarrows}} a \overset{f}{\longrightarrow} b$$

> We are now going to show that $fs = ft$, then invoke the definition of monic to conclude that $s = t$ which amounts to $x = y$ as we need. So we want to show that fs and ft do the same thing to every input — but the only input is $*$.

Now $fs(*) = f(x)$ by definition of s

$\qquad\quad = f(y)$ by hypothesis (i.e. using the thing we supposed)

$\qquad\quad = ft(*)$ by definition of t

thus $\quad fs = ft$ as functions (because they agree on every input).

But f is monic, so $s = t$, and so $s(*) = t(*)$. That is, $x = y$ as required.   □

We'll now check that being monic is part of being an isomorphism.

Proposition 15.6 *Let f be a morphism in any category. If f is an isomorphism then f is monic.*

Proof Let $f: a \to b$ be an isomorphism with inverse g.

Consider the usual fork diagram. We aim to show that if $fs = ft$ then $s = t$. We have the following situation.

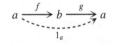

A category theorist might be satisfied at this point, by reading the diagram "dynamically". Here are the steps of that reading.

$$fs = ft \implies gfs = gft$$
$$\implies s = t \qquad \text{since } gf = 1_a$$

Thus f is monic as required. □

Reading a diagram dynamically means taking one part of the diagram at a time and moving through the diagram via some equalities. We will gradually build an understanding of how to do this.

Note we only used part of the definition of isomorphism, which is coherent with the fact that "monic" is a generalization of "injective", which is only part of the definition of bijection. This part is interesting enough that it gets a name.

Definition 15.7 Let $f: a \to b$ be a morphism in a category.

If there is a morphism g as shown here with $gf = 1_a$ then f is certainly monic (as in the proof above) and is called a *split monic*. Then g is called a *splitting, left inverse* or a *retraction* for f.

I confess I can never remember the convention on lefts and rights (it goes by algebraic notation rather than by arrows, which confuses me) so I find it safest to draw the diagram.

It is a slightly subtle point that not all monics are split. So we have a sort of Venn* diagram like this. Split monics are a particularly handy type of monic because they are stable under transfer to other categories. We'll look at this idea of stability in Chapter 20.

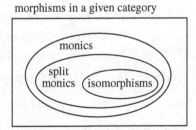

15.4 Epics: categorical surjectivity

Surjectivity is like injectivity but the other way round. The formal set-based (non-categorical) definitions don't look so much like "other way round" versions of each other, but with the categorical version we can see a very precise sense in which they are the other way round from each other: they are *duals*, which means that in the categorical definitions all we need to do is turn all the morphisms to point the other way. If we do this to the definition of monic we get the definition of epic, which is a categorical version of surjectivity.

── Things To Think About ──

T 15.5   Write out the definition of monic again but with every arrow pointing the other way. This is the definition of "epic". Now see if you can prove that in **Set** a morphism is epic if and only if it is surjective.

If I very literally turn all the arrows around in the definition of monic the diagram will become this.

$$m \xleftarrow[t]{s} a \xleftarrow{f} b$$

However, I prefer to keep the abstract shape but adjust the notation so that the morphism f still goes $a \rightarrow b$ and all arrows point to the right, as shown here.

$$a \xrightarrow{f} b \xrightarrow[t]{s} e$$

Here is the resulting definition; I will fully acknowledge that it's hard to remember which direction is which out of monic and epic unless/until you work with them a lot.

Definition 15.8   A morphism $f : a \rightarrow b$ in a category is called an *epimorphism* or *epic* if, given any diagram as above: $sf = tf \implies s = t.$

We're now going to show that the epics in **Set** are precisely the surjective functions. The proof proceeds analogously to the result for monics, the idea being to "translate" the properties of surjectivity into morphisms.

Proposition 15.9   *A function f is surjective if and only if it is epic in* **Set***.*

This proof is going to be helped along by a few preliminary ideas.

The idea is that when you have a function $a \rightarrow b$ the target set b splits up into the part that is "hit" by f and the part that isn't. Here's an intuitive picture. The function f maps all of a into b somewhere, possibly not landing on all of it.

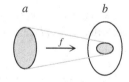

The part that the function f does land on is called the *image* of f.

Definition 15.10   The *image* of a function $f: a \longrightarrow b$ is a subset of b defined as follows: $\mathrm{Im}f = \{\, f(x) \mid x \in a \,\}$.

Here the vertical bar | indicates "satisfying", in the context of defining a set: the set consists of the things on the left of the bar, satisfying the conditions on the right. So this set is all elements $f(x)$; another, less succinct but perhaps more intuitive way of putting it, is: $\mathrm{Im}f = \{\, y \in b \mid \exists\, x \in a \text{ such that } f(x) = y \,\}$.

Now note that the set b then splits into two parts: the part that is hit by f and the part that isn't. I'll call the part that isn't hit the *non-image*, and write it NonIm. This is not standard, I just invented it for now. The set b is then the *disjoint union* of those two sets: $b = \mathrm{Im}f \sqcup \mathrm{NonIm}f$.

That is, everything in b is in the image or the non-image, and not both.

Now the situation we're looking at involves having two more functions that follow on from f, so they map the whole of b somewhere else. And the point is that if they are equal upon pre-composition with f then they must agree on the image of f. However, they could disagree on the non-image.

This picture is trying to capture that. The two further functions s and t both map the gray part of b to the same place, but map the outside to different places.

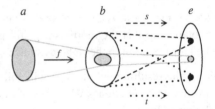

The essence of the proof is then that such an s and t can be different if and only if something exists in the non-image of f, otherwise there will be no space for them to be different while being the same when pre-composed with f. Incidentally I have implicitly used the following characterization of surjectivity.

A function $f: a \longrightarrow b$ is surjective if and only if $\mathrm{Im}f = b$.

We are finally ready to prove that a function is surjective iff it is epic.

Proof of Proposition 15.9   First we suppose that f is surjective and aim to show that it is epic.

So we consider functions s and t as in this fork diagram $a \xrightarrow{\ f\ } b \underset{t}{\overset{s}{\rightrightarrows}} e$
such that $sf = tf$, and want to show $s = t$.

> The idea is that f lands on all of b, so there's no extra room for s and t to be different given that they're the same when pre-composed with f.

Consider $y \in b$. Now f is surjective, so: $\exists\, x \in a$ such that $f(x) = y$.

$$
\begin{aligned}
\text{Thus} \quad s(y) \;&=\; sf(x) \quad \text{by substitution} \\
&=\; tf(x) \quad \text{by hypothesis} \\
&=\; t(y) \quad \text{by substitution}
\end{aligned}
$$

so $s = t$ as functions, as required.[†]

For the converse we are going to prove the contrapositive, that is, that if f is not surjective then it is not epic.

> The idea is that if f is not surjective then we can fabricate two functions s and t that agree on the image of f but disagree on the non-image. I'm going to take e to be a set of 3 objects so that there's a landing spot for where s and t are the same, and one landing spot each for where they're different. It doesn't matter what these three objects are so I just called them $1, 2, 3$.

Define a set $e = \{1, 2, 3\}$, and define functions $s, t : b \longrightarrow e$ as follows.

$$
s(y) \;=\;
\begin{cases}
1 & \text{if } y \in \mathrm{Im}f \\
2 & \text{if } y \in \mathrm{NonIm}f
\end{cases}
\qquad
t(y) \;=\;
\begin{cases}
1 & \text{if } y \in \mathrm{Im}f \\
3 & \text{if } y \in \mathrm{NonIm}f
\end{cases}
$$

First note that since f is not surjective, $\mathrm{NonIm}f$ is not empty, so $s \neq t$. However we are going to show $sf = tf$, that is: $\forall x \in a,\ sf(x) = tf(x)$.

Now, given any $x \in a$, we know $f(x) \in \mathrm{Im}f$ (by definition), so $s(f(x)) = 1$. Similarly $t(f(x)) = 1$. So $sf = tf$, but $s \neq t$, so f is not epic. $\qquad\square$

This proposition gives rise to what is probably my favorite category theory joke:[‡] "Did you hear about the category theorist who couldn't prove that a function was surjective? Epic fail."

In the previous section we showed that "isomorphism $\Longrightarrow$ monic", so we could now show the analogous result for epics.

─────────── **Things To Think About** ───────────

T 15.6 Look back at the proof of the result for monics and see if you can directly translate it into a proof for epics.

Proposition 15.11 *Let f be a morphism in any category. If f is an isomorphism then f is epic.*

[†] This wording might sound strange. We could just say "$s = t$ as required", but I put "as functions" in to remind ourselves that we have checked they are equal on each element, which means that the entire function s is equal to the entire function t.

[‡] I believe I first heard this joke from James Cranch.

In Chapter 17 we are going to talk about the fact that we can get results "for free" by turning all the arrows around, which is called duality. This means we will be able to say "and dually" for epics here, but it's not a bad idea to try writing it out directly anyway, by turning around all the arrows in the proof that isomorphisms are monic. Note that this will use the rest of the definition of isomorphism, the part we didn't use for the proof for monics. This part is named analogously.

Definition 15.12   Let $g\colon b \longrightarrow a$ be a morphism in a category.

If there is a morphism f as shown here with $fg = 1_b$ then g is certainly epic. Then g is called a *split epic* and f is called a *splitting*, a *left inverse* or a *section* for g.

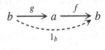

Here g is the "other side" inverse (left as opposed to right) from the split monic case, and we have an analogous Venn* diagram like this.

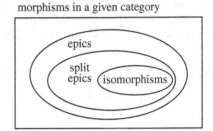

morphisms in a given category

The categorical definitions of monic and epic are perhaps less intuitive than the direct definitions of injective and surjective, but in a way they are more related to the idea of cancelation. We are quite used to the idea of "canceling" in equations of numbers. For example if you're faced with the equation $4x = 4y$ in ordinary numbers I hope you'd feel like "canceling" the 4 to conclude $x = y$. Formally we might say we're dividing both sides by 4, that is, we're using the multiplicative inverse of 4.

However, technically we're only using one side of the inverse, the part that says this:
$$\frac{1}{4} \times 4 = 1$$

We are not using the part that says this (although as multiplication is commutative it comes to the same thing):
$$4 \times \frac{1}{4} = 1$$

Sometimes (in worlds without commutativity) things are cancelable on *one* side but they don't have *both* sides of an inverse, and this is what happens with monics and epics. Given $fs = ft$ then f being monic says we can "cancel" it to conclude $s = t$. If f is epic we can cancel it from the other side, so that if $sf = tf$ we can deduce $s = t$.

Although this makes the definition match some existing intuition I will say that this is not the intuition I carry with me when I'm thinking about monics and epics in category theory, so although it might motivate the definition more I'm not sure it's a good thing to get hung up on. I'd rather stress that the abstraction has various points:

1. It enables us to make connections with more situations other than just sets and functions, by now looking for these properties in other categories.
2. It gives us a sense in which injective and surjective are precisely "the other way round" from one another, by categorical duality.

We're now going to finish thinking about the relationship with isomorphisms.

15.5 Relationship with isomorphisms

We have seen that all isomorphisms are both monic and epic. However, the converse isn't true: if f is a *split* monic and a *split* epic then it has an inverse on both sides which is precisely the definition of isomorphism, but a morphism could be monic and epic without necessarily being an isomorphism, if it doesn't split. To exhibit this we will need to look at a category of objects with more structure than sets, because in **Set** we know that

1. monic = injective
2. epic = surjective
3. isomorphism = bijective = injective + surjective

So in **Set** it happens that "monic + epic $\implies$ isomorphism". However, we can at least show that the splitting of epics is subtle, by considering infinite sets.

───── **Things To Think About** ─────

T 15.7 We are going to think about a function that is surjective but might not have a splitting. We'll warm up with finite sets and then do infinite sets.

1. Let $f\colon \{-1, 0, 1\} \longrightarrow \{0, 1\}$ be defined by $f(x) = |x|$. This function is surjective thus epic; can you find a splitting for it?
2. Let $f\colon \mathbb{Z} \longrightarrow \mathbb{N}$ be defined by $f(x) = |x|$ as above. This function is surjective thus epic; can you find a splitting for it?

Remember that for f to be a split epic we need g as shown here.

$$b \xrightarrow{\;g\;} a \xrightarrow{\;f\;} b$$
$$1_b$$

The idea is to try and make a function g going backwards, sending every element x to a *pre-image* of x under f, that is, an element that f maps to x. If f

maps several elements to the same place then we have a choice. So for the first example 0 has to go to 0 but 1 can go to 1 or −1.

Here are pictures of those two possibilities. In either case if you follow the dotted arrows from the beginning to the end, 0 and 1 each go back to themselves, they just take a slightly different route in the middle.

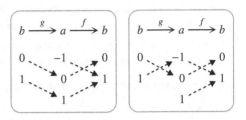

Now for the second example it's the same idea but we are doing the entire integers. So 0 has to go to 0, but any other n can go to n or $−n$. This means that to construct a splitting $\mathbb{N} \to \mathbb{Z}$ we need to make that choice for each natural number. In this case it's not hard: we can just decide right from the start that we're going to send everything to the positive version, say.

However, if we were in a situation without a way to make one global choice then we would have to make an individual choice for each natural number. In case you've heard of it, I'll briefly mention that this is exactly what the Axiom of Choice is about: the question of whether or not it is possible to make an infinite number of choices, even if in each case the choice is only between a finite number of things. The Axiom of Choice says that it is possible, and this axiom is *logically equivalent* to saying that all epics split. There are different versions of set theory that include or don't include the Axiom of Choice. You needn't worry about it if you don't want to, it's just good to be aware if you are implicitly invoking it to make an infinite number of choices.

Finally here I'd like to stress a difference between property and structure. The definition of monic/epic is a *property* of a morphism, whereas a splitting is a piece of *structure*, and that's a technically and ideologically important difference. A property is a bit nebulous, whereas structure is something that we can specify and more likely carry around with us to other categories. It's a bit like saying someone is beautiful (a property) rather than being specific about a feature they have, like a warm smile, or a generous heart.

15.6 Monoids

In the category **Mnd** of monoids and monoid homomorphisms things are a little more subtle because of the requirement of preserving structure. This means we have less freedom over where to send elements, and some things end up being determined for us.

For example, consider the monoid of natural numbers $\mathbb{N}$ under addition. We will try to define a monoid homomorphism f to some other monoid M.

- We know that 0 is the identity in $\mathbb{N}$ so it must go to the identity in M; we have no choice about that.
- We can send 1 anywhere we want.
- We know that $2 = 1+1$ in $\mathbb{N}$, and by the definition of monoid homomorphism we must have $f(1 + 1) = f(1) + f(1)$ so we have no choice about where 2 goes: it has to go to $f(1) + f(1)$.
- Something similar is now true for every other natural number. Every n is a sum of 1's and so we have no choice over where it goes: it has to go to the n-fold sum of $f(1)$'s.

What we've done here is based on the fact that $\mathbb{N}$ is the *free monoid* built up from a single element. We saw this in Section 11.2 when we started with one non-trivial arrow and built up a whole category from it, and we'll come back to it in Chapter 20 when we talk about free functors. For now I want to show that this forces more things to be epic than just the surjective homomorphisms.

Proposition 15.13 Let $f: \mathbb{N} \to \mathbb{Z}$ be the monoid homomorphism given by the inclusion (that is, $f(n) = n$). Then f is epic in **Mnd**.

─────────── **Things To Think About** ───────────

T 15.8 Have a think about monoid morphisms $\mathbb{Z} \to M$, in the same spirit as with $\mathbb{N}$ above. What choice do we have about where things go? If you're confused by M being a random monoid, try just taking it to be $\mathbb{Z}$.

Once you've worked that out, try and make monoid morphisms s, t as in the definition of epic, with $sf = tf$. What choice do we have over s and t?

$$\mathbb{N} \xrightarrow{f} \mathbb{Z} \underset{t}{\overset{s}{\rightrightarrows}} M$$

Mapping out of $\mathbb{Z}$ forces our hand in much the same way that mapping out of $\mathbb{N}$ does. Deep down this is because $\mathbb{Z}$ only differs from $\mathbb{N}$ by having additive inverses, but where those go is forced by preserving the monoid structure, as we saw when we defined group homomorphisms. So we know that when we define s and t the only choice we have is about where 1 goes, and the only way they can be different is if they do different things to 1.

Now if we also have $sf = tf$ that means s and t agree on the image of f. However, $1 \in \mathbb{Z}$ is in the image of f (because $f(1) = 1$) and so we must have $s(1) = t(1)$, which in turn forces $s = t$. This shows that f is epic even though it isn't surjective.

Note that f is injective and this is enough to ensure it is monic in **Mnd**. So in fact f is monic and epic, but it is not an isomorphism.

15.7 Further topics

Density

There are many examples similar to monoids where morphisms are epic without being surjective, on account of the action of a morphism being determined by just part of it. I'll include some examples here just in case you've heard of them, to make some connections in your mind; if you haven't heard of them then it won't make much sense to you, but we won't use it anywhere.

For example, with topological spaces a continuous map is entirely determined by its action on a "dense" subset. A basic example is if we consider a closed interval such as $[0, 1]$ — remember this means it's a portion of the real line including the endpoints 0 and 1. If we define a continuous function from here to somewhere else, once we've defined it on the *open* interval $(0, 1)$ (that is, without saying what it does to 0 and 1) there is no choice left about what it does to 0 and 1. A set A is called dense inside B if the "closure" of A is B; the closure of A is essentially the most economical way to enlarge A just enough to make it a closed set.

This is related to how we can construct the real numbers from the rationals, by taking the closure under limits, which amounts to throwing in the irrational numbers. In general B is the closure of A if the only thing B has extra is limits for sequences in A that really "want" to converge but didn't have a limit. Continuous functions preserve limits, so their action on B will be entirely determined by the action on A.

This sort of thing happens any time something has been constructed "freely" from some generators. If the morphisms in that category preserve the structure, then all you have to know is where the generators go, and then there's no choice of where the rest of the structure can go. You might also have seen this with vector spaces and basis elements. There's a concept of "limit" and "density" in category theory that mimics the situation of limits and density in analysis.

Subobjects

One special case of injective functions is subset inclusions. For example we have seen the function $\mathbb{N} \longrightarrow \mathbb{Z}$ which just takes each natural number and sends it to the same number as an integer. A subset inclusion is when A is a subset[†]

[†] A is a subset of B if all the elements of A are elements of B. It could be the whole of B or it could be empty, or it could just have some of the elements of B.

of B and all you do is map each element of A to "itself" in B. Now this is an un-categorical notion because we're talking about *elements* of sets, and moreover we're asking for elements of A to be *equal* to some elements of B.

There's no categorical reason that a subset inclusion is any better than any other kind of monic, and so in more general situations monics are used to indicate *subobjects*. This helps us account for the possibility that something might have different labels but apart from that be really like a sub-thing. This makes less sense with sets because they have no structure; it might seem odd to think of the set $\{1, 2\}$ as a subset of the set $\{a, b, c\}$ but it's a little less odd to think of a cyclic group of order 2 as a subgroup of a cyclic group of order 4 even if the elements have different names.

Higher dimensions

Monic and epic are categorical versions of injective and surjective functions, and when we go into higher dimensions we want to generalize injective and surjective functions to the level of *functors*, that is, the morphisms in the category **Cat**. We can try looking for monics and epics in that category, but it won't really give us "the right thing" because that is working at the wrong dimension. **Cat** is really trying to be a 2-category (as we mentioned before) with a more subtle type of relationship between morphisms. That means that we shouldn't be asking for morphisms to be equal, as we do in the definition of monic/epic. In Chapter 21 we will see that we want something more like "injective and surjective up to isomorphism" and that this gives the concepts of full and faithful functors.

16

Universal properties

The idea of characterizing something by the role it plays in context. Looking for extremities in a particular category, key to the whole idea of category theory.

16.1 Role vs character

Characterizing things by the role they play in a particular context is different from thinking about their intrinsic characteristics. We have seen a few examples of this already. We could describe 0 as a number representing nothing, which is describing it by an intrinsic characteristic. Or we could describe it as the unique number such that when you add it to any other number nothing happens. Not only does 0 play this role but, crucially, it is the *only* number that plays this role, so we can describe it by that role and all know we're talking about the same number.

In normal language we often use words that mix up whether we are talking about a role or an intrinsic characteristic. I mentioned some examples in Chapter 1, including pumpkin spice: I used to think it was spice made of pumpkins, when in fact it is a spice typically used in pumpkin pie. However, it is now used in all sorts of other things that don't involve pumpkin, including (infamously) pumpkin spice lattes.[†] A more pernicious example is "bikini body" which comes with the (largely sexist) idea that you need to be slim and toned (intrinsic characteristic) to wear a bikini, whereas, as some amusing memes put it, all you need to do to get a bikini body is get a bikini and put it on your body.

When those linguistic quirks are accidents of history rather than signs of misogyny I enjoy their quirks, but either way they make me think of category theory. I am intrigued by how deeply we have internalized some of this language so that we might not notice how contradictory it is deep down, but it can be confusing to people learning a new language, or moving to a different country.

[†] I hear that some brands of pumpkin spice latte do now have pumpkin as an ingredient. I'm sure if I don't include this footnote someone will write and tell me I made a mistake.

┌─────────────────── **Things To Think About** ───────────────────┐

T 16.1   Think about the following and how the words we use are mixing up role and intrinsic character, and in what way they do and don't make sense. Note that some are particular to American English.

1. baseball cap   4. pound cake
2. film   5. biscuit (from the French "bis cuit")
3. hang up the phone   6. fat-free half-and-half

└──┘

It interests me that food items often have quirks of language in them especially if they are traditional things that have been passed down through generations and possibly they mutated as they went along.

┌─────────────────── **Things To Think About** ───────────────────┐

T 16.2   Now for some more mathematical examples. We have seen how to characterize the number 0 by a role rather than a property. What about the number 1? −1? i? Do these characterizations specify the number uniquely?

└──┘

The number 1 is the unique number such that when we multiply it by any other number nothing happens. Formally we can say:[†]

$$\exists!\, a \in \mathbb{R} \ \text{such that} \ \forall x \in \mathbb{R} \ ax = x.$$

Every part of this definition is important and we must get all the clauses in the right order. If we just look at the end part "$ax = x$" we might be tempted to say this is true when $x = 0$. Now, "$ax = x$" is indeed true when $x = 0$, but we're not supposed to be looking for one value of x where this is true, we're supposed to be looking for one value of a that makes this true for *all* x. The only number that works is 1, so 1 is the unique number with this property.

This characterizes the number 1 as the *multiplicative identity*, and shows a way in which 1 is analogous to 0, as 0 is the *additive identity*. When we characterize things by universal properties one of the points is to find things that play analogous roles in different contexts.

We can characterize the number −1 as the additive inverse[‡] of 1, and that pins it down uniquely. However the imaginary number i is more subtle. It is defined as a square root of −1, that is, by it satisfying the equation $x^2 = -1$. However it is not the *unique* number with that property: $-i$ also satisfies the equation. This results in the curious fact that if we think about roles then we can't tell the difference between i and $-i$. This is a favorite weird conundrum

---

[†] Translation: there exists a unique a in the reals such that for all x in the reals $ax = x$.
[‡] That is: $1 + (-1) = 0$.

of mine in abstract mathematics. We have to arbitrarily pick one and call it i, and then the other one is $-i$. However, as we can't tell them apart until we've picked one, we also have no way to say which one we've picked. The whole thing should feel very disconcerting, maybe a bit quantum.

The question of how uniquely we can pin things down is an important question where universal properties are concerned. We may find several objects that all satisfy the property, but they will all be *uniquely isomorphic* so the category "can't tell them apart". Thinking categorically we should embrace not being able to tell them apart, even though there are several of them.

To me this is a lot like treating people equally even though they are different. It does lead to a linguistic conundrum where we're not sure whether to say "a" or "the" for something that is categorically unique but not literally unique. This can be frustrating to non-categorically minded people, or people who are not (yet) in the spirit of category theory. We are also liable to use the word "is" in a categorical way to say that something "is" something else if it is categorically so, rather than literally so.

Anyway, when we pin things down by a universal property, we look at the roles things play in context, rather than their intrinsic characteristics, and then we look at the extremities.

16.2 Extremities

Sometimes we look for things that play a certain role in context, and then we look for the one that is most extreme among all of those. For example, to find the *lowest common multiple* of two numbers we look at all the common multiples, and find the lowest one. This is what we do notionally anyway; in practice there are ways to calculate it.

Looking for extremities is one way to examine the inner workings of a situation and make comparisons. For example, if we look at the tallest mountain in the UK we find that it is really not very tall compared with the tallest mountain in the US. This gives us a rudimentary hint of the difference in general scale between the two countries.

However, we can also do it the other way round: we could start with an object we're trying to study, and then look for a context in which it is extreme. This then tells us something about the nature of that context. For example if I am ever the tallest person in the room then it shows it's a room full of rather short people, or children. Whereas if I'm the shortest person in the room it's not saying much; it might just be saying I'm in a room full of men, which happens quite often at math events.

In the cube diagrams from earlier, 30, 42 and rich white men occupy an analogous extreme position, in this case at the top because of how I've drawn the cubes.

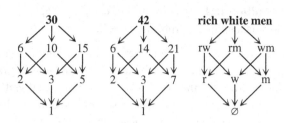

It is more rigorous to identify the extremity relative to the arrows in the category, rather than relying on the physical positioning on the page. We can say that 30, 42 and rich white men are each in a position where all arrows begin (with the direction for arrows that I've chosen here). By contrast we could look at the place where all arrows end: 1, and non-rich non-white non-men.

Things To Think About

T 16.3   What can we say about extremes in each of the following situations? Is there an extreme where arrows begin? Is there an extreme where arrows end?

1. The natural numbers expressed like this: $0 \longrightarrow 1 \longrightarrow 2 \longrightarrow 3 \longrightarrow 4 \longrightarrow \cdots$

2. The integers like this: $\cdots \longrightarrow -2 \longrightarrow -1 \longrightarrow 0 \longrightarrow 1 \longrightarrow 2 \longrightarrow \cdots$

3. A category with two objects and no arrows: • •

4. A category that goes round in circles: •⇄•

The natural numbers expressed in a line has all its arrows starting from 0. However, it has no ending place for the arrows as the natural numbers keep going "forever" — there is no biggest natural number. The integers keep going forever in both directions so they have no starting point or ending point.

In the example with no arrows the objects are completely unrelated so neither of them seems like an extreme. The category in a circle also doesn't seem to have an extreme but for the opposite reason — everything is very connected in a symmetrical way.

Evidently our description of extremes so far has been rather vague and intuitive. We are going to express it in a precise and rigorous way using the concept of initial and terminal objects.

16.3 Formal definition

We will now look at the formal definitions. This is our first example of a universal property: an initial object.

Definition 16.1   An object I in a category $\mathcal{C}$ is *initial* if for every object $X \in \mathcal{C}$ there is a unique morphism $I \longrightarrow X$.

The idea is that I has a universal property among all the objects of the category: if you compare it with any object, it "comes first". This sounds a little competitive, but the idea isn't about winning really, it's more about the object helping us out as a baseline or a sort of hook on which to hang things, as it enables us to induce morphisms to other places. Imagine you're trying to record the positions of various things and you have a tape measure. It would be really helpful to have a reference point, particularly if you can attach one end of your tape measure to it. That's something like an intuition about initial objects.

─────────────── **Things To Think About** ───────────────
T 16.4   Go back to the above examples and see if you can see formally why they do or don't have an initial object.

In this category of natural numbers 0 is initial. $0 \longrightarrow 1 \longrightarrow 2 \longrightarrow 3 \longrightarrow \cdots$

It might not look like there is a morphism to every other natural number, but remember that we have not drawn composite arrows in this diagram.

Formally: in this category an arrow $a \longrightarrow b$ is the assertion $a \leq b$, so we immediately know that any arrow $a \longrightarrow b$ is the unique such.[†] (An assertion just is or isn't true, so there is either one morphism or zero morphisms.) For any natural number n we know $0 \leq n$ so there is a morphism $0 \longrightarrow n$, again unique by construction.

> This is an example where we proved existence and uniqueness but uniqueness actually came first; that is we first prove there is "at most one" and then that there is "at least one". We often refer to something as being "unique by construction" when that happens, and it happens quite often with universal properties.

This category of integers
has no initial object. $\cdots \longrightarrow -2 \longrightarrow -1 \longrightarrow 0 \longrightarrow 1 \longrightarrow 2 \longrightarrow \cdots$

This is essentially because there is no smallest integer, but formally we need to prove that for all objects a, a is not initial: for any object a there is no morphism $a \longrightarrow a - 1$, which shows a is not initial.

For the next example (now with names for the objects) there is no morphism $a \longrightarrow b$ so a can't be initial, and there is no morphism $b \longrightarrow a$ so b can't be initial.

$\begin{matrix} a & \quad & b \\ \bullet & & \bullet \end{matrix}$

───────────────

[†] This wording might sound odd to you. There's an implied concept at the end, which has been omitted for brevity. So the sentence really says "any arrow $a \longrightarrow b$ is the unique such arrow (that is, unique arrow $a \longrightarrow b$)".

More generally in any category with more than one object and no non-trivial morphisms there is no initial object because there just aren't enough morphisms. To have an initial object we must *at the very least* have an object with a morphism going everywhere else.

Next here's the example that "goes round in circles". We need to be careful what we mean — if f and g are inverses we will get something different from if they're a circle that loops around and never comes back to the identity.

If the composite $g \circ f$ is *not* the identity then a can't be initial, because we have (at least) two different morphisms $a \longrightarrow a$: the composite $g \circ f$ and the identity 1_a. In fact we may have an infinite number of morphisms $a \longrightarrow a$; for example, if f and g satisfy no further equations then we will keep getting a new morphism every time we go round the loop.

On the other hand suppose f and g are in fact inverses. Now this category has exactly four morphisms: f, g, and the two identities 1_a and 1_b. All composites will be equal to one of those, as f and g compose to the identity both ways round. Then we can see that both a and b are initial.

- a is initial: there are unique morphisms $a \xrightarrow{1_a} a$ and $a \xrightarrow{f} b$
- b is initial: there are unique morphisms $b \xrightarrow{g} a$ and $b \xrightarrow{1_b} b$

Thus we have two initial objects. However, categorically speaking they're not *really* different. This is the idea of uniqueness for universal properties.

16.4 Uniqueness

There might be more than one initial object in a category, but the category "can't tell the difference" between them, in the following precise sense.

Proposition 16.2 *If I and I' are initial objects in a category $\mathcal{C}$ then they must be uniquely isomorphic, that is, there is a unique isomorphism between them.*

With proofs in higher mathematics I always think it's good to get the idea and then be able to turn the idea into a rigorous proof, rather than try and learn the proof. Also I personally just prefer understanding the idea and making it rigorous, rather than trying to read the proof and extract the idea.

The idea of this particular proof is to use the universal property of I and of I' in turn. We use the universal property of I to get a unique morphism to I', then use the universal property of I' to get a unique morphism back again, and then show that those are inverses.

Proof   Since I is initial there is a unique morphism $I \xrightarrow{f} I'$.
But also since I' is initial there is a unique morphism $I' \xrightarrow{g} I$.
We aim to show that these are inverses.

First consider the composite gf as shown.[†] $I \xrightarrow{f} I' \xrightarrow{g} I$
Since I is initial there is a unique morphism $I \longrightarrow I$.

But 1_I and gf are both morphisms $I \rightarrow I$ so we must have $gf = 1_I$. (See footnote about this notation.[‡]) Similarly $fg = 1_{I'}$, thus f is an isomorphism $I \rightarrow I'$ and is unique by construction.   □

> I hope you weren't put out by that "similarly" at the end. I and I' play the exact same role as each other in this proof so the part showing that $f \circ g$ is the identity is really no different from showing that $g \circ f$ is the identity, just swapping I for I' and f for g everywhere.

We can also show the converse.

Proposition 16.3   *If I' is uniquely isomorphic to an initial object I then it is also initial.*

Proof   Suppose I is initial, and that I' is uniquely isomorphic via f and g as shown on the right. To show I' is initial we consider any object X and exhibit a unique morphism $I' \rightarrow X$.

Now, since I is initial there is a unique morphism $I \rightarrow X$, say s, and so we have a morphism $I' \xrightarrow{g} I \xrightarrow{s} X$.

> To show it's unique the idea is that if we had another one it would produce another morphism from I as well, contradicting I being initial.

[†] As I mentioned before, we sometimes write composites as gf to make it shorter than $g \circ f$.
[‡] This unideal notation is a 1 with a subscript I, denoting the identity on I.

Consider any morphism $I' \xrightarrow{t} X$. This produces a composite $I \xrightarrow{f} I' \xrightarrow{t} X$. But I is initial so we know there is a *unique* morphism $I \rightarrow X$.

Thus $\qquad s = t \circ f \qquad$ as these are both morphisms $I \rightarrow X$

and so $\quad s \circ g = t \circ f \circ g \quad$ pre-composing by g

$\qquad\qquad = t \qquad\qquad$ since f and g are inverses.

Thus $s \circ g$ is a unique morphism $I' \rightarrow X$ and I' is initial as claimed. $\qquad\square$

Note that we did not need to use the fact that I and I' are *uniquely* isomorphic, as that uniqueness follows from I being initial. So in fact we have the following result.

Proposition 16.4   *Any object isomorphic to an initial object in $\mathcal{C}$ is initial.*

This categorical uniqueness for objects with universal properties is important as it means it makes sense to characterize things by universal property. If it were possible to have substantially different objects with the same universal property then it would be ambiguous to do that.

16.5 Terminal objects

So far we've only done the formal definition of universal property of objects at the "beginning" of everything. The property of being at the "end" is not really different, because it's the same idea just with all the arrows turned around. We will talk more about this in the next chapter, but it's another situation in which turning all the arrows around gets us some things for free.

―――――――――― **Things To Think About** ――――――――――

T 16.7   In the examples earlier we thought about the category made from the natural numbers with an arrow $a \rightarrow b$ whenever $a \leq b$. What if we made a category with $a \rightarrow b$ whenever $a \geq b$ instead? What does the diagram of this category look like, and what happened to the initial object?

Here we're not expressing any different information about the natural numbers, just expressing the same information in a different way.

Here's the original category of natural numbers $\qquad 0 \rightarrow 1 \rightarrow 2 \rightarrow 3 \rightarrow \cdots$

and the new one: $\qquad\qquad\qquad\qquad\qquad 0 \leftarrow 1 \leftarrow 2 \leftarrow 3 \leftarrow \cdots$

Visually it is apparent that we no longer have a starting point for all the arrows, but an ending point. This is the universal property that is dual to initial objects, and is called a terminal object.

T 16.8   The definition of terminal object is the same as the definition of initial object, but with all the arrows turned around. See if you can write down the definition, and see that the uniqueness results immediately follow.

For the definition we just turn round the arrows in the definition of initial object, but like for monics and epics I'll also flip the diagram so that the arrows still point right, and change the letters so that we have a T for terminal.

Definition 16.5   An object T in a category $\mathcal{C}$ is *terminal* if for every object $X \in \mathcal{C}$ there is a unique morphism $X \longrightarrow T$.

The following uniqueness result can be proved as for initial objects, but with all the arrows pointing the other way. I've stated it in a slightly different form from before, because for initial objects we built this up bit by bit but now I'm saying it all at once. This form of stating a result is dry[†] but quite common, as we often want many different equivalent ways of saying the same thing.

Proposition 16.6   *Let T be a terminal object in a category $\mathcal{C}$ and T' any object. The following are equivalent:*

1. *T' is a terminal object in $\mathcal{C}$.*
2. *T' is uniquely isomorphic to T.*
3. *T' is isomorphic to T.*

What we did for initial objects was: first we showed (1) implies (2). Then we showed (3) implies (1). But (2) implies (3) by definition and so we have a loop of implications showing that the three statements are logically equivalent.

T 16.9   Try turning all the arrows around in the cube diagrams for the factors of 30, the factors of 42, and privilege. Observe that the initial object becomes the terminal object in each case. We'll address this formally in the next chapter.

16.6 Ways to fail

Not all categories have terminal and initial objects.

Our original category of natural numbers has an initial object but no terminal object: $0 \longrightarrow 1 \longrightarrow 2 \longrightarrow 3 \longrightarrow \cdots$

The version with the arrows the other way has a terminal object but no initial object: $0 \longleftarrow 1 \longleftarrow 2 \longleftarrow 3 \longleftarrow \cdots$

[†] By "dry" I mean terse and expressionless, like a dry writing style.

As I've said before, instead of just declaring yes or no to whether something is true, it's illuminating to classify the different ways in which things can fail.

─────────────── **Things To Think About** ───────────────

T 16.10   Try drawing some small pictures of categories that fail to have a terminal object or fail to have an initial object. Try to isolate one issue at a time that causes the categories to fail, like when we looked at failure to be an ordered set. Also, as in that exercise, it doesn't matter what the objects of the category are; what matters is the structure of the arrows. Remember that anything that fails to have a terminal object can be entirely turned around to give something that fails to have an initial object. In some cases when you turn the arrows around you get the same thing though; make a note of when that happens as that's interesting too.

We can think about pictures and what it means for something to have no beginning or end. Or we can think about the formal definition, which involves the phrase "there exists a unique morphism", meaning there is exactly one. As usual, there are two ways to fail to have exactly one: by having more than one, or by having fewer than one (i.e. none).

That is, some categories fail to have a terminal/initial object because they have too many morphisms, and some fail because they have too few.

Going on forever

The example of the natural numbers showed us that a category can fail to have a terminal or initial object by "going on forever" in one or other direction. In that case the problem is there is no object with enough morphisms.

This category has no terminal object because no object has enough morphisms going *to* it.
$$0 \longrightarrow 1 \longrightarrow 2 \longrightarrow 3 \longrightarrow \cdots$$

Everything only has morphisms going to it from one side, not from everywhere. This is because there is no "biggest" natural number: the thing that is trying to be a terminal object is infinity, but that isn't a natural number.

In fact we can try to add in a terminal object. That turns out to be one way of fabricating something like infinity to add in to the natural numbers.

We would need an extra object, say ω, and a morphism to it from every other object, so it might look a bit like this.[†]

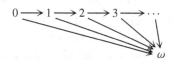

In this diagram I have, unusually, drawn some arrows that are composites.

───────────────

[†] The symbol ω is the Greek letter omega, which is often used to represent infinity when it's an actual object, as opposed to ∞ which is more of a concept.

- We could omit the arrow from 0 to ω as it is this composite: $0 \longrightarrow 1 \longrightarrow \omega$
- We could omit the arrow from 1 to ω as it is this composite: $1 \longrightarrow 2 \longrightarrow \omega$

$\vdots$

The trouble is we can't keep omitting arrows forever because then we won't have any arrows, so at some point we're going to have to start drawing arrows to ω. But it's most consistent to start at the beginning, and just draw them all.

This way of adding a terminal object that's missing is quite in the spirit of higher mathematics: we find ourselves in a situation that is lacking something, and that lack is preventing us from doing something we want, so we construct a new world in which we've added that thing in. This is how we get from the natural numbers to the integers and then to the rationals, the reals, and the complex numbers. In category theory this is the concept of "adding structure freely", which means you don't change anything about the structure you already had, and you don't impose any unncessary conditions on the new structure you add in. We'll come back to this in Chapter 20.

Branching

Another way to fail is by having a branching point. In that case there might be a beginning or end but not a place where *everything* begins or ends.

This example has an initial object, a, but no terminal object. The objects b and c are both ending points but because the arrows branch out, there are "too many" ending points.

Formally the problem is that b has no arrow to it from c, so b can't be terminal. But c has no arrow to it from b, so c can't be terminal. (And a has no arrow to it from either b or c.) So this is a situation where there are too few morphisms.

This one is the *dual* — I have just flipped the directions of the arrows. Thus now a is terminal but there is no initial object.

Going round in circles

We have already seen a situation of "going round in circles" that had no initial object, provided that the arrows were not inverses.

Those circles could be bigger "circles" as shown here; as long as the cycle does not compose to the identity there will be neither an initial nor a terminal object.

These examples are all self-dual, that is, if we turn all the arrows around the diagram is the same overall diagram (just with labels in different places).

Parallel arrows

This example has a pair of parallel arrows, meaning there are too many morphisms for *a* to be initial and too many for *b* to be terminal. This is another self-dual example.

Disconnectedness

If a category is "disconnected" then there are too few arrows for anything to be initial or terminal. A category is disconnected if it separates into disjoint parts with no morphism joining them together at any point.

This example has no non-trivial morphisms, but more than one object, so no object can be terminal or initial.

This example has two disconnected parts with non-trivial morphisms in each. The objects *a* and *x* are "locally initial" in their respective regions, but there is no morphism between the parts so neither can be initial overall. Likewise for *b* and *z*, with respect to terminal objects.

Empty category

The empty category has no terminal or initial object because it has no object!

Things To Think About

T 16.11   Is it possible for an object to be terminal and initial at the same time?

Here are some trivial situations in which an object is both terminal and initial.

* If the category has only one object and one (identity) morphism then the object is both terminal and initial.
* If the category has more than one object but they're all uniquely isomorphic then every object is both terminal and initial.

The second example is a useful and important type of category that we'll come back to. For now we're going to build up to seeing that the category of groups has an object that is both terminal and initial, in a more profound way.

16.7 Examples

Most of the examples we've been looking at have been abstract ones where I've just drawn categories with no particular meaning, to demonstrate structural features. Now let's look at what terminal and initial objects are in some of the examples of categories of mathematical structures that we've seen.

Sets and posets

In the category of sets and functions we've actually already seen the terminal and initial objects.

─────────────── **Things To Think About** ───────────────

T 16.12   What are the terminal and initial objects in **Set**? Think about what set of outputs ensures that there is only one possible function *to* it, no matter what set of inputs you have (and the other way round for initial objects). A more formal but less intuitive approach is to think about the formula for the number of possible functions from one set to another, and consider what set would force that formula to give the answer 1.

You may prefer thinking the intuitive way or through the formula, but it's good to be able to do both to check them against each other. For the formula, we saw that if we have n inputs and p outputs then the number of possible functions is p^n. Now observe that $p^n = 1$ if and only if $p = 1$ or $n = 0$.

Let's think about the outputs first. Intuitively the condition $p = 1$ means we have only one output, so only one option for where any inputs go, thus only one possible function. Thus any set with only one element is terminal: if a set B has only one element, then for any set A there is exactly one function $A \longrightarrow B$.

Note that it doesn't matter what the single element is. It could be the number 1, the number 83, an elephant, or anything else. Uniqueness here is the fact that all of the one-element sets are uniquely isomorphic to one another.

There is always a unique bijection[†] between any one-element set and any other, as illustrated here.

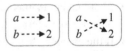

By contrast, a pair of 2-element sets has two bijections between them, so they are isomorphic but not *uniquely* isomorphic.

If we're talking about a one-element set we often declare the single element to be an asterisk $*$ to remind ourselves that it has no meaning. The resulting one-element set is then written $\{*\}$.

Let us now think about initial objects. We saw that $p^n = 1$ if $n = 0$, and this condition means there are no inputs: the set of inputs is the empty set. We saw in Section 12.1 that this means there is always exactly one function out: it's the "empty function", which is vacuously a function as there are no inputs.

As for uniqueness of this initial object, we usually consider that there is only one empty set, as sets are defined by their elements. That is, sets A and B are

───────────────

[†] Remember a bijection is a perfect matching up of objects in the two sets; formally it is injective and surjective.

usually declared equal if and only if they have the same elements, which is the case if they both have *no* elements. Thus this initial object is actually *strictly* unique. In category theory we say something is "strictly" the case to contrast it with being the case "up to isomorphism".

The terminal and initial sets are not the same. However, although they are opposite extremes categorically, they might not feel like opposite extremes as sets: they're both very small. This shows that categorical intuition may need to be developed slightly differently from other intuition.

Incidentally the terminal set is characterized by what happens when we map *into* it, but it's also a very useful set for mapping *out* of.

Things To Think About

T 16.13 What does a function $\{*\} \to A$ consist of?

Defining a function $\{*\} \to A$ consists of choosing one output, as we just need to know where the element $*$ is going to go. If we write 1 for the 1-element set what we're saying is: a function $1 \to A$ "is" just an element of A.

As often happens in category theory, the "is" in the above statement needs some interpretation. Informally this means "specifying a function $1 \to A$ is logically equivalent to specifying an element of A". If you find this a little unsatisfactory your instincts are good; there are various ways to state it more rigorously but it's good categorical practice to become comfortable with thinking of it as stated above with the word "is".

> Aside on foundations: this logical equivalence is how we can translate between category theoretic and set theoretic approaches to math. The category theoretic approach takes functions as a basic concept, rather than elements of sets, but the above shows that we can use functions out of a 1-element set to get back the concept of elements of sets, while still remaining categorical.

The basic principle of the empty set and the 1-element set can guide us to looking for initial and terminal objects in categories of sets-with-structure. It is at least a place to start looking.

Things To Think About

T 16.14 For these types of set-with-structure see if an empty one is initial, and if a 1-element one is terminal: posets, monoids, groups, topological spaces.

This is true of posets and topological spaces, but is not entirely true of monoids and groups, as we'll see in the next section. The key is that the unique function from any poset to the 1-element one is definitely order-preserving, as every-thing is just mapped to the same place. Similarly for the unique function from the empty poset to any other poset: there is nothing to order in the empty poset, and so the order is vacuously preserved.

Likewise for spaces, the intuition is that the unique function from any space to the one-element space is continuous because everything lands on the same point, thus everything ends up close together and nothing is broken apart. The unique function from the empty space to any other is vacuously continuous because there is nothing to map, so it is vacuously all kept close together.

Monoids and groups

Monoids and groups work a bit differently from sets, posets and spaces because they can't be empty: the definition says there must be an identity element. Indeed, the smallest possible monoid or group is the one with a single element, the identity. This is called the trivial monoid/group as it is the most trivial possible example.

─────────────── **Things To Think About** ───────────────

T 16.15   Try investigating group homomorphisms to and from the trivial group, based on what we know about functions to and from the 1-element set. Mapping *to* it, we know there's only one possible function; is it a homo-morphism? Mapping *from* it, there's one possible function for each element of the target group; which ones are homomorphisms?

For the part about mapping *into* the trivial group, the point is we know that the trivial group has the terminal set as its underlying set, but is it terminal as a group? The answer is yes: the unique function to it (from anywhere else) will have to map everything to the identity, and that is definitely a homomor-phism. The axioms are all equations in the target group, which only contains the identity element, so all equations are true (the two sides can only equal the identity). So the trivial group is terminal in the category **Grp**.

Now for the part about mapping *out of* the trivial group. When we're doing ordinary functions out of a 1-element set we know that we can map that single element anywhere in the target set. However, for group homomorphisms we don't have a free choice: a group homomorphism must preserve the identity element, and the single element in the source group is the identity, so it has to go to the identity in the target group. Any function picking out some other element of the target group will not be a homomorphism.

This shows there is only one possible homomorphism from the trivial group to any other, that is, the trivial group is initial in **Grp**, as well as being terminal. This is quite an interesting situation so gets a name.

Definition 16.7   A *null object* in a category is one that is both terminal and initial.

Inspired by the example of sets, terminal objects are often generically written as 1, and initial objects as 0. Thus a null object arises when $0 = 1$ in a category. This slightly shocking equation alerts us to the fact that a category with a null object behaves differently in some crucial ways. Arguably the null object for groups is one of the things that makes the category of groups such a rich and interesting place in which to develop algebraic techniques, and that is in turn why so many branches of math have advanced by putting the word "algebraic" in front of them and making some sort of connection with group theory.

Inside a poset

In a poset we can generically call the ordering "less than or equal to" as the conditions on a partial ordering make it behave like that whatever it actually is. Then an initial object corresponds to the minimum element, if there is one, and a terminal object corresponds to the maximum element.

─────────── **Things To Think About** ───────────

T 16.16   The minimum is an element that is less than or equal to every element. How does this correspond to an initial object?

In a poset, an arrow $a \longrightarrow b$ is the assertion $a \leq b$. A minimum, say m, is defined by being less than or equal to every element, meaning there is an arrow from m to every element. For m to be initial each of those arrows must also be unique, which is true because in a poset there is always at most one arrow $a \longrightarrow b$ for any pair of objects a, b. The fact that a maximum is terminal is analogous.

Here is a poset with an initial object but no terminal object. The element at the bottom is less than or equal to all elements, so it is initial. The ones at the top are "locally maximal" but not maximal overall, so are not terminal.[†]

─────────── **Things To Think About** ───────────

T 16.17   Turn all the arrows around in that diagram. What is the situation with terminal and initial objects now?

[†] When we say "locally" we mean something like "this is true if you zoom in to a small enough region". We could define "local maximum" formally but I hope you can see the intuition.

If we turn all the arrows around then the bottom object (physically on the page) becomes terminal and there is no longer an initial object.

Things To Think About

T 16.18   In the diagram of generalizations of types of quadrilateral in Section 5.5 what was initial and what was terminal? What about in the diagrams for factors of numbers, and cubes of privilege?

In the diagram of quadrilaterals "square" was initial and "quadrilateral" was terminal. This was because everything was a type of quadrilateral, and squares turned out to be special cases of everything.

In the lattice/cube/poset for factors of 30 it depends which way we draw our arrows. With the directions shown on the right, 30 is initial and 1 is terminal. It might look like there is more than one arrow from 30 to 5, for example, as we could go via 10 or via 15; however, the diagram commutes and so both ways around that square are equal.

The moral of that story is that a terminal object has exactly one morphism to it from any object, but the one morphism could be expressed as a composite in more than one way.

In the corresponding diagram for privilege, rich white men are at the top, so are initial if we are drawing arrows according to *losing* one type of privilege. People with none of those three types of privilege are terminal; we previously drew the diagram with the empty set in that position, but if we put the words in, it is non-rich non-white non-men who are terminal in that particular context.

16.8 Context

One of the important things we can do with universal properties is move between different contexts and make comparisons: we can compare universal objects in different categories, or we might observe something being universal in one category but not another. We did some comparing when we looked at the initial objects in various categories of sets-with-structure.

We compared them with the initial object in **Set** and asked whether they "matched". We could think of this in terms of a schematic diagram as shown on the right.

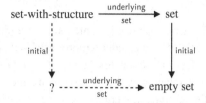

The question then is whether we can go round the dotted arrows and still get to the empty set. For posets and topological spaces we could, but for monoids and groups the dotted arrow took us to a 1-element set, not the empty set. We will talk about this more when we talk about functors preserving structure.

A slightly different but related question is about objects being universal (having a universal property) in one category but not another. In the case of sets compared with groups, the empty set could not become an initial object in the category **Grp** because there is no such thing as an "empty group"; a group must at least have one element, the identity. In the diagram of factors of 30, 30 is initial because we're only considering factors of 30. If we consider factors of 60 or 90, say, then 30 is there but will no longer be initial. The number 1 will still be terminal in all those categories, but if we allowed fractions then something more complicated would happen at the bottom of the diagram.

For the categories of privilege we could look in the diagram for "rich, white, male", but we could then restrict our context to women and consider the three types of privilege "rich, white, cisgender".

In the first case rich white women are not at the top.

However, in the context of just women they *are* at the top, as shown here.

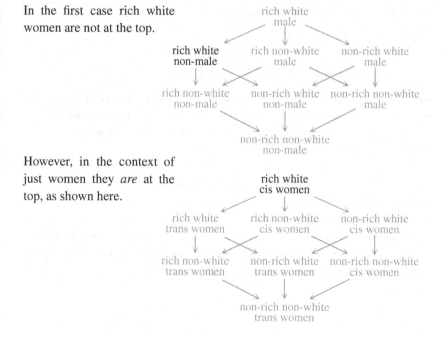

Thus rich white cis women are not initial in the broader category, but they are initial if we consider a more restricted category.

I wrote about this in *The Art of Logic* as a structure that has helped me understand why there is so much anger towards white women in some parts of the feminist movement, because they only think of the first diagram and of the sense in which they are under-privileged compared with white men. This will be especially true if they tend to mix only with white people. They then forget how privileged they are relative to all other women. As with the anger of poor white men, I think it is more productive to understand the source of this anger rather than simply get angry in return.

I think it is fruitful to think about categories in which we have different universal properties so that we gain understanding of what it's like to play different roles relative to context. I can pick three types of privilege that I have: I am educated, employed and financially stable. I can also pick three types of privilege that I don't have: I am not male, I am not white, and I am not a citizen of the country I live in. I think it would be counter-productive to fixate on either of those points of view, but it is illuminating to be able to keep them both in mind (and others) and move between them.

Similarly category theory should not be about getting fixed in one category. Expressing things as categories is only the start. What's more important is moving between different categories and thereby seeing the same structure from different points of view, which is why we will soon come to functors, which are morphisms between categories.

16.9 Further topics

Mac Lane (in)famously wrote "All concepts are Kan extensions". I would say that all concepts are universal properties — the question is just *in what context*? Furthermore, all universal properties are initial objects *somewhere*. So all concepts are initial objects.

Other universal properties
There are other universal properties that we study in category theory, and we will look at some of them in Chapters 18 and 19. They can all be expressed as initial objects *somewhere*, but if a concept is very common and illuminating it can be more practical to isolate its universal property in its own right, rather than always referring to it as an initial object in some more complicated category.

Sometimes the initial objects are to do with freely generated structures, such as with monoids where the initial object, $\mathbb{N}$, was necessarily the monoid freely generated by a single element 1. Freely generated algebraic structures are related to the use of functors to move between categories. A freely generated structure is typically produced by a particular type of functor, a "free functor", which in turn has a universal property expressed via a relationship with a "forgetful functor". Those relationships are called *adjunctions*; we will not get to those in this book but I thought I'd mention the term in case you want to look them up.

Preservation by functors

I have mentioned the fact that moving between categories to change points of view is very important, and we'll come back to this in Chapters 20 and 21 on functors, the "morphisms between categories". When we do so we will ask questions about whether terminal and initial objects are preserved, that is, if we start with an initial object in one category and apply a functor, will the result still be initial in the target category of the functor?

Characterizing by universal property

In the categorical way of thinking, if a construction is somehow natural, canonical, or self-evident, it should have a universal property making that "naturalness" precise. This is a way of making our intuition rigorous. Otherwise the idea of something being natural is very flimsy because what seems natural to one person will not be at all natural to another, even among mathematicians. Exhibiting a satisfying universal property for something is satisfying in its own right, aside from it being a handy tool we can use. It can be a sort of abstract validation of our ideas.

This type of math is not about getting the right answers, and so it is not enough just to check the validity of your logic to make sure your answer is correct. This type of math is more about defining new abstract frameworks that are fruitful. If your definition is logically sound that's only the first step. If it's useful for something that's another step. But in category theory if it has a good universal property, then that shows it has internal logic of its own, and that it's not just something utilitarian that we cooked up to serve a purpose, but rather, that it grows organically from the strong roots of abstract mathematical principles.

17

Duality

We can regard all arrows in a category as pointing the other way, and this gives us the dual category. One advantage is that we immediately get a dual version of every construction and every theorem.

17.1 Turning arrows around

When we draw an arrow ⟶ normal convention has it that we are indicating a direction, in this case moving towards the right. This may be something to do with the shape of actual arrows (as in the weapon) and that this is an abstract diagram of one of those, but I imagine that most of us have not seen very many arrow-weapons in real life. I've seen a few in museums but that's it.

This assocation of direction with the diagram is just a convention. We could take the convention to be the other way without changing anything about the structure of the situation, we'd just risk confusing everyone.

Duality in category theory is about doing just this: turning all the arrows around without really changing the structure of the situation. It also risks confusing everyone, but when the concepts in question are very abstract it's sometimes clear that the choice of direction is arbitrary.

One example we saw early on, in Chapter 5, was the diagram of types of quadrilateral. We were thinking about generalizations and drew each arrow to represent relaxing a condition and thereby moving from a special case to a more general one.

Here is an example of one of those arrows, and two different ways we can read the same information.

> square $\xrightarrow{\text{generalization}}$ rhombus
>
> *"A rhombus is a generalization of a square."*
> *"A square is a special case of a rhombus."*

These two readings are not expressing any different information, they're just expressing it the other way round.

We could also draw it the other way round, like this.

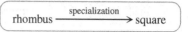

rhombus —— specialization ——→ square

──────── **Things To Think About** ────────

T 17.1   Here is an example of a family relationship we saw in Chapter 5.

$A \xrightarrow{\text{is a parent of}} B$

If we turn this arrow around what will the relationship be? Is it the opposite?

If we turn around the parent arrow we get this: $B \xrightarrow{\text{is a child of}} A$

In both the generalization and parent examples it might seem that the two directions are opposites of each other, but opposites are a hazy concept. I suppose if we asked someone in normal life "What's the opposite of a parent?" they would probably say a child, but we could also say it's a non-parent. Duality, by contrast, is a precise concept.

──────── **Things To Think About** ────────

T 17.2   What relationships do we get if we turn these arrows round?

1. $a \xrightarrow{\text{is a factor of}} b$ 3. $a \xrightarrow{\text{gain of one type of privilege}} b$

2. $a \xrightarrow{\leq} b$ 4. $a \xrightarrow{\sim} b$ (indicating an isomorphism)

If we turn around the arrow for "is a factor of" we get: $b \xrightarrow{\text{is a multiple of}} a$

For $\leq$ we get $\geq$, which is not exactly the "opposite" as both versions include the = part. $b \xrightarrow{\geq} a$

For the third example, gain of privilege becomes this: $b \xrightarrow[\text{of privilege}]{\text{loss of one type}} a$

For the last example, we get the same relationship: $b \xrightarrow{\sim} a$

So far all we've done is turn around individual arrows, but if we then look at the effect on structures, more interesting things happen.

──────── **Things To Think About** ────────

T 17.3   Take a diagram of factors of 30, turn every arrow around, and then physically turn the new diagram upside-down. What do you notice if you compare this with the original diagram? Which numbers play analogous roles? What are the relationships between the pairs of numbers x, y where the role of x in the first diagram is analogous to the role of y in the last?

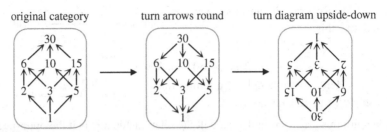

original category   turn arrows round   turn diagram upside-down

Probably the most immediately obvious thing is that the last diagram has the same shape as the first one — it's a cube, so upside-down it's still a cube. This means every object has an analogous object in the other diagram. 1 and 30 have switched places (they're opposite corners of the cube), which means that the terminal object and initial object have switched places.

We then see that we have the following analogous numbers.

upper row	lower row
6 is analogous to 5	2 is analogous to 15
10 is analogous to 3	3 is analogous to 10
15 is analogous to 2	5 is analogous to 6

It turns out these two tables give us the same information, just stated a different way round. Moreover each pair of analogous numbers multiplies together to make 30. That is a more standard school-level way of thinking about factors: in pairs that multiply together to make the number in question.

All of these features are about duality and are a way we can glean more understanding from situations by looking at them the other way round.

17.2 Dual category

We have made many references to this idea of turning arrows around but now we are finally going to present it formally. The basic concept is that given any category, there is a related category where we simply consider every arrow to be facing the other way. The information isn't different, just the presentation, but it's a remarkably powerful concept. We use the notation $\mathcal{C}^{\mathrm{op}}$, pronounced "C op". The "op" stands for "opposite".

Definition 17.1 For any category $\mathcal{C}$ the *dual (or opposite) category* $\mathcal{C}^{\mathrm{op}}$ is given as follows.

- objects: $\mathrm{ob}\,\mathcal{C}^{\mathrm{op}} = \mathrm{ob}\,\mathcal{C}$
- morphisms: given objects a, b, we have $\mathcal{C}^{\mathrm{op}}(a, b) = \mathcal{C}(b, a)$
- identities are the same
- composition is reversed (which I'll explain below).

In order to explain the last point, I first want to give a little more intuition about what this definition means. This can all get very confusing so I'm going to draw arrows in $\mathcal{C}^{op}$ with a little circle on them, like this: $\longrightarrow\!\!\circ\!\!\longrightarrow$

The definition says an arrow $a \xrightarrow{\;\;f\;\;}\!\!\circ\, b \;\in\; \mathcal{C}^{op}$

is given by an arrow $b \xrightarrow{\;\;f\;\;} a \;\in\; \mathcal{C}$

The "meaning" of the arrow is no different, just the direction we're regarding it as having, formally. Now for composition in $\mathcal{C}^{op}$ I am going to write $\odot$.

We need to consider this configuration in $\mathcal{C}^{op}$

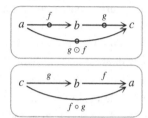

The diagram this corresponds to in $\mathcal{C}$ is this:

So it makes sense for us to define composition in $\mathcal{C}$ by: $g \odot f = f \circ g$.
This is what I mean by "composition is reversed" in the definition.

--------------------------- **Note on type-checking** ---------------------------

Here, "makes sense" means that the composite $f \circ g \in \mathcal{C}$ has the correct source and target to be a candidate for giving the composite $g \odot f \in \mathcal{C}^{op}$. This is called *type-checking*, which is a term that is more prevalent in computer science. When you ask a computer to handle a variable, you first have to tell it what type of variable to expect: is it going to be a number, a letter, a string, and so on. If you then ask the computer to perform a procedure that produces a variable that is not of the type the computer was expecting then the computer will get confused as your procedure "doesn't type-check".

Category theory puts so much structure in place in advance that often a large part of a construction or proof is just a question of type-checking. Sometimes we're guided in our arguments or definitions because there's only one thing that will type-check, as in the above situation. We couldn't even try and make $g \odot f = g \circ f$ because it doesn't type-check: f and g are not composable that way round in $\mathcal{C}$, they are only composable the other way round. Type-checking can pre-empt a lot of muddy thinking. It's a bit like starting cooking by getting out all your ingedients, measuring them into little bowls, and lining them up neatly like they seem to on television cooking shows. I'm usually too lazy/impatient to do that but I have to admit that if I can be bothered to do it, it makes the actual cooking process more calm and organized.

So far I have written out all the arrows in $\mathcal{C}^{op}$ with circles on them, but in general I prefer to completely avoid drawing arrows in the dual category. Apart from in the definition itself, it's rarely necessary, and I usually find it more confusing than it's worth. Usually we use dual categories as either a thought process or a way of formalizing the idea of functors that flip the direction of arrows, as we'll see in Chapter 20.

We say "dually" to mean "now repeat this argument but in the dual category". But rather than draw the arrows in the dual category, we flip the arrows in the argument, as we'll see in the following examples. The idea is that given any construction or argument that we have made in a category $\mathcal{C}$, there is a dual version which consists of doing the exact same thing but in the category $\mathcal{C}^{op}$. The dual version is often indicated by the prefix "co-". So for example:

- we define products and then the dual concept is coproducts
- we define limits and then the dual concept is colimits
- we define equalizers and then the dual concept is coequalizers
- we define projections and then the dual concept is coprojections.

This gives rise to all sorts of silly jokes involving either sticking the prefix "co-" onto a word in normal English, or re-interpreting words that start with "co" in normal English but not as a prefix, such as "countable" (co-untable: what does "untable" mean?). Here are a couple of jokes. If you get them then you're getting the idea of duals.

1. What do you call someone who reads category theory papers? A co-author.
2. A mathematician is a machine for turning coffee into theorems...

     ...A co-mathematician is a machine for turning co-theorems into ffee.

The last joke depends on the fact that the dual of a dual takes us back to the original concept.

Proposition 17.2   *Let $\mathcal{C}$ be any category. Then* $(\mathcal{C}^{op})^{op} = \mathcal{C}$.

I experienced a little shudder there because I wrote an *equality* between categories! This is two levels too strict: the correct notion of sameness between categories, which we aren't ready to define formally yet, is equivalence. Isomorphism is a little less nuanced, and equality is dire. And yet I think this really is an equality because it was defined via some equalities of sets, which is already one level stricter than morally correct.

17.3 Monic and epic

We informally pointed out that the definition of epic is just the same as monic with the arrows turned around, but here is what that means formally.

Proposition 17.3 *A morphism f is epic in C if and only if it is monic in C^op.*

Maybe this should be the definition of epic, but given that we defined epic directly, this is duly a proposition. Also the word "it" needs some care: the morphism f is a morphism in both C and C^{op} because the morphisms in the dual category are the same as those in C, just regarded the other way round.

This gives us a model for what "dually" means: it means we take the concept in question, that we have studied in C, and now study it in C^{op}, but still draw the arrows in C. The following steps take us from the definition of monic to the dual definition of epic.

The definition of a monic in C involves this shape of diagram: $m \rightrightarrows a \longrightarrow b \quad \in C$

So a monic in C^{op} involves a diagram like this: $m \overset{s}{\underset{t}{\rightrightarrows}} a \overset{f}{\rightarrow} b \quad \in C^{op}$

Now, following the principle of only drawing arrows in C and not in C^{op}, we draw this as: $m \overset{s}{\underset{t}{\leftleftarrows}} a \overset{f}{\leftarrow} b \quad \in C$

Finally, as we prefer arrows to point right, we flip it physically on the page, giving the usual diagram for the definition of epic: $b \overset{f}{\longrightarrow} a \overset{s}{\underset{t}{\rightrightarrows}} m \quad \in C$

Things To Think About

T 17.4   What is the dual of an isomorphism? That is, if f is an isomorphism in C^{op} what does that tell us about it in C?

The diagram on the right shows f as an isomorphism, with inverse g. If we turn the arrows around it's the same diagram.

So if f is an isomorphism in C^{op} then it is also an isomorphism in C. In fact that is exactly the same information. If a concept gives the same information whether it's in C or in C^{op} then it is called *self-dual*.

Another dual situation we saw in Chapter 15 on monics and epics was the proof that every isomorphism is monic: we rather briefly said the proof that every isomorphism is epic follows "dually".

Things To Think About

T 17.5   Can you state this formally using the dual category?

The formal version of the dual proof goes like this. We have proved that if f is an isomorphism in C then it is monic in C. But this is true in any category, so it is also true in C^{op}. Now if f is an isomorphism in C then it is an isomorphism in C^{op}, thus it is a monic in C^{op}, but this precisely says that it is epic in C.

This is an example of what I call a BOGOF[†] on theorems.

―――――――――――― **Duality BOGOF principle** ――――――――――――

Any concept in a general category can be considered, in particular, in $\mathcal{C}^{op}$, and then translated back into a concept in $\mathcal{C}$ with no extra proof or argument needed. This is true of definitions and proofs. We just say "and dually".

Here's an example of how we get to use the BOGOF principle.

―――――――――――― **Things To Think About** ――――――――――――

T 17.6   Suppose we have morphisms $a \xrightarrow{f} b \xrightarrow{g} c$ in $\mathcal{C}$.

1. Prove that if f and g are both monic then the composite gf is monic.
2. What does the dual give us for free?

Proof   To show that gf is monic we need to consider a diagram as on the right, and show that $gfs = gft \implies s = t$.

$$m \underset{t}{\overset{s}{\rightrightarrows}} a \xrightarrow{f} b \xrightarrow{g} c$$

So suppose $gfs = gft$. Since g is monic we know that $fs = ft$. Since f is monic we know that $s = t$ as required.

The dual tells us that if f and g are both epic then gf is epic.   □

Note that if we really translated this result into $\mathcal{C}^{op}$ we might get rather confused about the direction of the composite, which in $\mathcal{C}^{op}$ would be written fg. This is one of the reasons I prefer not to write down arrows in the dual category. Another reason (less utilitarian and more ideological) is that I find that the *point* of considering the dual is to think about new diagrams in $\mathcal{C}$, not pretend you're in a different category temporarily.

Here's another BOGOF example.

―――――――――――― **Things To Think About** ――――――――――――

T 17.7   Given morphisms $a \xrightarrow{f} b \xrightarrow{g} c$ in $\mathcal{C}$, prove that if gf is monic then f must be monic but g need not be. [Hint: the first part follows by working your way through the definitions. For the second part you could try finding a counterexample in **Set** using injective[‡] and non-injective functions.] What does the dual give us for free?

―――――――――――――――――――――――――――――――――――――

[†] Buy One Get One Free (also called BOGO in the US).
[‡] Reminder: injective means "no output is hit more than once".

Proof   Again we are considering the following diagram:

$$m \underset{t}{\overset{s}{\rightrightarrows}} a \xrightarrow{f} b \xrightarrow{g} c$$

Suppose that gf is monic. To show that f is monic we suppose $fs = ft$. But this certainly means $gfs = gft$ and then as gf is monic we can deduce that $s = t$ as required.

To show that g need not be monic let us consider this in **Set** and find functions f and g such that gf is injective but g is not.

> The idea is that f has to be injective because if it mapped two different things to the same place then gf would be doomed to do the same, whereas g doesn't need to be injective on the non-image[†] of f, because the non-image of f never gets in on the action of gf. So we just need b to be big enough to have some non-image of f.

Consider the sets and functions shown on the right. The composite gf is monic (in fact it's an isomorphism) but g is not, so this is a counterexample. Dually, if gf is epic then g must be epic but f need not be.   □

$$a \xrightarrow{f} b \xrightarrow{g} c$$
$$0 \dashrightarrow 0 \dashrightarrow 0$$
$$1$$

The BOGOF dual part perhaps requires some care because of the directions.

Here is a line-by-line comparison of the argument in $\mathcal{C}^{\mathrm{op}}$ and how it translates to $\mathcal{C}$.

$\in \mathcal{C}^{\mathrm{op}}$	$\in \mathcal{C}$
$a \xrightarrow{f} b \xrightarrow{g} c$	$a \xleftarrow{f} b \xleftarrow{g} c$
if the composite is monic	if the composite is epic
then f must be monic	then f must be epic

I find that the best way to avoid confusion is to detach my brain from the letter names of the functions and think of their position instead: if a composite is monic then the *first* arrow you travel along must be monic, and if a composite is epic then the *last* arrow you travel along must be epic.

───────── **Things To Think About** ─────────

T 17.8   Although it's not logically necessary, can you find a counterexample in **Set** to exhibit the fact that gf can be epic even if f isn't epic? Look back at the example for monics.

The example we gave for monics also works for epics. The composite gf was epic (it was an isomorphism) but f was not epic.

These proofs involving monics, epics and composites are things I really enjoy again and again no matter how many times I do them. It's not for their

───────────────

† Reminder: the image is "all the outputs that are actually hit", and non-image means "outputs that are not hit".

usefulness, it's not particularly because of what they illuminate, it's not because of profound applications or relevance to my life. I just find them deeply satisfying. It's the sense of fitting logical pieces together perfectly. This is why I loved category theory on impact, the very first time I studied it. It was the purest logical joy I had encountered. Loving it for its illuminating and applicable aspects came much later.

17.4 Terminal and initial

Another situation in which we thought about turning arrows around was when we started with the definition of initial object and then "turned the arrows around" to make the definition of terminal object.

──────────────── **Things To Think About** ────────────────

T 17.9   Can you make that correspondence between terminal and initial objects precise using the formal concept of $\mathcal{C}^{op}$?

Proposition 17.4   *Let $\mathcal{C}$ be a category. A terminal object in $\mathcal{C}$ is precisely an initial object in $\mathcal{C}^{op}$.*

Note that as with epics, I might prefer this to be the *definition* of terminal object, but as we already defined them directly this is a result that follows. It's another of those situations where we can choose which is the definition and which is the result. Anyway, this is what we mean by saying that terminal objects are the dual of initial objects. Indeed some people call them final objects and cofinal objects.

Every universal property we look at (well, also all the ones we don't look at) will come with a dual version, and from now on we will typically say that first, and then unpack what the dual means when translated back into $\mathcal{C}$.

17.5 An alternative definition of categories

I want to finish this chapter by pointing out where this duality in categories really comes from, or rather, by presenting a slightly different point of view on the definition of category that makes the concept of duality jump out at us. There are many different ways of presenting the underlying data for a category, emphasizing different aspects. Depending on how you present the underlying data, you then have to present the structure and the properties slightly differently too, but I want to focus on the data for now.

We previously defined a (small) category $\mathbb{C}$ to have

- a set of objects, and
- for every pair of objects a, b a set of morphisms $\mathbb{C}(a, b)$.

However, we never really talked about the "set of all morphisms", just the set of morphisms from any given object to another. Instead we could say there is

- a set of objects C_0, and
- a set of morphisms C_1

and now we need to do a bit of extra work to say that each morphism goes from one object to another. We express this using functions.

The idea is that every morphism has a source object, so we can make a function as shown on the right.

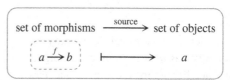

The small dotted box is to indicate that the morphism f is one element in the set of "inputs" here, as the set of inputs is the set of morphisms. It is mapped to the "output" a.

Likewise we have a function sending each morphism to its target object, as shown here.

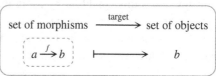

So, going back to the notation of the sets C_0 and C_1 we have two functions like this.

$$C_1 \overset{s}{\underset{t}{\rightrightarrows}} C_0$$

I hope it is now clear that there is symmetry between s and t. Each one is merely a function from C_1 to C_0, and flipping the morphisms of the category $\mathbb{C}$ just consists of swapping the names s and t here.

This different way of describing the underlying data for a category is particularly important for certain kinds of generalization, including into higher dimensions. This diagram of two sets and two functions between them is sometimes called a graph or a directed graph, not to be confused with the sorts of graph you are made to sketch at great length at school. The idea here is that we could draw out the information contained in this "graph" as a diagram of actual arrows pointing from and to their designated sources and targets. It's called directed because the source and target functions give direction to the arrows. In category theory we rarely consider undirected graphs, so we're more likely to call this a graph and leave it to be specified when it's *not* directed.

─────── **Things To Think About** ───────

T 17.10   The next step in turning this into a full definition of category would be to work out how to express identities and composition. Try making a function that takes an object a and produces the identity on it as output. What conditions must this function satisfy to ensure that the identity 1_a has the correct source and target?

To express identities we need a function as shown here, with "id" for identity.

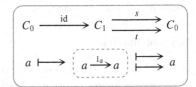

In order to be sure that the identity on a really has a as its source and target, we need to apply the source and target functions after the identity function, and know that we get a back again.

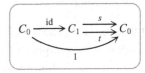

We can express this condition by using the diagram on the right, and asking that the composite functions $s \circ \text{id}$ and $t \circ \text{id}$ are both the identity function 1 (which sends every element to itself).

We're not ready to develop this into a full definition of categories yet as we need to look at another universal property first, in order to define composition. We'll see that in the next chapter.

18

Products and coproducts

The slightly more advanced universal properties of products and coproducts, together with examples in various of the categories we've seen.

Products and coproducts are our next universal property, slightly more complicated than terminal and initial objects. I have mentioned that every universal property is an initial object *somewhere*, and indeed products and coproducts in a category $\mathcal{C}$ are "just" terminal and initial objects in a "souped-up" category based on $\mathcal{C}$. However, it is often more helpful to translate a universal property back into $\mathcal{C}$ rather than thinking about that other souped-up category all the time. This is like when we thought about dual concepts by placing a concept in $\mathcal{C}^{op}$, but we translated the concept back into $\mathcal{C}$. I think it's like when you make a paper snowflake by folding up a piece of paper and cutting out a pattern — the fun part is opening out the paper again.

We will start with products, and then see the dual concept, coproducts.

18.1 The idea behind categorical products

In basic math the product of two numbers a and b is a number $a \times b$ that has a certain relationship with the original two numbers. When introducing this to children we might start by investigating $2 + 2$, and then $2 + 2 + 2$, and then $2 + 2 + 2 + 2$, and we might say that the first one is "two times two" and the next is "three times two" and the next is "four times two" and so on. Then there's the tricky question of why 4×2 is the same as 2×4, when the first is $2 + 2 + 2 + 2$ and the other is $4 + 4$.

We might then get them to play with some small objects and line up four rows of two in a grid. Then if you rotate the grid you see two rows of four instead.

So $a \times b$ can be thought of as a lots of b, or b lots of a, or the number of things in a grid of a rows and b columns, or the number of things in a grid of b rows and a columns. Some people argue about which view of multiplication is "best"

but I think the best thing (if we must make it a competition) is to understand all points of view and their relationships with each other.

Expressing multiplication of numbers by repeated addition does not generalize very well, either to more complicated types of number (like fractions or irrational numbers) or to more complicated types of object such as shapes, and this is where category theory comes in. We're going to look for a general type of multiplication that we can attempt in any category. It won't necessarily be *possible* in every category, just like the fact that some categories have terminal and initial objects and some don't. Some categories have products and some don't. It's pretty handy when they do.

For categorical products we will start with two objects a and b in our category, and produce a third object $a \times b$ which deserves to be called a product of a and b by virtue of certain relationships it has with a and b. The question is what those relationships are.

If we think about our 2 by 4 grid of objects above, what is the relationship between this grid and the numbers 2 and 4? We might say there are 2 rows and 4 columns, but that sort of presupposes an understanding of what rows and columns are. We might say its height is 2 and its width is 4, but that also presupposes concepts of height and width. Here is something we could say that's (perhaps) more fundamental.

Imagine projecting shadows of our objects onto a wall. If we project in one direction (and perfectly line everything up) then we will see only 2 things, and if we project in the other direction we'll see 4.

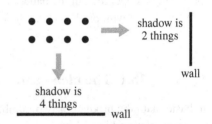

This is pretty much the idea of how we formalize products in category theory.

18.2 Formal definition

Given objects a and b we are going to think about configurations of arrows as shown on the right; the two arrows are like the two arrows in the shadow diagram above.

But we don't want any old thing that will do this: we want a universal one. This means we want it to be somehow "canonical", or inherent, or essential.

──────────── **Things To Think About** ────────────

T 18.1 Can you think of some irregular configurations of objects whose shadows as above still produce 2 things in one direction and 4 in the other?

Here are some configurations that produce 2 and 4 object shadows in sneaky, non-canonical ways.[†]

The full definition of product in a category demands that we have a diagram with two "projection" morphisms to a and b, and moreover that the diagram is universal among all such diagrams, in the following sense.

Definition 18.1 Given objects a and b in a category, a *product* for them is:

an object $a \times b$ equipped with morphisms p and q as shown below	such that given any diagram of this form	there is a unique morphism $x \xrightarrow{k} a \times b$ making the following diagram commute.[‡]

This is a standard format for a universal property. The second diagram is abstractly the same shape as the first, I've just drawn it differently physically to emphasize how they fit together in the third diagram; the second is often not drawn as we might just say "given any such diagrams…". The third diagram says that the first is universal among all diagrams of the same shape, so once we understand what that means we might not draw that one either.

The morphism k in the third diagram is a type of comparison between the universal situation (the first diagram) and the non-universal one (the second diagram). Note that asking for the diagram to commute means that the triangle on each side commutes. Written out in algebra this means $\boxed{pk = f \text{ and } qk = g}$ but I think it's really better to think of it geometrically in the diagram.

Finally note that the uniqueness of the morphism k needs to be taken together with the "such that" clause afterwards: it might not be the only morphism $x \to a \times b$, but it has to be the only morphism that makes both of those side triangles commute.

[†] This reminds me of the image on the cover of Douglas Hofstadter's spectacular book *Gödel, Escher, Bach: An Eternal Golden Braid* featuring a carved three-dimensional object which miraculously manages to project the shadows G, E, B, and also E, G, B. If you don't know it, I urge you to look it up, partly just because it's an amazing book and partly because the image on the cover is impossible to describe in words.

[‡] The ! symbol signifies "unique".

Terminology

This sort of situation is going to come up repeatedly so it's useful to have some terminology.

- The morphisms p and q are called *projections*.
- The morphism k is called a *factorization*. We are factorizing f and g.
- There is a general type of diagram called a *cone* which consists of a *vertex* and the morphisms from that vertex to everything else. In this case the vertex is $a \times b$ and "everything else" is just the objects a and b.
- The product is the *universal cone*, universal among all cones of this shape.
- We say the universal property *induces* the factorization k.

18.3 Products as terminal objects

Products are in fact a special type of terminal object. It's just that a product in $\mathcal{C}$ isn't a terminal object in $\mathcal{C}$, it's a terminal object in some more complicated category based on $\mathcal{C}$, which I like to call a "souped-up" category. We already saw this principle for terminal objects themselves: terminal objects in $\mathcal{C}$ are initial objects in $\mathcal{C}^{op}$. We are generally going to see that we can make a more complicated category based on $\mathcal{C}$, look for a simple structure in there, and then translate it back to a structure directly in $\mathcal{C}$.

For products, we need a category whose objects are whole diagrams of the shape in question. That is, given objects a and b whose product we want to define, we make a new category whose objects are diagrams of the form shown here.

Note that a and b stay the same, but we have a choice of x, f and g.

Now we need to define morphisms, for example between the objects shown here. Remember that the diagram in each box counts as one object in this "souped-up" category.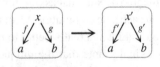

A morphism is going to look like the diagram in the definition of product. It consists of a morphism $x \xrightarrow{k} x'$ making the diagram on the right commute. We can think of it as a "structure-respecting map" where "respect" means "make everything commute".

Identities, composition and axioms are "inherited" from $\mathcal{C}$, which makes sense because morphisms in this category are just some particular morphisms in $\mathcal{C}$.

This is an example of when it's almost more important to understand what the question is than what the answer is. For identities we just need to observe that the diagram on the right commutes.

This shows that the morphism 1_x gives a valid morphism from this object to itself:

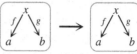

For composition we need to consider the type of scenario shown on the right.

This means we have the situation shown in the first diagram on the right. What we need to know is that the composite $k' \circ k$ is a valid morphism from the beginning to the end, which means we need to show that the second diagram commutes.

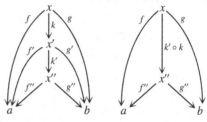

This is true because if you stick together commutative diagrams you get commutative diagrams. That's rather vague, but the thought process typically goes as follows.

We want to show that the first diagram on the right commutes, so we "fill it in" with the dotted arrows as shown in the second diagram. (This is where it would really help to be a video.) Then because all the small areas commute, we know that the larger diagram commutes.

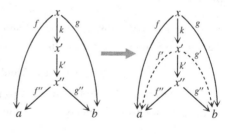

If we wrote it out in algebra it would look like this:

$$
\begin{aligned}
f'' \circ k' \circ k &= f' \circ k &&\text{as the bottom left triangle commutes}\\
&= f &&\text{as the top left triangle commutes}\\
g'' \circ k' \circ k &= g' \circ k &&\text{as the bottom right triangle commutes}\\
&= g &&\text{as the top right triangle commutes}
\end{aligned}
$$

I hope you can see that the method using diagrams is much more illuminating while containing all the information, plus a little more.

Proving things in category theory involves a lot of trying to show that some large diagram commutes by "filling it in" in this way. This is what is known as "making a diagram commute".[†] Personally I find it really fun and satisfying in a way that strings of symbols are not (to me). I appreciate being able to invoke geometric intuition. Moreover, a string of symbols omits the source and target information for the morphisms and so loses a possibility for type-checking. For example, in the above argument it is possible to write the symbols $k \circ k'$ by mistake, although those morphisms are not composable that way round. Whereas if you draw them out as arrows, then as long as you get their sources and targets right you can't physically draw the composite the wrong way round.

─────────────── **Things To Think About** ───────────────

T 18.3   Can you see how terminal objects in our "souped-up" category correspond to products in $\mathcal{C}$?

A terminal object in this category is:

an object   such that for any object   there is a unique morphism

     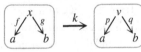

But such a morphism k in the souped-up category is a morphism $x \xrightarrow{k} v$ in $\mathcal{C}$ making the diagram on the right commute, which is exactly the unique factorization we need for the definition of product.

Note that k is both a morphism in $\mathcal{C}$ and, because it makes the appropriate diagram commute, a morphism in the souped-up category.

─────────────── **Things To Think About** ───────────────

T 18.4   As the product is a terminal object, we get some uniqueness from the uniqueness of terminal objects. Can you unravel that and say what it is in $\mathcal{C}$?

We know that terminal objects are unique up to unique isomorphism, and products are a terminal object *somewhere*, so products must also be unique up to

─────────────────────

[†] Some more pedantic people object to this term on the grounds that we're not making it commute, we're showing that it commutes.

unique isomorphism. However we have to be careful what that means: it's only going to be unique in the *somewhere* category, not in $\mathcal{C}$.

That is, in our souped-up category of diagrams, if the two objects shown here are both terminal then there must be a unique isomorphism between them *in that category*.

This means there is a unique isomorphism $v \xrightarrow{k} v'$ making the diagram on the right commute. Importantly it is not just a unique isomorphism between the vertices v and v'. There might be many isomorphisms $v \to v'$, but there will only be one that makes the diagram commute.

This is the idea of uniqueness for all universal properties. A product in a category is not just an object $a \times b$: it's an object together with the projection morphisms, so uniqueness has to be up to unique isomorphism respecting those.[†] Finally note that we write the product of a and b as $a \times b$ although many different (isomorphic) objects could be regarded as $a \times b$. We'll see this now with examples in **Set**.

18.4 Products in Set

You might well have seen some products in **Set** in your life, even if you didn't realize it. Every time we draw a graph with an x-axis and a y-axis we're using a product. If each axis is the real numbers $\mathbb{R}$ then the 2-dimensional plane is a product $\mathbb{R} \times \mathbb{R}$, also written $\mathbb{R}^2$. Using axes in this way is named "cartesian" after Descartes, and this type of product is generally called a "cartesian product".

--- **Things To Think About** ---

T 18.5 Can you see how $\mathbb{R}^2$ is a categorical product $\mathbb{R} \times \mathbb{R}$ in the category **Set**? See if you can work out what the projections are, and how the universal property works. Remember that an object of the cartesian plane is a point given by a pair of coordinates (x, y).

To show that $\mathbb{R}^2$ is a categorical product first we need to exhibit the projections, that is, give the whole diagram on the right. The idea here is that one of the copies of $\mathbb{R}$ in the "feet" is the x-axis and the other copy is the y-axis.

[†] This didn't arise with terminal and initial objects because there weren't any projection morphisms, so "respecting those" was vacuously satisfied.

The projections p and q then project onto the x- and y-coordinate respectively, as specified here.

The universal property then says: given any diagram as in the curved part on the right, we have a unique factorization k as shown. (Again, the ! symbol signifies "unique".)

This says that for any set A, a pair of functions f and g as shown amounts to a function producing points in the cartesian plane, as we can interpret f and g as producing an x- and y-coordinate respectively. This is an important interpretation of the universal property. Whenever we use the phrase "given any... there exists a unique..." we're saying, among other things, that there's a bijective[†] correspondence between the first type of thing and the second type of thing. For products it means we have a bijective correspondence between:

- pairs of individual morphisms to the "feet", and
- a single morphism to the vertex.

The fact that the bijection is produced via composition with a factorization makes it even better, and we'll come back to the meaning of that. I think understanding this "meaning" of the universal property is at least as important as grasping the technical procedure.

--- **Things To Think About** ---

T 18.6   See if you can formally construct the factorization, using the ideas above, and prove that it is unique. As usual, to show uniqueness we can suppose there is another one, say h, and demonstrate that we must have $h = k$.

For the above example with $\mathbb{R}^2$ we construct the unique factorization k from f and g as follows: given any object $a \in A$ we need to specify the point $k(a) \in \mathbb{R}^2$. That point has its x-coordinate produced by f and its y-coordinate produced by g. Formally it's this: $k(a) = (f(a), g(a)).$

Now we can *sense* it's the unique one making the diagram commute because we had no choice about how to do it. We knew that k had to make the two sides of the diagram commute; the left-hand triangle says doing k and then taking the x-coordinate has to give the same result as doing f, and similarly for the right-hand triangle doing k and taking the y-coordinate has to be the same as doing g. That is more or less the whole proof.

[†] Reminder: a bijective correspondence is one that pairs things up perfectly.

Proposition 18.2 *The function k defined above is the unique morphism making the diagram defining the universal property commute.*

Proof Consider any morphism k making the diagram commute. By definition of p and q, we know $k(a) = (pk(a), qk(a))$.

- We know that the left-hand triangle commutes so $pk(a) = f(a)$.
- Similarly from the right-hand triangle we get $qk(a) = g(a)$.
- Thus $k(a) = (f(a), g(a))$.   □

This always reminds me of an Etch-a-Sketch: you have one dial controlling your x-coordinate, and another dial controlling your y-coordinate, and then if you're very well coordinated you can draw pictures of anything at all in the 2-D drawing space, by controlling each coordinate separately. It is just one example of general cartesian products in **Set**. Given any two sets A and B we can make a set $A \times B$ that is something like making "A-coordinates and B-coordinates".

Things To Think About

T 18.7   See if you can come up with a definition of what this product set is.

Definition 18.3 Let A and B be sets. Then the *cartesian product* $A \times B$ is the set defined by $\boxed{A \times B = \{(a, b) \mid a \in A,\ b \in B\}.}$

The elements (a, b) are called *ordered pairs*, and there are functions p and q as shown on the right called *projections*, sending an element (a, b) to a and b respectively.

We saw ordered pairs in Section 15.2 when we looked at the example of Lisa Davis in New York. We were thinking about pairs "name and date of birth", so we could say the set A was the set of names, the set B was the set of dates of birth, and the function in question was from the set of people to the cartesian product $A \times B$.

Things To Think About

T 18.8   Let $A = \{a, b\}$ and $B = \{1, 2, 3\}$.

1. Write out all the elements of $A \times B$. Can you lay them out on the page in a way that shows how it's related to products of numbers?
2. Can you think of some places in normal life where this sort of thing arises?

Here are the elements of $A \times B$. I have laid them out in a grid to relate it back to the grids I drew at the start of the chapter for multiplication of numbers. I've also put in "axes", to make it look even more like coordinates.

b	$(1, b)$	$(2, b)$	$(3, b)$
a	$(1, a)$	$(2, a)$	$(3, a)$
	1	2	3

This is now very similar to how squares on a chessboard are typically labeled for reference, with the numbers 1–8 in one direction and the letters A–H in the other. It's also how maps used to have grids on them back when we used to look up a street in an index — it would say something like "p.27, 4A" and we'd then have to look in that square to try and find the street we wanted.

It's also why I like drawing a grid of a hundred numbers from 0 to 99 (as in Chapter 3) rather than 1 to 100, because that way we basically get to see the two digits of the numbers as coordinates. Whereas if you start at 1 then the rows won't have a consistent first coordinate; for example the row starting 11 will end with 20, changing the first digit.

———— Things To Think About ————

T 18.9   Prove that the cartesian product together with its projections is a categorical product of A and B. This is exactly like the proof we did for $\mathbb{R}^2$ but just with general sets A and B instead. Can you also prove that $B \times A$ is a categorical product of A and B? What is the unique isomorphism with $A \times B$?

Proof   Consider morphisms f and g as in the curved part of the diagram on the right; we need to show that there is a unique factorization k as shown. Now, in order for the diagram to commute we must have $k(x) = (f(x), g(x))$ for any $x \in X$, and this definition makes k a unique factorization.

This shows that $A \times B$ is a categorical product as claimed.   □

Note that we have streamlined this argument somewhat, basically saying "if there were a function making these triangles commute, it would have to behave like this, and oh look, there is a function that does that" which means, again, we did the uniqueness part in advance. This will be even more noticeable when we do this for structure-preserving maps like group homomorphisms, because we will say "if there is a group homomorphism here it has to do this, and now we check that that is in fact a group homomorphism".

Note also that we have given up drawing the diagram in stages. In the absence of books with diagrams that grow in front of our eyes (otherwise known as a video), we often draw a single diagram and read it "dynamically". The dotted arrow is understood to come later, induced by the rest of the structure.

Now let's think about the cartesian product $B \times A$. This is a set of ordered pairs but ordered the other way round: $\boxed{B \times A = \{(b, a) \mid a \in A, \ b \in B\}.}$

There is then an isomorphism of sets switching between those two expressions, as shown here.

$$A \times B \longrightarrow B \times A$$
$$(a, b) \longmapsto (b, a)$$

This is the unique isomorphism between those two sets *as products*; there are many other isomorphisms between them as sets. We will now specifically examine uniqueness for products of sets.

18.5 Uniqueness of products in Set

We previously saw that any object isomorphic to an initial object is also initial (and likewise for terminal objects). We have also seen that uniqueness of products is more subtle: the unique isomorphism between products is only unique *such that* the diagram involving the projections commutes. We will now investigate this for products in **Set**, to see it working in practice.

─────── **Things To Think About** ───────

T 18.10   Let $A = \{a, b\}$ and $B = \{1, 2, 3\}$ again as in the earlier example, and let $C = \{1, 2, 3, 4, 5, 6\}$.

1. Show that C is isomorphic to the cartesian product $A \times B$. How many isomorphisms are there?
2. Show that each isomorphism uniquely equips C with the structure of a categorical product of A and B. This means that it is equipped with projection maps and has a universal property.
3. What is the content of saying that $A \times B \cong C$ non-uniquely, but that when C has the structure of a product it is uniquely isomorphic to $A \times B$?

Proof   The cartesian product $A \times B$ has 6 elements, and so does C, so we know they are isomorphic because finite sets are isomorphic if and only if they have the same number of elements. There are many isomorphisms however. We did some counting of isomorphisms in Section 14.5 and we can do something similar here to find that the number of possible bijections is $6 \times 5 \times 4 \times 3 \times 2 \times 1$ which is written as 6! and known as "6 factorial".[†]

Now, given an isomorphism $C \xrightarrow{j} A \times B$ we can compose with p and q as shown to get *putative* projections for C. ("Putative" means that they are candidates for being the projections but we can't say they definitely work until we check some more stuff; however usually when we say this it means that it is going to turn out to be right.)

─────────────────

[†] If we construct a bijection $C \longrightarrow A \times B$ then we could first pick where 1 goes, and we have 6 possibilities, and then we pick where 2 goes, for which there are 5 remaining possibilities, and so on. You may have seen this if you have ever studied how to count permutations of n elements; a permutation is precisely a bijection from a set to itself.

We now need to consider f and g as in the diagram on
the right, and show that there is a unique factorization k as
shown. In this diagram the parts in gray are to remind us
how pj and qj are constructed, and the $\sim$ sign is to remind
us that j is an isomorphism. All the arrows are functions.

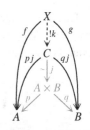

> The thing to think now is: what ways are there to get from X to C? We can get from
> X to $A \times B$ from the latter's universal property, and then we can go "backwards" up
> j as it is an isomorphism. We now make that formal.

The universal property of $A \times B$ induces a unique factorization $X \xrightarrow{!h} A \times B$
making a certain diagram commute, and we can then use the inverse j^{-1} of j
to make a morphism $X \xrightarrow{!h} A \times B \xrightarrow{j^{-1}} C$.

> We now have to check some things, but if you're thinking categorically you will
> already *feel* that this must be correct. Eventually you wouldn't need to see a proof
> like this beyond this point. See remarks on "abstract nonsense" after the proof.

We now have to check two things: that this really is a factorization (i.e.
the relevant diagram commutes) and that it is the unique one. To show that
the diagram commutes consists of a "diagram chase" where we follow arrows
around a diagram progressively. It's quite hard to show this in a static diagram
on a page, but I will try. Again this is really all part of one diagram, but we're
reading it "dynamically", working our way through it as shown. At each stage
we're really just considering the black arrows and not thinking about the gray
ones at the moment. Then we notice that the shaded region commutes so we
can move across it, which takes us to the black arrows in the next diagram.

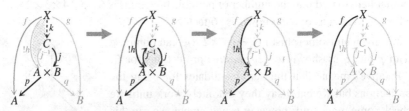

Written out in algebra, it would look like this:

$$
\begin{aligned}
pjk &= pjj^{-1}h & \text{by definition of } k \\
&= ph & \text{by definition of } j^{-1} \\
&= f & \text{by definition of } h
\end{aligned}
$$

Each of the equals signs corresponds to one move across a large gray arrow to the
next diagram. The explanation in words to the right corresponds to the gray shaded
area in each diagram, which is telling us how we can move to the next diagram and
thus to the next line of the algebra. I hope you can see how the algebra corresponds
to the dynamic reading of the diagram, and that you can become comfortable with
the diagrams, and then become comfortable with a single diagram read dynamically. This is a key technique in category theory.

This shows that k is indeed a factorization; we still need to show it is the
unique one.

The idea here is that k is inextricably linked to h by construction, and h is unique
so k "must be" unique too. To make this formal we can assume there's a different
one and show that this would produce a different version of h too.

The key is the diagram on the right, showing how we constructed the projections for C and the factorization k.

If you read the diagram dynamically you can see that
k is a factorization for the product C precisely if jk is a
factorization for the product $A \times B$. Thus if k' is also a factorization for the product C then jk' is also a factorization
for the product $A \times B$.

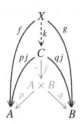

But we already know that the factorization for $A \times B$ is unique, thus we must
have $jk = jk'$, and then since j is an isomorphism it follows[†] that $k = k'$.   □

Note on "abstract nonsense"

Many category theory proofs proceed in this way, especially where universal properties are involved: you need to exhibit a piece of structure satisfying
some properties, and once you've found a putative structure that type-checks,
you *know* it's going to satisfy the properties because of how you produced the
structure. In the above proof to produce the factorization we used exactly the
information available: the universal property of $A \times B$, and the isomorphism
with C. There was no other way to get any sort of morphism $X \longrightarrow C$; it was
almost just type-checking that led us here. This is quite typical of category theory proofs and can sometimes lead people to think that they have no content,
giving rise to the description "abstract nonsense". This probably started as an

---

[†] If we wrote this out fully in algebra it would look something like this. Suppose k' is also a
factorization, that is $(pj)k' = f$ and $(qj)k' = g$. But this means $p(jk') = f$ and $q(jk') = g$. But
there is a unique morphism h such that $ph = f$ and $qh = g$, and so $jk' = h$. But we already
knew $jk = h$, so $jk' = jk$, and since j is an isomorphism we can apply j^{-1} on the left of both
sides of the equation to get $k' = k$. [I find the dynamic diagram much more enlightening.]

insult to category theory and then was reclaimed by some category theorists as an affectionate way of describing this process. I don't like using it, even in affectionate jest, because I don't like the word "nonsense". I think this is abstract profundity. The framework beautifully slots everything into place for us.

Finally I'll stress that there may be many isomorphisms $C \dashrightarrow A \times B$ as sets (as we saw with the small example producing 6! isomorphisms), but only one isomorphism as products. This means that once C has been equipped with the structure of a product, there is only one isomorphism with $C \dashrightarrow A \times B$ respecting that structure, that is, commuting with the projections.

There is something perhaps a little hazy in the notation of products in **Set**, as both categorical products and cartesian products are likely to be notated $A \times B$. However, cartesian products are just one particular construction of categorical products in **Set**, a particularly common one. One consequence is we end up having to be rather careful about things like associativity.

──────────── **Aside on associativity** ────────────

I will just mention this briefly: cartesian product is not strictly associative, for a slightly pedantic reason. For the two different bracketings we get:

$$(A \times B) \times C = \Big\{ ((a,b),c) \ \Big| \ a \in A, \ b \in B, \ c \in C \Big\}$$
$$A \times (B \times C) = \Big\{ (a,(b,c)) \ \Big| \ a \in A, \ b \in B, \ c \in C \Big\}$$

Both ways round involve the same information, that is, ordered triples of elements, but written in a slightly different way. If you feel that the difference between these two things is just an annoyance then I agree, and it shows that we're working at the wrong level here. We are trying to ask about this equation: $\boxed{(A \times B) \times C = A \times (B \times C).}$ But this is an *equality* between objects in a category: we should be asking about isomorphisms. The two sets above are certainly isomorphic, in a particularly contentless way, but we need one more dimension to express that rigorously. It is the question of "weak associativity".

Finally note that for categorical products it's not just asking for the wrong level of sameness: it doesn't even make logical sense to pose the question, because the products $(A \times B) \times C$ and $A \times (B \times C)$ are only *defined* up to isomorphism. If something is defined up to isomorphism you can't ask if it equals something else — or, if you do, you have to be rather careful what you mean. In the case of categorical products the only thing we could mean is that each side of the "equation" canonically has the structure of the product on the other side. And to complicate things further, there is another three-fold product that is not derived from the binary products. For cartesian products it is this:

$$A \times B \times C = \Big\{ (a,b,c) \ \Big| \ a \in A, \ b \in B, \ c \in C \Big\}.$$

T 18.11   Try investigating the categorical three-fold product that corresponds to this. It will have three projections instead of just two.
└──┘

This all becomes much more crucial when we move into higher dimensions. The more dimensions we have, the more crucially subtle it all becomes.

Terminology

It is not always possible to find products of objects in a category — there might be no diagrams with the desired universal property. If a category has enough structure that we can take products of *any* two objects in the category, then we say that the category *has binary products*. If it has binary products then by induction we can build up to *n*-fold products for any *n*, so we say that the category *has finite products*.

So far we have looked in quite some detail at products in **Set**. We will now take a look at products in some of the other categories we've seen.

18.6   Products inside posets

Recall that a poset is a category with at most one morphism between any two objects, and that this means all diagrams commute. So in a poset the definition of product simplifies: we don't have to check any commutativity or uniqueness for the factorization.

T 18.12   Recall the poset of factors of 30 on the right. Show that 6 is the categorical product of 2 and 3. This seems "obvious", but don't get too complacent: what is the categorical product of 6 and 10 in this category? Can you generalize this to posets of factors of a general *n*?
└──┘

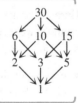

To show that 6 is a categorical product we need to exhibit projections and a universal property. We can see from the diagram that it has putative projections as shown here.

For the universal property we need to consider any number in the diagram that has morphisms to 2 and 3. The only number that does (other than 6) is 30, and we duly do have a factorization as shown here.

It might seem coherent that the categorical product of 2 and 3 is the "ordinary" product, but let's now look at the categorical product of 6 and 10. There is only one candidate for product now because there is only one object with projections to 6 and 10, and that is 30; as it is the only one it must be universal.

> The "souped-up category" in which we're looking for a terminal object has only one object and only one (identity) morphism, so that object is definitely terminal.

There are now (at least) two ways we could proceed, to see how to understand and generalize this. They sort of correspond to experiment and theory: we could experiment with other posets of factors to see if we can see a pattern, or we could go through the definition of product in this particular context to understand what it produces.

―――――――――――― **Things To Think About** ―――――――――――

T 18.13   Here are some suggestions for either of those thought processes.

1. If you want to explore other numbers, I suggest the diagram for factors of 36. What is the categorical product of 4 and 6 in there? What about 6 and 9? 4 and 9? 2 and 4?

2. If you want to go ahead and explore by theory, remember that a morphism $a \longrightarrow b$ in this category (the way I've drawn it here) is the assertion that a is a multiple of b. Write out the definition of product using that definition in place of each morphism, and see what you get.

Here is the diagram for the poset of factors of 36. Remember we can work this out from the prime factorization $36 = 2 \times 2 \times 3 \times 3$ which tells us we have

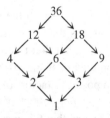

- two dimensions, one for 2 and one for 3, and
- paths of length two in each direction, as each prime factor appears twice in the factorization.

The categorical products can more or less be seen as the lowest point in the diagram that has a morphism to each object in question. So the pairs of numbers I asked about have categorical products as shown in this table.

		categorical product
4	6	12
6	9	18
4	9	36
2	4	4

If we now work through the definition of product in this context, we see that for the product of a and b we need:

1. A diagram of the form shown here, which in this category just means that v is a multiple of both a and b.

2. Given any such diagram with vertex x, we need a morphism $x \rightarrow v$. Here is a "translation" of what that means in this particular category.

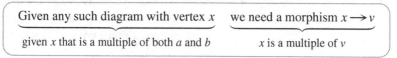

These two thoughts together, translated, give exactly the definition of lowest common multiple. You might think that the lowest common multiple is usually defined to be the lowest among all common multiples, but in fact it follows that any other multiple is not only larger but is also a multiple of the lowest one, and this is the definition that is often used in more abstract mathematics.

In a general poset we might generically call the ordering $\geq$ and in that case the definition of categorical product gives us this:

1. an object v such that $v \geq a$ and $v \geq b$, and
2. for any object x such that $x \geq a$ and $x \geq b$, we must have $x \geq v$.

This is exactly the definition of *least upper bound* or *supremum*. The first point says that v is an upper bound for a and b, and the second point says that if x is any upper bound, it is greater than or equal to v. Least upper bounds are important in analysis and careful study of the real numbers, as the existence of least upper bounds is a way of distinguishing the reals from the rationals.[†]

18.7 The category of posets

We will now zoom out, and instead of looking inside a poset expressed as a category, we will look at the category **Pst** of posets and order-preserving maps.

When looking for a universal property for sets-with-structure, a typical approach is to start with the universal construction on the underlying sets, and then see if there's a natural way to extend the structure on them. We are making use of the system of interactions below. (The squiggly sign is an isomorphism

---

[†] A set of rational numbers might have a least upper bound that is irrational, so that the set doesn't have a least upper bound in the rational numbers.

sign on its side, and there's a question mark as the isomorphism is the thing we're wondering about.)

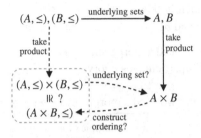

The dotted box contains the product we are trying to understand, and we can speculate that perhaps the underlying set of the product is $A \times B$, the product of the underlying sets. So we dream up a "sensible" ordering on $A \times B$ and see if we can exhibit a universal property.

We sort of try this method "optimistically" and find that it works in some individual cases. The idea of category theory is then to make the method into a precise theory and prove it abstractly so that we can use it in more complicated cases without having to rely on optimistic trial-and-error.[†]

We could start by trying this for $\mathbb{R}^2$, since although we talked about $\mathbb{R}^2$ as just a product of sets earlier, $\mathbb{R}$ is naturally a poset with the ordering $\leq$. Is there a natural way to order $\mathbb{R}^2$? One way is to do it *lexicographically*, that is, like in a dictionary: you look at the first coordinate first, and look at the second coordinate second. So everything starting with x-coordinate 1 will be less than everything with x-coordinate 2, regardless of the y-coordinates.

Another way is to consider both coordinates at once and declare this:

$$(x_1, y_1) \leq (x_2, y_2) \iff x_1 \leq x_2 \text{ and } y_1 \leq y_2.$$

In this case *both* coordinates have to be $\leq$ in order for the pair to count as $\leq$.

Things To Think About

T 18.14   Consider the points in $\mathbb{R}^2$ shown on the right. Try drawing a minimal diagram of arrows between these points, according to the lexicographic ordering and then according to the other ordering we described above. Draw an arrow $a \longrightarrow b$ whenever $a \leq b$, but as usual there is no need to draw identities or composites.

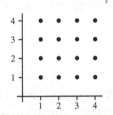

Which of your two pictures do you think looks more like a product? Remember, in the definition of a product $A \times B$, A and B play the same role, so in the product picture there should be some sort of symmetry.

[†] The abstract explanation for this approach involves limits in categories of algebras for monads. This is beyond our scope but I mention it in case you want to look it up later.

According to the lexicographic ordering we know that everything in the column $x = 1$ comes before everything in the column $x = 2$ which comes before everything in the column $x = 3$ and so on. Here are the two diagrams.

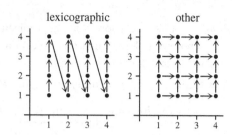

I hope you agree that there's something more satisfying about the second diagram. That's a rather subjective point though; a non-subjective point is that the second one is symmetrical in the x and y values, where the first one isn't.

That is, if we switch the x- and y-axes in the second diagram it will stay the same, but if we switch them in the first diagram it will become the one on the right, which is in fact called the *colexicographic* ordering.

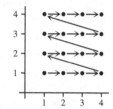

Of course, neither the feeling of satisfaction nor the thought about symmetry is enough to prove that the "satisfying" ordering makes that thing into a categorical product.

─────── **Things To Think About** ───────

T 18.15   What do we need to do to show that this makes $\mathbb{R}^2$ into a product of posets, not just a product of sets?

To show that this is a product we need to exhibit the universal property *in the category of posets* not just in **Set**. We can piggy-back on the universal property in **Set**, but we need to show these things in addition:

1. the projections are now order-preserving functions, not just functions, and
2. when we're inducing a unique factorization, if all the functions we start with are order-preserving then the factorization is also order-preserving.

We might as well state this in generality now, not just for $\mathbb{R} \times \mathbb{R}$.

Proposition 18.4   *Let $(A, \leq)$ and $(B, \leq)$ be posets. Then we can define an ordering $\leq$ on $A \times B$ by* $\boxed{(a_1, b_1) \leq (a_2, b_2) \iff a_1 \leq a_2 \text{ and } b_1 \leq b_2}$ *and this makes $(A \times B, \leq)$ into a categorical product $(A, \leq) \times (B, \leq)$ in* **Pst**.

One might "abuse notation" and simply say that given posets A and B the product $A \times B$ has ordering as above.

Proof   First we show that the projections are
order-preserving. Recall that the projections are the
functions p and q given on the right.

$$A \xleftarrow{p} A \times B \xrightarrow{q} B$$
$$a \longmapsfrom (a,b) \longmapsto b$$

Now to show p and q are order-preserving we need to show

$$(a_1, b_1) \leq (a_2, b_2) \implies \underbrace{p(a_1, b_1) \leq p(a_2, b_2)}_{\text{i.e. } a_1 \leq a_2} \text{ and } \underbrace{q(a_1, b_1) \leq q(a_2, b_2)}_{\text{i.e. } b_1 \leq b_2}$$

and this is true by definition of the ordering.

> The top parts of the boxes are what we need to show, and the bottom parts together
> are the definition of the ordering; the top and bottom of each box is equivalent by
> the definitions of p and q.

Next we need to consider the unique factorization as in
the diagram on the right. We know there is a unique factorization k in **Set**; we need to show that if f and g are
order-preserving then k is too.

Recall that k is defined by $k(x) = (f(x), g(x))$
so to show that k is order-preserving we need to show

$$x_1 \leq x_2 \implies \underbrace{(f(x_1), g(x_1)) \leq (f(x_2), g(x_2))}_{\text{i.e. } f(x_1) \leq f(x_2) \text{ and } g(x_1) \leq g(x_2)}$$

But this is precisely the definition of f and g being order-preserving.   □

Note that this proof doesn't just show that the necessary things are true, it also
shows us how the parts of the structure correspond precisely to the things that
we need. However, this is a type of proof where I didn't go in with any great
"feeling" about how the proof was going to proceed. Rather, I just followed
the definitions through, more or less by type-checking. The part that involved
"feeling" was in deciding what the appropriate ordering on $A \times B$ would be.

─────────────── **Things To Think About** ───────────────

T 18.16   Do products work like this for totally ordered sets? That is: if we
take the product of two tosets regarded as posets, will the result be a toset? Try
it for a pair of very small tosets, for example two copies of $0 \longrightarrow 1$.

Tosets (totally ordered sets) do not work so neatly. My favorite (or perhaps
least favorite) example of this is Tupperware boxes. Rectangular Tupperware
boxes have a length and a width, which is like a pair of coordinates (x, y). We

could order them by either of those things but it's much more useful to be able to stack them, and we can only do that if one of them has *both* its length and its width smaller than the other. Typically this not the case and I end up with a whole load of boxes that won't stack, because the product ordering is only a partial ordering, not a total ordering. At least, that's how I think of it.

For a more formal example: the product of the very small poset $0 \longrightarrow 1$ with itself has four elements, ordered as shown (I have drawn in the diagonal to remind us that it is there, but dotted as it's redundant in the diagram).

This is not a totally ordered set. It is visibly not in a straight line; more formally we could observe that there is no arrow between $(0, 1)$ and $(1, 0)$.

There are some trivial cases in which a product of tosets will be a toset (if either of the original tosets is empty or has only one element) but in general it will not be a toset. Note that this is an advantage of the lexicographic or colexicographic orderings: we take some tosets and definitely produce a total ordering on the product of the underlying sets. This is useful if we're doing something like making a dictionary and really need a total ordering. However, to get the total ordering we have to sacrifice the universal property. In fact it is quite often the case that a universal structure is impractical in some way and that a non-universal structure will be more useful for practice, where the universal one is more useful for theory.

───────────────────── **Note on "in general"** ─────────────────────
In mathematics, "in general" means something very precise. In normal life it means something more like "most of the time" but in mathematics it means that this is what happens in the absence of special properties. So for tosets, the product is sometimes a toset but only for some special cases of toset. To take an extreme example we might say "$\frac{1}{x}$ is not defined for numbers in general, but only when $x \neq 0$". In this case it is defined for almost all numbers, but that still doesn't count as "in general".

────────────────────────────

The diagram we drew for the product of two copies of the poset $0 \longrightarrow 1$ was reminiscent of coordinates, but if we rotate it slightly we can see it as a kind of blueprint for the posets of privilege, specifically the interaction between any two types of privilege.

$$(1,1)$$
$$(0,1) \qquad (1,0)$$
$$(0,0)$$

For example if we consider an individual poset representing the privilege that white people have over black people, and another for the privilege that male people have over female people, the product poset gives us the square as shown below. It is one face of the cube of privilege.

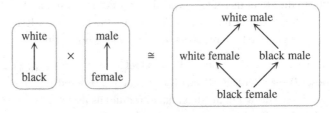

In this case the first projection ignores a person's gender and the second projection ignores a person's race.

There are two forms of antagonism that I see happening in this and analogous situations. One is that those with just one of the two types of privilege tend only to think about the type they lack rather than the type they have. For example, white women only compare themselves with white men and forget how much more privileged they are than black women. The other antagonism is between people who are incomparable in this poset, in this case white women and black men, as they each fight against their own oppression.

This doesn't always happen: some people who lack one type of privilege are then able to empathize with everyone who lacks one (or more) types of privilege, by performing an isomorphism of categories in their head (whether they think of it like that or not).

18.8 Monoids and groups

When we made a product of posets $(A, \leq)$ and $(B, \leq)$ we took the product $A \times B$ of the underlying sets and then developed a partial ordering on it based on the individual orderings. We can do something analogous for monoids and groups, but now we start with groups $(A, \circ)$ and $(B, \circ)$, and instead of making an ordering on $A \times B$ we need to make a binary operation based on the individual binary operations on A and B.

As with posets we can try this first on $\mathbb{R}^2$ as that may be a more familiar context. So far we have treated $\mathbb{R}$ as a set, and then as an ordered set. We can also treat it as a monoid or a group, under addition. The idea for the categorical product $\mathbb{R} \times \mathbb{R}$ is to come up with a definition of addition for ordered pairs, like $(x_1, y_1) + (x_2, y_2)$, based on how we add individual numbers.

Things To Think About

T 18.18 What pair of coordinates could be the answer to $(x_1, y_1) + (x_2, y_2)$? If you're stuck, think about these coordinates not as points, but as an instruction to move a certain direction horizontally and a certain direction vertically.

If we think of (x, y) as telling us to move x units horizontally and y units vertically then it might be more obvious how to add these things together.

The expression $(x_1, y_1) + (x_2, y_2)$ says to move x_1 units horizontally and y_1 vertically, then x_2 horizontally and y_2 vertically, as shown on the right.

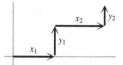

The total result is that we have gone $x_1 + x_2$ units horizontally and $y_1 + y_2$ units vertically, so we seem to be saying $(x_1, y_1) + (x_2, y_2) = (x_1 + x_2, \; y_1 + y_2)$.

Things To Think About

T 18.19 Check that this binary operation makes $\mathbb{R}^2$ into a group. What is the identity and what are inverses?

We call this doing addition "pointwise" or "componentwise" as we do it on each component separately, just like when we did posets we defined the ordering on the product componentwise. When we do anything pointwise, typically axioms all follow as they hold in each component individually. That is the case here. In particular the identity is $(0, 0)$, and the inverse of (x, y) is $(-x, -y)$.

Now let's generalize this.

Things To Think About

T 18.20 Can you generalize our approach to defining addition on $\mathbb{R}^2$, to define categorical products of monoids in general? For the proof, you could try taking our proof about products of posets as a blueprint.

Proposition 18.5 Let $(A, \circ, 1)$ and $(B, \circ, 1)$ be monoids. We define a binary operation on $A \times B$ by $(a_1, b_1) \circ (a_2, b_2) = (a_1 \circ a_2, \; b_1 \circ b_2)$.
This makes $A \times B$ into a monoid with unit $(1, 1)$, and it is the categorical product of the original two monoids.

The idea of the proof is similar to our proof for posets. We know that we have projections and a universal property in **Set**, and we need to show the following:

1. the projections are now monoid homomorphisms, not just functions, and
2. the factorization induced in **Set** by a pair of monoid homomorphisms is also a monoid homomorphism (see below for further explanation of this).

I strongly believe that after a certain point in this proof it's much easier to
write your own proof than try and decipher the symbols involved. I think many
(or most or all) mathematicians read research papers this way — reading just
enough to try and work out the proof for themselves. But here it is anyway.

Proof   First we show that the projections are homomorphisms. As usual the projections are the functions p and q given on the right.

$$A \xleftarrow{p} A \times B \xrightarrow{q} B$$
$$a \longmapsfrom (a,b) \longmapsto b$$

We will deal with p; then q follows similarly. To show that p is a homomorphism we need to show that it respects the binary operation and the unit.

> For the binary operation we need to show $p((a_1, b_1) \circ (a_2, b_2)) = p(a_1, b_1) \circ p(a_2, b_2)$.
> We basically unravel the definitions on both sides and show they're the same. When
> writing it out formally it might be better to start on the left and work our way to the
> right, even though in rough we might work on both sides at the same time.

For the binary operation:

$$
\begin{aligned}
p((a_1, b_1) \circ (a_2, b_2)) &= p(a_1 \circ a_2,\ b_1 \circ b_2) &&\text{by definition of } \circ \text{ in the product} \\
&= a_1 \circ a_2 &&\text{by definition of } p \\
&= p(a_1, b_1) \circ p(a_2, b_2) &&\text{by definition of } p
\end{aligned}
$$

For the identity $p(1, 1) = 1$ as needed. So p is a homomorphism, and so is q.

Next we need to consider the unique factorization as in the
diagram on the right. We know there is a unique factorization k in **Set**; we need to show that if f and g are homomorphisms then k is too.

> Again, we basically just unravel both sides of the definition of homomorphism and
> find that they're the same.

Recall that k is defined by $k(x) = (f(x), g(x))$. Now:

$$
\begin{aligned}
k(x_1 \circ x_2) &= (f(x_1 \circ x_2),\ g(x_1 \circ x_2)) &&\text{by definition of } k \\
&= (f(x_1) \circ f(x_2),\ g(x_1) \circ g(x_2)) &&\text{since } f \text{ and } g \text{ respect } \circ \\
&= (f(x_1), g(x_1)) \circ (f(x_2), g(x_2)) &&\text{by definition of } \circ \text{ in the product} \\
&= k(x_1) \circ k(x_2) &&\text{by definition of } k
\end{aligned}
$$

so k is a homomorphism. Finally we show that k preserves identities:

$$
\begin{aligned}
k(1) &= (f(1), g(1)) &&\text{by definition of } k \\
&= (1, 1) &&\text{since } f \text{ and } g \text{ each preserve identities}
\end{aligned}
$$

which is the identity in the product as required.   □

The analogous result is true for groups but it's actually slightly easier as we don't have to check for the preservation of identities.

18.9 Some key morphisms induced by products

The standard morphism we've seen induced by a product as shown on the right is often written (f, g), based on its construction in **Set**.

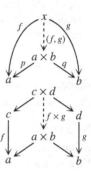

Another useful morphism induced from a product is shown here. This one is really induced from some composites: we use the composites from $c \times d$ all the way to a and b via f and g respectively to produce a factorization often called $f \times g$.

Things To Think About

T 18.21   See if you can work out what $f \times g$ does in **Set**.

Given functions $C \xrightarrow{f} A$ and $D \xrightarrow{g} B$, the function $f \times g$ has the effect shown on the right.

$$C \times D \;\to\; A \times B$$
$$(c, d) \;\mapsto\; (f(c), g(d))$$

I mention this here as we will need maps of this form later, when we talk about products of categories in Section 21.3.

18.10 Dually: coproducts

We have spent quite a long time on products. We are now going to look at the dual concept, which is called a coproduct. Remember "look at the dual" means we look at products in $\mathcal{C}^{op}$ and then translate the concept back into $\mathcal{C}$.

Definition 18.6   A *coproduct* in a category $\mathcal{C}$ is a product in $\mathcal{C}^{op}$.

Things To Think About

T 18.22   Can you "translate" this definition back to a concept directly in $\mathcal{C}$?

Unraveled definition of coproduct

We just need to turn around every arrow in the definition.

Let a and b be objects in a category $\mathcal{C}$. A *coproduct* of a and b
is a universal diagram of the shape shown on the right.

> Eventually the aim is to be comfortable enough with the general idea of universal
> properties that this succinct wording is enough, but I'll spell it out here as we do
> need to be careful about some directions.

The universal property says: given any diagram of the same
shape, as in the curved arrows on the right, there is a unique
factorization k as shown.

Note that the factorization k now goes *from* the universal vertex v *to* the vertex
x that is not universal, where for products it was the other way round. The key
here is not to think of that direction exactly, but to think of the fact that we
are "factorizing" the non-universal diagram: expressing it as a composite of
the universal one and k. Then some type-checking will ensure that the unique
factorization map is pointing in the appropriate direction.

We have some terminology analogous to the product situation:

- p and q are called *coprojections* or *insertions*.
- Instead of a cone, this is now a *cocone*. In a cone we have projections pointing away from the vertex, and in a cocone we have coprojections pointing to
  the vertex. The coproduct is then the *universal cocone*.
- The coproduct is written $a \coprod b$, $a \sqcup b$, or $a + b$ for reasons we'll see.

┌──────────────────── **Things To Think About** ────────────────────┐

T 18.23   A coproduct is an initial object *somewhere*; can you work out where?
It's another kind of "souped-up" category.

└───┘

A coproduct for a and b is an initial object in a category whose
objects are cocones over a and b, that is, diagrams of the form
shown on the right. Morphisms are morphisms between the vertices (the top objects) making everything commute.

┌──────────────────── **Things To Think About** ────────────────────┐

T 18.24   What results immediately follow by duality with products?

└───┘

As coproducts in $\mathcal{C}$ are products in $\mathcal{C}^{\mathrm{op}}$ we have dual versions of everything that
is true for products. So coproducts are unique up to unique isomorphism (mak-
ing cocones commute), there is symmetry in the variables, and there are n-fold
versions. Now let's see what coproducts are in various example categories.

18.11 Coproducts in Set

For a coproduct of sets A and B, we're looking for a way of producing a canonical set with functions from A and B *into* it. It turns out to be the *disjoint union*. Disjoint unions have a technically slightly obscure looking definition, but the idea is that you stick the two sets together and ignore the fact that some of the objects might have been the same. You can think of it as painting all the elements of A red and all the elements of B blue to force them to be different, and then taking the "normal" (not disjoint) union.

In case this isn't clear, let's consider these sets: $A = \{a, b, c\}$, $B = \{b, c, d\}$. When we take the normal union we get this: $A \cup B = \{a, b, c, d\}$ This is because the normal union says that if there are elements in the intersection (that is, in both sets) we don't count them twice.

The *disjoint* union is something like this: $A \sqcup B = \{\text{red: } a, b, c, \ \text{blue: } b, c, d\}$

Or perhaps I could use fonts instead of colors: $A \sqcup B = \{a, b, c, \mathbf{b}, \mathbf{c}, \mathbf{d}\}$

For the normal union we can't tell how many elements there will be unless we know how many there are in the intersection; for the disjoint union we just add the number of elements in A and B together (that's if they're finite; if they're infinite then the union or disjoint union will also be infinite).

We can picture the disjoint union as just sticking the sets side by side, as shown here.

$A \sqcup B$

The technical definition of disjoint union is: $A \sqcup B = (A \times \{0\}) \cup (B \times \{1\})$. We won't really need it here but it's quite a clever definition and you might like to think about what it's doing.

─────────────── **Things To Think About** ───────────────

T 18.25   Can you show that the disjoint union is a coproduct in **Set**? The first step is just understanding what it is we need to show, and that's a good step even if you can't then show it. You can do it for the specific example above, or for disjoint unions in general; you can use the formal definition if you've understood it, or just get the idea via pictures.

We need to exhibit the coprojections and then verify the universal property.

Here's a picture encapsulating the idea of
the coprojections. Note that technically we
shouldn't say "$A \sqcup B$ is a coproduct" without saying what the coprojections are, but
we often do when the coprojections can be
understood to be the "obvious" inclusions.

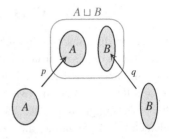

Personally I find that this picture is much more illuminating than the formal
definition of the coprojection functions, but it's important to be able to turn the
idea into a rigorous definition if you want to become a rigorous mathematician.
However, the idea is also important, and as the rigorous part is much easier to
look up elsewhere than the idea, I'll leave it at the picture here.

It remains to check the universal property. Again I think the idea is possibly
more important than the formality. The idea is that to define a function *out of*
the disjoint union, exactly what you have to do is define it on the A part and
also define it on the B part. Those are completely separate issues, so this is the
same as defining a function out of A and a function out of B separately. Thus
for any set X we have a bijective correspondence between

1. pairs of functions $A \xrightarrow{f} X$ and $B \xrightarrow{g} X$, and
2. functions $A \sqcup B \xrightarrow{k} X$.

Moreover it's not "any old" bijection: the correspondence is produced by composition with the coprojections.

This is an important way of thinking about universal properties. In general it
says there is a bijective correspondence between (co)cones and factorizations,
so that we can equivalently encapsulate the information of any (co)cone as a
single morphism to (or from) a universal one. It's very handy because a cone
generally involves many morphisms, and we can now encapsulate it as one
morphism together with a "reference" cone.

This is crucially the idea behind the higher level abstract definition of universal property that we can't yet do as we haven't seen natural transformations.
It is the root of what category theorists mean when they say that a bijection is
"natural" when they use "natural" technically rather than informally.

18.12 Decategorification: relationship with arithmetic

We sometimes use the notation + for coproducts because coproducts of sets
correspond to addition of ordinary numbers in a way that we'll now discuss.

When we first introduce addition to small children we might give them a number of objects, and then some more, and see how many there are altogether. So to do 3 + 2 we give them 3 objects and then 2 more objects and then take the disjoint union of those two sets; we just don't say it like that, typically. The coprojection functions might consist of us physically sliding some small objects into place, as in this picture.

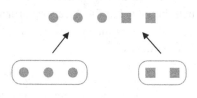

In fact, when we do this physical addition, I think we are doing profound categorical mathematics with children; likewise if we do physical multiplication in grids that we saw earlier on. We are showing them deep structures behind arithmetic, which then get flattened out if we degenerate into drilling "number facts" in the slightly later years in school.

I often say that numbers are an abstraction because we make them by forgetting a lot of details about objects in the real world. However another point of view is that they're a "de-categorification",[†] regarding isomorphic sets as strictly the same. For example there are infinitely many one-element sets, and they're all isomorphic in the category **Set**. If we regard them as strictly the same there is just one of them and this is, essentially, the number 1. Likewise there are infinitely many two-elements sets and they're all isomorphic, and we make them strictly the same and call that thing the number 2.

Then addition such as $a + b$ is defined by taking any two sets that were turned into the numbers a and b, finding their categorical coproduct, and then turning those into numbers again. This is what counting with little children is like, when you give them objects and ask them to just see how many there are. Multiplication is defined by taking categorical products. This shows that addition and multiplication are dual to each other, which is a curious point of view given that multiplication is also repeated addition.

The characterization as repeated addition ends up being something that can be proved using universal properties together with the fact that every set is a coproduct of one-element sets. Many other "laws" of arithmetic can also be proved by categorifying them, using universal properties in **Set**, and then decategorifying again. This includes commutativity, associativity, and the distributive law of multiplication over addition $a(b + c) = ab + ac$. I think it's a shame that we tend to dismiss "physical arithmetic" as a mere stepping stone to abstract arithmetic.

[†] This was coined in J. Baez and J. Dolan, Categorification. *Higher category theory* (Contemporary Mathematics, No. 230) 1–36, American Mathematical Society, 1998.

18.13 Coproducts in other categories

We will now look at coproducts in the other categories where we've seen products, but only briefly. A general principle is that universal properties involving cocones (rather than cones) are much harder to construct for "sets with algebraic structure".[†]

─────────────── **Things To Think About** ───────────────

T 18.26   1. What are coproducts in a category of factors of n? Follow through what we did for products but do it dually. Then try and generalize to any category that is a poset.

2. In the category of posets and order-preserving maps, we found products by taking the product of the underlying sets and then putting an ordering on it. Can we do coproducts like this? Does it work for totally ordered sets?

3. Coproducts of monoids are hard: disjoint union doesn't work. Why not?

Inside a poset

In the poset of factors of n we saw that products are lowest common multiples. Coproducts are then highest common factors. In more general posets we saw that products are least upper bounds; coproducts are then greatest lower bounds. The fancy word for a least upper bound is *supremum* and the fancy word for a greatest lower bound is *infimum*.

In the category of posets

We can take a disjoint union of two posets A and B and it will still be a poset: everything in the A part will simply be incomparable to everything in the B part. I hope at this point I can leave it to you to check the universal property.

However, as for products if we take the coproduct of two tosets A and B, the result is in general not a toset: the elements of A will be incomparable to the elements of B, and this is not allowed in a toset. As with products it will work in some special cases (if either of the original sets was empty).

In the category of monoids

In the category of monoids if we take the disjoint union of A and B as sets then the result is not in general a monoid — we have a binary operation on the A part and a binary operation on the B part but we do not have a way of doing $a \circ b$ with an element from each side.

───────────────

[†] In case you're interested in looking it up: this statement can be made precise in terms of algebras for monads.

Sorting this out is somewhat complicated. We don't want the answer to $a \circ b$ to be anything that already exists because that would be non-canonical, so we "throw in" an extra element to be the answer. Essentially this consists of throwing in all ordered pairs $a \circ b$ for any $a \in A$ and $b \in B$.[†] But then what if we try and do the binary operation on one of those new elements together with another element?

For example $\boxed{a' \circ (a \circ b)}$ would need to be the same as $\boxed{(a' \circ a) \circ b}$ by associativity, and this is still of the form "element of A $\circ$ element of B".

However $\boxed{(a \circ b) \circ a'}$ can't be reduced in general as we don't know that $\circ$ is commutative. Likewise $\boxed{(a \circ b) \circ (a' \circ b')}$ and so on — in fact we could end up with infinite strings of "stripy" elements, that is, alternating between elements of A and elements of B. These are what we have to construct to make the coproduct of monoids.

If we insist that the binary operation is commutative then things are much simpler. Then in any long string of a's and b's the a's can all commute past the b's and be gathered together, for example $\boxed{a \circ b \circ a' \circ b' = (a \circ a') \circ (b \circ b')}$.

So if we restrict to the category of *commutative* monoids everything in a coproduct can indeed be expressed as "element of A $\circ$ element of B".

You might notice that this is oddly similar to the elements of the product $A \times B$ apart from in notation, as the elements of the product are

$$(\text{element of } A, \text{element of } B).$$

This is the same information, just with a different symbol in between the elements. This means that in the category of commutative monoids or (more typically) commutative groups, the coproduct and the product are the same. This is unusual, but is a similar phenomenon to the fact that the terminal and initial objects are the same. It is one of the things that makes the category of commutative groups so special. Commutative groups are also called Abelian groups after the mathematician Abel.[‡]

There are some unhelpful issues of terminology around universal properties. One is that coproducts of groups are typically called "free products" in group theory, and products are called "direct products" (there is something else called a "semi-direct product"). It took me ages to work out that the free product of groups was actually the coproduct and I wish someone had told me. Some people think it's pedagogically better if you work these things out for yourself, but I was the kind of student who was convinced that I must be confused and

---

† We're used to writing ordered pairs as (a, b) but writing them $a \circ b$ is the same information just presented differently; we'll come back to this in a moment.

‡ This gives rise to the ridiculous joke: What's purple and commutes? An Abelian grape.

doing something wrong if I thought of something that hadn't been told to us, so I was afraid to ask, for fear of being thought stupid. One solution to this is to encourage all students to be more arrogant and believe in their own brilliance; another is to explain more things and make sure you never make anyone feel stupid if they ask a question.

18.14 Further topics

Topological spaces

The product of topological spaces follows the same principles as above but is somewhat more technically difficult, and we haven't actually done the technical definition of topological space in the first place. However the idea is the same: we take the cartesian product $A \times B$ of underlying sets, and then try and put a topology on it in a canonical way. As with posets there are several different possibilities, but it's more subtle to see which is the categorical product as it's not just a simple case of symmetry; it's really about universality. The technicalities of what is called the *product topology* are a little subtle, especially if you want the possibility of infinite products, but the ideas are quite vivid in low dimensions at least: the product of two shapes is the result of waving one shape in the shape of the other, and imagining what higher-dimensional space you "sweep out". For example the product of a square and a line is a cube, because if you move a square in a line you sweep out a cube in the air.[†] The product of a circle with a circle is a torus (like the surface of a bagel), which I like to demonstrate by taking a slinky (essentially a circle) and wrapping it round on itself in a circle. You can even make a vertical cut in a bagel and then insert the bagel into the slinky-torus.

Anyway, when we work categorically then we don't really need to know exactly what the formal definition is. We just need to know that the categorical product exists, and then we work with its universal property rather than its definition.

In the category of categories

We can take products and coproducts of categories, and perhaps you can at this point guess what they are. We'll come back to it after we've spent more time with functors and the category **Cat** of (small) categories, where these products and coproducts live.

[†] This does depend on some details such as the line being the same length as the edges of the square, and the "sweeping" motion being exactly at right-angles to the face of the square.

Further analogies with numbers

There are more things we can do inspired by the analogy with products of numbers. We have already taken a passing look at commutativity (the symmetry between a and b in the categorical product $a \times b$) and associativity. Other things we could ask include:

- Is there an identity? (Often yes. What might it be?)
- Are there inverses? (Often no. Why?)
- Do we have something like unique factorization into primes? If so what are the "primes"?

The last one is quite well studied for groups and is particularly straightforward for Abelian groups, where there is a "fundamental theorem of Abelian groups" rather like the fundamental theorem of arithmetic (which says that every natural number can be uniquely expressed as a product of primes).

For the first question we can show that terminal objects are a sort of weak unit for products, in that $a \times 1$ is canonically isomorphic to a for any object a (and dually for coproducts and initial objects). This is then a prototype example of a "monoidal category" which is the categorification of a monoid, that is, it is a monoid but up one dimension. A monoid is a set with a binary operation; a monoidal category is essentially a category with a binary operation. However, in order to make the notion appropriate for its dimension, associativity and unitality on objects does not hold strictly as this would involve equalities between objects. Instead there are canonical isomorphisms similar to the ones we've seen for products, but as they are not necessarily induced by a universal property we impose some axioms on them to make sure they behave almost as well as the ones induced by a universal property.

We will touch on this in the last chapter, on higher dimensions.

19

Pullbacks and pushouts

These are more advanced universal properties, showing more of the general features that were missing from the special cases we've seen so far.

The universal properties we have seen so far are terminal/initial objects, and products/coproducts. As we mentioned at the end of the last chapter, universal properties are all universal cones over some starting data. Our starting data so far has not been very general: it has not included any morphisms. For terminal/initial objects the starting data was trivial, that is, it was empty. For products/coproducts it was a pair of objects. In this chapter we are going to do a first example where the starting data is a diagram including some morphisms. This gives all the ideas needed for universal cones over general diagrams, although the formal definition of the full generality is beyond our scope.

Pullbacks and pushouts are dual to each other although neither of them is called "co" of the other. Pullbacks are also sometimes called "fibered products" especially in some fields where they really are fibered products. As this term indicates, they're a bit like products but restricted, or refined, in a sense we'll discuss.

This chapter is a culmination of our progressive exploration of universal properties. As these chapters have been a build-up, the material in this chapter is probably harder than what has gone before. However, the next chapter will begin a fresh topic, so if you feel confused in the current chapter I am optimistic that you will be able to pick up again in the next chapter, and come back to this one at a future point.

19.1 Pullbacks

Pullbacks can be thought of as refinements of products, intuitively as well as technically. Here is the intuitive idea for pullbacks.

A product of two sets takes all the possible ways of making pairs, with one element from each set. But perhaps we want to make pairs that are coherent in some way, perhaps pairing up people who are the same height to be dancing

partners, or pairing pants/trousers with shirts that are the same color: we do
that using a pullback.

If we were just pairing any old pants with any old
shirts then our starting data would just be two sets:
our set of pants, and our set of shirts. However, to
match colors we need to take into account the functions shown on the right.

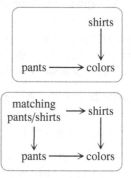

Then, we're looking for pants and shirts that are
mapped to the same place in the set of colors,
which gives us a commuting square as shown here.

But it's not just any old commuting square on the right, it's a *universal* one,
giving all matching pants–shirts pairs, and nothing else.

This is an example of a *universal cone*. Terminal objects and products are
also universal cones, but we didn't express those quite so explicitly. Here's
how it works for products.

When we take products, the data we start
with is just a pair of objects, as shown in the
first diagram below. A cone over it is shown
in the middle, and the scheme for its universal property on the right.

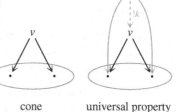

In a general limit, the starting diagram can be any diagram, involving arrows
as well as objects.

For example, given the diagram on the left
below, a cone over it is shown in the middle,
and the scheme for a universal property on
the right.

(The ellipse is just
to make the cone
look like a cone.)

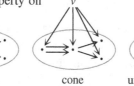

A cone is an object v (the "vertex") together with morphisms to every object in
the starting diagram, such that everything commutes. To be universal it must
then be universal among all cones of that shape: given any such cone with

vertex x, say, there must be a unique factorization $x \longrightarrow v$, as shown in the third diagram. As before, being a factorization means it makes everything commute.

To express this rigorously in full generality much more formalism is needed, as we can't always simply draw these diagrams. Sometimes we want to take limits over infinite diagrams that thus can't be drawn. We will see how to express the notion of "diagram" formally in the next chapter using functors, but that's as far as we'll go. Here is the definition of a pullback, via universal cones.

Definition 19.1 A *limit* over a diagram is a universal cone over it. A *pullback* in a category is a limit over a diagram of the shape shown on the right.

Admittedly we haven't exactly given a formal definition of "universal cone", only some pictures and the general idea. We'll now unravel what this means for the definition of pullback.

The first step is to understand what a cone over this diagram is. It will help us to name our objects, which I've done below.

A priori [†] a cone is then a vertex v together with morphisms to a, b and c making everything commute, as shown on the right.

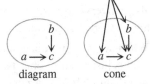

diagram   cone

However, "everything commutes" means that the morphism $v \longrightarrow c$ is determined by being the composite $\boxed{v \longrightarrow a \longrightarrow c}$, or indeed $\boxed{v \longrightarrow b \longrightarrow c}$ so it is redundant for us to specify it.

So this cone is actually just a commutative square — if we omit the arrow $v \longrightarrow c$ and physically move v down, it becomes this:

The universal property then says: given any square[‡] with vertex x, say, as shown, there must be a unique factorization as shown on the right. I have named the arrows, for convenience.

Note that the conditions on being a factorization are the same as for a product — all the triangles involving projections must commute. I hope you will become happy with reading the condition from the diagram, but in algebra

---

[†] This means what we know in advance because of the general definition, before we sit down and think about the specifics of this particular situation.

[‡] As usual in category theory a shape with four arrows counts as a square even if it's not geometrically drawn as a square.

that condition says $\boxed{pk = f \text{ and } qk = g}$. No extra conditions arise from extra complications in the starting diagram; those only affect the notion of cone.

I have used some of the same notation here as for products, because this is a bit like a more restricted notion of product. The projections p and q are then a lot like the projections for a product.

To indicate a pullback we often put a little corner sign at the top left, as shown here. It goes where the vertex of the universal cone is.

Definition 19.2   We say that a category *has pullbacks* when it has pullbacks over all diagrams of the relevant shape (that is, as shown on the right).

We will now look at some examples, as usual starting with sets and functions.

19.2 Pullbacks in Set

Given functions s and t as shown, we want to define a set which is like the canonical way to complete the diagram into a commuting square. It's going to turn out to be like a restricted product, so we will denote it $A \times_C B$.

$$
\begin{array}{c}
B \\
\downarrow t \\
A \xrightarrow{\ s\ } C
\end{array}
$$

The idea is that if we didn't have the morphisms to C, the canonical thing to do would be to take the product $A \times B$, which is the set of all ordered pairs. However, to make sure the diagram commutes we take the "fibered" product, which means we restrict to those ordered pairs (a, b) such that a and b are mapped to the same place in C. This is a subset of the product $A \times B$.

Formally we define this: $\boxed{A \times_C B = \{(a, b) \in A \times B \mid s(a) = t(b)\}.}$

As this is a subset of the product $A \times B$, it inherits projection functions onto the A component and B component respectively.

Proposition 19.3   *The square on the right is a pullback, where p and q are the projections inherited from $A \times B$.*

─────────── **Things To Think About** ───────────

T 19.1   See if you can prove this for yourself. Try the method for products of sets in the previous chapter, constructing a factorization by following through the definitions — there is no choice of what to do here.

When I say there's "no choice" of what to do, I mean there aren't really any decisions to make. Sometimes category theory proofs are like an automatic conveyor belt that carries you along if you can just manage to step on it.

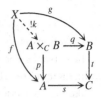

Proof   Consider a commuting diagram of sets and functions as in the outside of this diagram. We need to show there is a unique factorization as shown.

> We need to say what $k(x)$ is for each $x \in X$. It has to be an ordered pair, and the commutativity of the two triangles means the A component has to be $f(x)$ and the B component has to be $g(x)$, just like for the product; we then have to check that's a valid element of $A \times_C B$. Again, this does uniqueness first, and existence afterwards.

Commutativity of the triangles means that if a factorization k exists, it must be given by $k(x) = (f(x), g(x))$. (This is called "unique by construction".)

We need to check that this is an element of $A \times_C B$, that is, that the two components map to the same element of C. This follows by the commutativity of the outside of the diagram (see below). Thus k is indeed a factorization and is unique by construction, so the diagram is a pullback as claimed.   □

> I hope that last part will make sense to you if you follow things around the diagram. Written in algebra, the condition we need to check is $sf(x) = tg(x)$ which is exactly the commutativity of the outside.

Note that when we write the vertex of the pullback as $A \times_C B$ it only tells us what the corner object C is, without reference to the morphisms s and t. In some situations those are very natural, but I think it's good to be aware that the notation is missing something.

────────────────── **Note on terminology** ──────────────────

Sometimes we talk about pulling back morphisms along each other. So we might say that p is the pullback of t along s, and that q is the pullback of s along t. Although the definition is symmetrical in s and t sometimes the situation isn't. Sometimes one of the starting morphisms is more interesting and the other is in more of a supporting role.

────────────────── **Things To Think About** ──────────────────

T 19.2   1. See if you can show that there is always a canonical morphism from the pullback $a \times_c b$ to the product $a \times b$, if they exist. (Use the universal property of the product to induce it.)

2. Show that this canonical morphism is always monic. This shows that the pullback is always something like a restricted version of a product.

3. Show that if c is the terminal object the pullback $a \times_c b$ is a product $a \times b$.

We won't rely on those thoughts for anything, so I'm just going to leave them there. Before we move on, I do want to show you my favorite square.

The diagram on the right is officially my favorite square, for reasons we're about to investigate. Here every arrow is the subset inclusion function.[†]

$$A \cap B \longrightarrow B$$
$$\downarrow \qquad\qquad \downarrow$$
$$A \longrightarrow A \cup B$$

─────────────────── **Things To Think About** ───────────────────

T 19.3   See if you can show that the above square is a pullback in **Set**. (This is half of why it's my favorite square.)

We know from our previous construction that the pullback should consist of pairs $(a, b) \in A \times B$ such that a and b land in the same place in $A \cup B$. But the only way for elements a and b to do that in this case is if they are the same, and in the intersection.

Note that this is not exactly a rigorous proof, but if we turned it into one we still wouldn't have directly checked the universal property of the square shown: we'd have shown that the diagram shown is suitably isomorphic to the construction of the pullback we did earlier. At that point we'd need to invoke a theorem saying that everything isomorphic to a pullback is a pullback. I hope you can *feel* that this is true, from the sense we've developed of universal properties so far. To be sure it's true, in the most abstractly satisfying way, we will now re-frame pullbacks as terminal objects in a more complicated category related to $\mathcal{C}$.

19.3 Pullbacks as terminal objects somewhere

We have said that all universal properties are terminal objects *somewhere*, and showed that products in $\mathcal{C}$ are terminal objects in a category "souped-up" from $\mathcal{C}$. We can do this for pullbacks too. The idea is as for products: we start with a diagram and want to find a universal cone for it, so we make a category of all cones, and the terminal objects in there are the universal cones.

─────────────────── **Things To Think About** ───────────────────

T 19.4   Suppose we want to define pullbacks for the diagram shown on the right. See if you can make a "souped-up" category of cones over the diagram, in which terminal objects are pullbacks for the diagram.

───────────

[†] Remember $A \cap B$ is the intersection, that is, the set of all objects in both A and B, and $A \cup B$ is the union, where you take all the elements of A and all the elements of B and throw them in a set together (and if anything is in both sets you just take it once).

We make a souped-up category as follows. Objects are commutative squares as shown, and morphisms are given by a morphism k making the diagram on the right commute. Terminal objects in this category are then pullbacks over the original diagram.

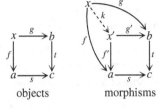

objects   morphisms

Note that for products we started with just the objects a and b, without the object c and morphisms s and t. If you erase c, s and t in the above diagrams, you get back the souped-up category for defining products; this new category just has an extra condition of commutativity with s and t.

We can now immediately deduce that pullbacks are unique up to unique isomorphism, where the isomorphism in question is unique making the whole diagram commute as in the diagram for a factorization.

19.4 Example: Definition of category using pullbacks

Before we move on to the dual of pullbacks I'd like to take a short aside to show one important use of pullbacks: to finish the alternative definition of category that we started talking about in Section 17.5.

This approach started with a set C_1 of morphisms and a set C_0 of objects, equipped with functions giving the source and target of each morphism as shown on the right.

$$C_1 \underset{t}{\overset{s}{\rightrightarrows}} C_0$$

We showed how to express identities in this context, but we couldn't do composition because it needs pullbacks. The idea is that composition is a function:

set of composable pairs of morphisms $\xrightarrow{\text{composition}}$ set of morphisms

$a \xrightarrow{f} b,\ b \xrightarrow{g} c$ $\longmapsto$ $a \xrightarrow{g \circ f} c$

A "composable pair" of morphisms looks like an ordered pair (f, g) but it's not *any old* ordered pair: the two morphisms have to meet in the middle in order to be composable. That is, the target of f has to be the same as the source of g.

———————— Things To Think About ————————

T 19.6   See if you can express that as a pullback in **Set**.

We take the pullback shown on the right; it's a pullback
in **Set** so can be defined as follows:

$$C_1 \times_{C_0} C_1 = \left\{ (f, g) \in C_1 \times C_1 \mid t(f) = s(g) \right\}$$

That is, an element is a pair of morphisms of the category we're defining, where
the target of one is the source of the other: a composable pair.[†]

Composition is then a function $\boxed{C_1 \times_{C_0} C_1 \longrightarrow C_1}$ with some conditions.

———————— Note on binary operations ————————

The fact that composition is defined on the pullback rather than the product is
what makes it a *partial* binary operation rather than a binary operation: it is
partial as it is only defined on part of the set of ordered pairs.

A binary operation on a set A is a function $\boxed{A \times A \longrightarrow A.}$

Our function for composition is defined on a pullback, which is a subset of the
product, not the whole product.

We still need to make sure that the composition function gives something that
behaves like composition. We are aiming for this: $a \xrightarrow{f} b \xrightarrow{g} c = a \xrightarrow{g \circ f} c.$

First we have to make sure that the source and target of the composite are correct, that is, that the
conditions on the right hold.

$$\boxed{\begin{aligned} s(g \circ f) &= s(f) \\ t(g \circ f) &= t(g) \end{aligned}}$$

We can express this by the commutative diagram
on the right. Note that s and t appear twice each
in the equations, so they appear twice each in the
diagram too.

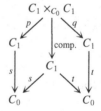

———————— Things To Think About ————————

T 19.7   See if you can "chase" a composable pair (f, g) around this diagram to
see how it gives the condition we want. It can sometimes be helpful to draw the
action on elements around the outside of the original diagram, or if that makes
a mess, draw a separate diagram with the same shape, but with elements at the
vertices instead of the sets C_0, C_1 and so on.

———

[†] Note that $t(f)$ isn't a composite; it's the result of applying the function t to the element $f \in C_1$.

Here is the diagram chase; I hope you will gradually become used to doing this for yourself so that the commutative diagram is enough. Eventually you may find that you don't even need to write out the diagram to chase elements around, but you can just do it by following the commutative diagram around in your head.

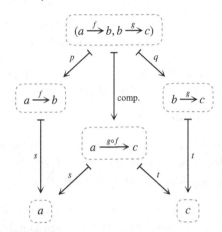

Things To Think About

T 19.8   See if you can now express the axioms for associativity and identities in commutative diagrams.

What we have started to do here is express the definition of a category categorically, if that doesn't sound too self-referential. That is, we have expressed it entirely using objects and morphisms in the category **Set**. One big payoff for having done that is that we can then pick it up and put it in other categories and see what happens. This gives us the notion of *internal category* and we'll come back to it in the last chapter, on higher dimensions.

19.5 Dually: pushouts

We will now move on to the dual of a pullback, which is called a pushout.

Things To Think About

T 19.9   Write down the dual concept of a pullback. That is, if we have a pullback in $\mathcal{C}^{op}$ what does this translate to as a structure in $\mathcal{C}$?

Definition 19.4   A *pushout* is a universal cocone for a diagram of the shape shown in the first diagram below. That is, it is a vertex v and morphisms as shown in the second diagram, such that given any diagram as in the third diagram, there is a unique factorization as shown in the fourth diagram.

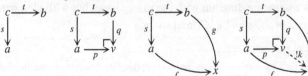

We indicate a pushout using the notation shown: a small corner marking the vertex of the universal cocone.

Note that as this is a cocone the factorization goes *out* of the universal vertex. In a cocone we have coprojections, that is, morphisms from the diagram in question to the vertex. Then the direction of the factorization sorts itself out by type-checking — it is a factorization of the non-universal cocone, a way of expressing the non-universal one as a composite of the universal one and the factorization morphism. The arrow can then only go one way.

───────────────── **Things To Think About** ─────────────────

T 19.10   What do we immediately know about pushouts, dually to what we know about pullbacks?

Dually to pullbacks we immediately know the following things:

• Pushouts are initial objects somewhere.
• Pushouts are unique up to unique isomorphism.
• Pushouts are canonically related to coproducts.

19.6 Pushouts in Set

Recall from Section 19.2 that I said the square on the right is my "favorite square". We have so far seen part of the reason; we can now see all of it.

$$
\begin{array}{ccc}
A \cap B & \longrightarrow & B \\
\downarrow & & \downarrow \\
A & \longrightarrow & A \cup B
\end{array}
$$

───────────────── **Things To Think About** ─────────────────

T 19.11   We showed that this is a pullback; can you now show that it is a pushout as well?

Proof   We need to exhibit the universal property, as in the diagram on the right. So we start by considering the outside, that is, a set X together with functions f and g making the outside commute.

$$
\begin{array}{ccc}
A \cap B & \xrightarrow{q} & B \\
{\scriptstyle p}\downarrow & & \downarrow{\scriptstyle t} \quad \searrow{\scriptstyle g} \\
A & \xrightarrow{s} & A \cup B \\
 & {\scriptstyle f}\searrow & \quad \dashrightarrow{\scriptstyle !k} \\
 & & X
\end{array}
$$

> As for when we did the pullback property, it may help to think about what it means for the outside to commute. If we start with an element $y \in A \cap B$ and chase it around, we see that commutativity means $f(y) = g(y)$, that is the functions f and g have to agree on the intersection $A \cap B$.

We need to construct a unique factorization. Now, the commuting condition for a factorization k amounts to the conditions on the right (since s and t are subset inclusions).

$$\forall a \in A, \quad k(a) = f(a)$$
$$\forall b \in B, \quad k(b) = g(b)$$

The question now is: can we use this as a *definition* of k? When defining a function on a union it works to define it on the individual parts as long as the definition agrees on the intersection, and that's true here as we know f and g agree on the intersection. We refer to this as the function being "well-defined".[†] We just need to say that formally.

So we define k by cases as shown on the right; k is well-defined because the outside of the diagram commutes, so if $z \in A \cap B$ then $f(z) = g(z)$ as needed.

$$k(z) = \begin{cases} f(z) & \text{if } z \in A \\ g(z) & \text{if } z \in B \end{cases}$$

Thus we have defined a factorization k and it is unique by construction. □

This is my favorite square because it is a pullback *and* a pushout, and it encapsulates some very fundamental principles in a way that I find much more structurally compelling than the usual Venn diagram (although I concede that the Venn diagram is better for actually putting things in).

One thing we can do with the pullback notation is stress that something that is not *a priori*[‡] the intersection is in fact isomorphic to the intersection.

For example, we can state that the commuting square of subset inclusions on the right is a pullback. The content of that statement is that the multiples of 10 are *precisely* those numbers that are multiples of both 2 and 5.

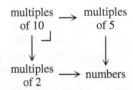

[†] This doesn't just mean we've defined it well, it means we haven't created an ambiguity in the definition, which is what would happen if our definition on individual parts did not agree on the intersection.

[‡] This means that we don't know in advance that it's the intersection, as we didn't define it to be the intersection, but we can then deduce that it is (isomorphic to) the intersection.

I saw a Venn diagram meme yesterday that showed the diagram on the right and said "Believe it or not, it's OK to be in all three."

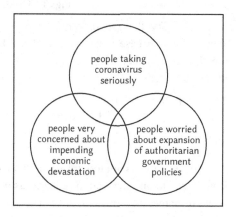

I would go a step further and say that the intersection of all three *exactly* consists of sensible people. That is, that all sensible people are in the intersection, and that everyone in the intersection is sensible.

Well, I admit that the last part might be more of a stretch (it's probably possible to be sensible about these issues and perhaps not at all sensible about something completely unrelated).

But if I really did think that the intersection was precisely the sensible people, I could express it as a universal property. It would be a 3-fold generalization of the pullback/pushout square above, and might look something like the diagram on the right.

$$A \cap B \cap C$$
$$A \qquad B \qquad C$$
$$A \cup B \cup C$$

Incidentally I think it is important to note that people define "authoritarian government policies" in different ways, and if someone doesn't think an illness is really dangerous, they might well think that a drive to vaccinate everyone is authoritarian, rather than protective.

Anyway these intersection/union situations are a particular kind of pushout in **Set**, and as we did with pullbacks we will now look at more general pushouts in **Set**. The idea is that the union is a way of sticking together two sets where we match them up along the elements they have in common. We could stick together two sets even if they have no elements in common, by defining the place we're going to match them up ourselves. This is how general pushouts work in **Set** and is actually quite a lot harder to define than pullbacks.

Consider the diagram of sets and functions on the right. To make a cocone we need to stick A and B together in such a way that C ends up in the same place (that's the commutativity).

To make it universal we have to do it in the most economical way possible, that is, we shouldn't stick anything together that doesn't need to be stuck together.

T 19.12   Try these two examples of sets and functions as shown. In each case make a set that is like sticking A and B together (union) but where you match up $s(c)$ with $t(c)$ everywhere, for each $c \in C$.

$$C \xrightarrow{t} B$$
$$s \downarrow \qquad$$
$$A \qquad$$

1. $A = \{1, 2\}$ s t
   $B = \{x, y\}$ $c \mapsto 1$ $c \mapsto x$
   $C = \{c\}$

2. $A = \{1, 2\}$ s t
   $B = \{x, y\}$ $c \mapsto 1$ $c \mapsto x$
   $C = \{c, d\}$ $d \mapsto 2$ $d \mapsto x$

The first example is not too unlike the intersection/union situation; in fact, it is isomorphic to an intersection scenario as the morphisms s and t are injective. If we re-named the elements then it would be just like an intersection/union situation. For example we could rename the elements of B to be 2 and 3 and then the element of C could be 2.

If we don't do that, we need to make a new set by starting with the disjoint union $A \sqcup B$ and then "identifying" $s(c)$ with $t(c)$ to ensure that the putative pushout square will commute. That is, we identify 1 with x. This means that we stick them together, or declare them to be the same object.

We could picture it as shown on the right. The large box with the dotted line represents $A \sqcup B$, with A on the left and B on the right. The long equals sign shows that we have identified the elements 1 and x.

So the resulting set has 3 elements: $1=x$, 2, and y.

The second one is more complicated. We need to identify $s(c)$ with $t(c)$, so we identify 1 and x. We also identify $s(d)$ with $t(d)$, that is, 2 and x. This gives the picture here.

We now see that, as a consequence, 1 and 2 also end up being identified with each other. So the resulting set has only two elements: $1=x=2$, and y.

This process is called *quotienting out by an equivalence relation*. Technically, however, the relation we're starting with isn't necessarily an equivalence relation, so actually we first have to use it to *generate* an equivalence relation, and then we quotient out by it. In my informal argument above, the relation we start with is the thing I've drawn with long equals signs. The fact that 1 and 2 became identified via x was not part of the original relation, but was a consequence of generating an equivalence relation.

We start with the following relation on $A \sqcup B$: $\boxed{s(z) \sim t(z) \quad \forall z \in C}$

┌─────────────────── **Things To Think About** ───────────────────┐
T 19.13   In what ways does this relation fail to be an equivalence relation in
the above two examples? (Assume that the definition is symmetric.)
└──┘

This relation is definitely not reflexive, as nothing is related to itself; however, it's easy to add that condition into a relation. We can also straightforwardly define it to be symmetric. Transitivity is more interesting: we have $1 \sim x$ and $x \sim 2$, but we don't know in advance that $1 \sim 2$, so transitivity fails.

Generating an equivalence relation consists of taking the "smallest" equivalence relation that satisfies the relation we already have. In this case it essentially consists of "closing" the relation under transitivity.

For the pushouts of sets we're not so concerned with what that equivalence relation is, as much as with what set we get when we quotient out by it. Quotienting out by a relation is where we identify everything that is related. Technically there are various ways to do this[†] but they're all isomorphic as sets, and so from the point of view of pushouts it really doesn't matter which one we do as the category will treat them all as "the same".

┌─────────────────── **Things To Think About** ───────────────────┐
T 19.14   Can you show that our second construc-
tion above (depicted on the right) is a pushout in
Set? Remember that "$1=x=2$" is a single element,
so the set at the bottom right has only two elements.
The coprojections p and q send everything to itself,
wherever it appears in that weird-looking set.
└──┘

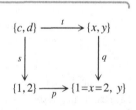

I think the important thing here is the idea: we made the set at the bottom right-hand corner by sticking elements together as economically as possible to make a commuting square. So it "should" be universal. However, for it to be rigorous mathematics we must also establish this formally.

Proof   We need to exhibit the universal property, as shown on the right. So we start by considering the outside, that is, a set X together with functions f and g making the outside commute. We need to construct a unique factorization k. (Again, the ! symbol signifies uniqueness.)

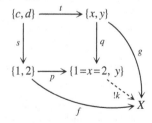

[†] If you're interested you could look up "equivalence classes" or "representatives of equivalence classes".

The conditions for a factorization say that the
two triangles in the above diagram must commute. The right-hand one tells us k must act as
shown on the right.

$$
\begin{array}{ccc}
 & k & \\
1=x=2 & \mapsto & g(x) \\
y & \mapsto & g(y)
\end{array}
$$

We must now check that the left-hand triangle commutes; this says we must
have $k(1=x=2) = f(1) = f(2).$ [†]

This is true because the outside of the diagram commutes, so in particular

$$
\begin{aligned}
f(1) &= fs(c) && \text{by definition of } s \\
&= gt(c) && \text{by commutativity} \\
&= gt(d) && \text{by definition of } t \\
&= fs(d) && \text{by commutativity} \\
&= f(2)
\end{aligned}
$$

> Note that I worked this out helped by deep belief that the quotient on the set in
> the bottom right is the exact right thing to make this pushout work, and thus that
> whatever we need to be true must be true. Believing that something is true before
> proving it can be dangerous as it can result in leaps of faith rather than rigor, but if
> the belief comes from deep structural understanding it can guide us towards rigor.

So the above definition of k makes the diagram commute and is unique by
construction, so the square is a pushout as claimed.   □

I honestly think that was quite ugly and in fact the proof in generality makes
more sense.[‡]

But I would also like to point out that these things just *are* a bit ugly. Quotienting out by relations is an ugly construction. We're sort of forcing things to
become the same as each other and it is typically a messy thing to do because
it comes with consequences. The more structure we have around, the more
consequences we end up with, and the messier it gets. Here we don't have any
structure, just elements, and it's already pretty messy.

Here's the construction of pushouts for general sets.

[†] If this confuses you, "chase" the elements 1 and 2 round the diagram individually, and write
down what it means for each one to land in the same place in X by both routes.

[‡] Actually with some more advanced understanding of category theory you'll see that the example is quite contrived. The function s is an isomorphism and some general theory then tells us
the pushout square is trivial: q is also an isomorphism and p does essentialy "the same" as t. I
still think it's a good exercise to examine the construction in a very small example.

Proposition 19.5 *The diagram on the right has a pushout given by $A \sqcup B \ / \sim$ where the relation $\sim$ is generated by:*

$$s(c) \sim t(c) \;\; \forall c \in C$$

$$C \xrightarrow{\;t\;} B$$
$$s \downarrow$$
$$A$$

> The notation $A \sqcup B \ / \sim$ means we are taking the disjoint union and quotienting out by the relation given.

Note that this construction makes the relationship with the coproduct explicit: we in fact start by taking the coproduct and then we quotient it. That quotient map is a function $\boxed{A \sqcup B \longrightarrow A \sqcup B \ / \sim}$ which is in fact the unique factorization induced by the universal property of the coproduct.

—————————— **Note on defining a function on a quotient** ——————————

At one level this proof is not going to be completely rigorous as we haven't really defined this quotient completely rigorously. However, really the one thing you need to understand about quotients is how we define functions *out* of them. We are essentially going to be defining a function out of a quotient, like this:

$$V/ \sim \; \longrightarrow \; X.$$

The quotient $V/ \sim$ consists of the elements of V at root, but some of them will be identified under the relation $\sim$. To define a function to X we can start by simply defining a function on *all* the elements of V, that is, a function $\boxed{V \xrightarrow{\;k\;} X.}$

We then show it is *well-defined*, that is, that it respects the relation $\sim$ so that if we then use this function on the quotient instead of on V no weird incompatibilities will happen. More precisely, we need to check that related elements in V are mapped to the same place by k, that is $\boxed{v \sim v' \;\implies\; k(v) = k(v').}$

This works whether we start with an equivalence relation, or a relation that we will then use to generate an equivalence relation.

————————————————

Proof   We need to exhibit the universal property, as shown on the right. So we start by considering the outside, that is, a set X together with functions f and g making the outside commute. We need to construct a unique factorization k.

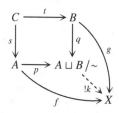

The conditions for a factorization tell us that the conditions on the right must hold, so we define a function k by those conditions.

$$\left(\begin{array}{l} \forall a \in A \quad k(a) = f(a) \\ \forall b \in B \quad k(b) = g(b) \end{array} \right)$$

We must show that this respects the relation $\sim$ (generated by $s(c) \sim t(c)$ for all $c \in C$). Now, given $c \in C$:

- $s(c) \in A$ so $ks(c) = fs(c)$ by definition of k.
- $t(c) \in B$ so $kt(c) = gt(c)$ by definition of k.
- The outer diagram commutes so $fs(c) = gt(c)$.

Thus k does respect the relation, so defines a function on the quotient.[†] Thus we have defined a factorization and it is unique by construction. □

Things To Think About

T 19.15   Go back to our proof of the universal property for the specific case (T 19.14) and see if you can see how that proof relates to our general one above.

19.7 Pushouts in topology

Pushouts in spaces are a key way of glueing spaces together to make more complex spaces. The idea is that when we (physically) glue things together we choose an area on each thing where the glue will go, and then those areas get identified.

For example here's a pushout square glueing two strips of paper together to make a longer strip. It's an example of my "favorite square" but this time in the category **Top** of topological spaces and continuous maps.

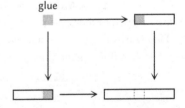

General pushouts are again more complicated, but these easier ones are extremely useful for building up spaces in topology by sticking them together.

[†]   You might prefer this sentence to say "Thus k does respect the relation, so it defines a function on the quotient". Mathematical writing often omits the "it" though. It streamlines the sentence, but it might just be a habit from the old days of desperately saving space in print journals.

The above example of making a longer strip was not that interesting but here is a pushout that sticks two intervals together to make a circle. The two copies of the interval end up on one half of the circle each, and the endpoints get glued together.

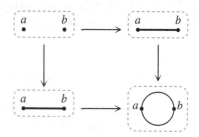

Remember that we are working in topology so the *geometry* of the situation doesn't matter. That is, it doesn't matter that the straight lines become curved; all that matters is how they are attached.

Pushout constructions in **Top** enable us to construct things that are very difficult to picture. For example, we could consider glueing a disk (a filled-in circle) to the edge of a Möbius strip.

A Möbius strip is the shape produced by taking a strip of paper and glueing the ends together with a twist.

This shape famously has "only one side" as the front and back of the paper have been joined up. It also only has one edge, because the "right and left" edges of the strip of paper have been joined up. As it only has one edge, this means its edge (or boundary) is topologically a circle. Thus we can abstractly (though not physically) glue a disk onto that boundary.

The square on the right is the pushout in question, where the morphisms s and t are each the inclusion of the circle into the boundary of the target shape. The shading is to show that the top right shape is a solid disk, whereas the top left shape is just the boundary and a hole in the middle. (I haven't shaded the Möbius strip as it was too hard.)

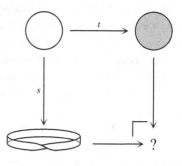

This is extremely hard to imagine visually or physically, but it is one possible construction of a very interesting space called the projective plane. In fact, many interesting spaces can be made by cutting disks out of things and glueing a Möbius strip into the circular hole that was made. This is helps with the *classification of surfaces*, where all possible 2-dimensional surfaces are organized up to homotopy equivalence. All the so-called non-orientable surfaces can be made from a sphere by cutting out a finite number of holes and glueing a Möbius strip onto each one. (The orientable ones are all the toruses with a finite number of doughnut-holes.)

19.8 Further topics

Pushouts of groups

We already saw that coproducts are much more complicated for monoids and groups than for sets and posets. This is because of the binary operation: it means that after making the disjoint union of the underlying sets, we must then generate new elements produced by the binary operation.

To find pushouts of groups we have to start with the coproduct of groups, which we saw in the previous chapter is the "free product", and then we have to do something analogous to quotienting out by an equivalence relation. In group theory this is provided by the theory of quotient groups. The idea is that one can "divide" a group by a special kind of subgroup, called a normal subgroup. We have actually seen an example of this although we didn't put it that way: the integers modulo n can be constructed as a quotient group where you start with the integers and then quotient by the subgroup consisting of all the multiples of n.

Pushouts of groups are related to those of topological spaces in a key way.

Pushouts in algebraic topology

Algebraic topology is about making relationships between spaces and groups. One way is via a functor that we'll mention in the next chapter, which assigns to every topological space its *fundamental group*. This gives us a way of studying spaces via the rich and well-established theory of groups.

A big problem is then actually calculating what the fundamental group is as a group, rather than just as an abstract definition. Pushouts help us do this because of a wonderful theorem called Van Kampen's theorem. This tells us that under certain circumstances we can express a space as a pushout and then calculate its fundamental group as a pushout of groups. The circumstances are a little stringent and, for example, would not help us in the situation above where we glued two intervals together to make a circle.

The pushout method works much better if instead of using groups we use groupoids. Recall that these are a "one dimension up" version of groups: groups are categories with a single object and every morphism invertible, whereas groupoids have every morphism invertible but no restriction on the objects.

In some ways the correspondence between spaces and groupoids works better than the one with groups, but there is much less long-standing groupoid theory than there is group theory. Thus there is still great benefit to making the effort to produce a group rather than a groupoid. However, as we move into higher dimensions the balance of advantage starts to shift, as we will see later.

More general universal properties

We have hinted at a more general theory of universal properties as we have built up more complicated versions:

1. We started with terminal and initial objects. These are universal (co)cones over empty diagrams: no objects and no morphisms.
2. We then saw products and coproducts. These are universal (co)cones over discrete diagrams: no morphisms.
3. Finally we saw pullbacks and pushouts. These are universal (co)cones over diagrams with just two morphisms pointing in opposite directions.

Another fairly straightforward type of universal property that we might have done next is equalizers and coequalizers.

These are universal (co)cones over a parallel pair of arrows, as shown here. You might like to try unraveling those definitions.

In general, universal cones over a diagram are called *limits* and universal cocones are called *colimits*, so all the universal properties we've been seeing have been limits and colimits. Pullbacks and pushouts are a good prototype for how limits and colimits generally work. In fact all finite limits (that is, limits over finite diagrams) can be constructed from combining finite products and pullbacks.

This result is usually stated in terms of products and equalizers; I just said it with pullbacks as we haven't really looked at equalizers. Equalizers and pullbacks are interchangeable in the presence of products, that is, we can make equalizers out of pullbacks and products, and we can also make pullbacks out of equalizers and products. It is perhaps more intuitive to state the result about limits in terms of products and equalizers, as products deal with multiple objects and equalizers deal with multiple morphisms between the same objects, and that is really the two features that a diagram can have.

In the next chapter we will, among other things, see how to express general diagrams other than just by drawing them.

20

Functors

We now change direction a little and address the concept of morphisms between categories. It is the "sensible" definition, using the principle of preserving structure.

20.1 Making up the definition

In Section 13.5 we briefly introduced the idea of a functor. The idea is that functors are the structure-preserving maps between categories. If you can't remember what the definition was at this point, I hope you might feel you could start making it up for yourself using the idea of preserving structure. I think this is the best way to get math into your brain: understand the principles so that you can create it for yourself. This is why I think that structural math shouldn't involve memorization, though some people do a bait-and-switch and claim that by memorization they mean "put into your memory". I think that's not what memorization typically means, and that we should keep separate the idea of rote memorization (without meaning) and the sort of process I'm calling structural, where you deeply understand a structural principle so that it becomes part of your consciousness.

The idea of a functor is that it's a function on underlying data, preserving the structure. In fact, now that we have expressed the definition of category in two slightly different ways I will also express the definition of functor in two slightly different ways.

Definition by homsets

The first definition we saw of (small) category has the underlying data as

- a set of objects
- for every pair of objects a set of morphisms, called a homset.

In this case the definition of functor consists of a function on sets of objects, and a function on every homset.

290

Definition 20.1 Let $\mathbb{C}$ and $\mathbb{D}$ be (small) categories. A *functor* $F\colon \mathbb{C} \to \mathbb{D}$ is given by

- a function $F\colon \mathrm{ob}\,\mathbb{C} \to \mathrm{ob}\,\mathbb{D}$, and
- for all objects $x, y \in \mathbb{C}$, a function $F\colon \mathbb{C}(x, y) \to \mathbb{D}(Fx, Fy)$

satisfying the following conditions called *functoriality*:

- on identities: for all objects $x \in \mathbb{C}$, $\boxed{F(1_x) = 1_{F_x}}$ and

- on composites: for all morphisms $x \xrightarrow{f} y \xrightarrow{g} z \in \mathbb{C}$, $\boxed{F(g \circ f) = Fg \circ Ff.}$

However, it is often stated in the "more general" way below, which does not specifically refer to functions, so that the definition works for categories of all sizes (whereas functions can only be invoked on *sets* of things). It is also a little more like how we typically use functors.

> You might be perplexed by the way the word "associates" is used in this more general definition. It's not really how we use it in normal English, but is how we use it in math. You might want to start by reading the definition below, and then read the footnote[†] for further unraveling.

─────────────── **More general definition of functor** ───────────────

Let $\mathcal{C}$ and $\mathcal{D}$ be categories. A *functor* $F\colon \mathcal{C} \to \mathcal{D}$ associates:

- to every object $x \in \mathcal{C}$ an object $Fx \in \mathcal{D}$,
- to every morphism $x \xrightarrow{f} y \in \mathcal{C}$ a morphism $Fy \xrightarrow{Ff} Fy \in \mathcal{D}$

satisfying functoriality as above.

─────────────────────────────

We could think of functoriality in terms of plane ticket prices in the following way. Suppose we have one category consisting of journeys by plane, so a morphism from A to B is a route from A to B by plane (possibly with layovers). Composition is then just doing one journey followed by another. We also have a category of numbers, and we could try and map journeys to numbers by their prices as an attempt at a functor F. The question of functoriality is then this: suppose you're flying from A to B to C. If you buy the ticket from A to B and

───────────────

[†] A more natural phrasing in normal English would be "A functor F associates an object Fx to every object x" but in math we turn it round so that the object that comes first in the logic also comes first in the sentence. The other slight oddity is that we say "associates to" rather than "associates with". Here "associate" is an active verb, perhaps more like "assign". What this definition really says is "Given any object $x \in \mathcal{C}$, the functor F produces an object $Fc \in \mathcal{D}$", and similarly for morphisms.

then the ticket from B to C as two separate journeys, does it cost the same as buying the whole ticket from A to C as one journey? The following is a schematic diagram depicting this situation, which means the triangular-headed arrows aren't morphisms in any particular category, but rather, they are processes.

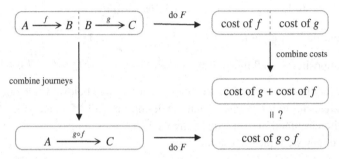

There are two entries for the bottom right corner because *a priori* we don't know if they're the same. One comes from going around the top-and-right edges of the square, and the other comes from going around the left-and-bottom. I would say the answer is that they are not in general the same. Airline ticket pricing is somewhat obscure (especially for international routes), and I would typically not expect anything as simple as adding the prices together.

Definition by graphs

We saw another definition of category, where the underlying data is a diagram of sets and functions as shown on the right. $C_1 \underset{t}{\overset{s}{\rightrightarrows}} C_0$

Things To Think About

T 20.1   What do you think the underlying data for a functor should be in this context?

In this context, the underlying data for a functor is a pair of functions F_1 and F_0 as shown on the right, making the diagram "serially commute".

$$C_1 \underset{t}{\overset{s}{\rightrightarrows}} C_0 \quad F_1 \downarrow \quad \downarrow F_0 \quad D_1 \underset{t}{\overset{s}{\rightrightarrows}} D_0$$

This is something we sometimes say about diagrams involving parallel arrows. If a diagram "serially" commutes it means that only the sub-diagrams[†] involving corresponding arrows commute. In the diagram above there is a square

---
[†] By "sub-diagram" I mean a part of the diagram in question.

involving just the top arrow s of each parallel pair (together with the vertical arrows), and there is a square involving just the bottom arrow t of each parallel pair. Each of those two squares is required to commute, but nothing mixing up an s and a t is required to commute.

The definition of category and functor by graphs is of interest for abstract reasons, as we'll see briefly in the last chapter, Chapter 24 on higher dimensions. However, the full definitions are a little beyond our scope.

20.2   Functors between small examples

We'll start by looking at our usual examples of mathematical structures expressed as categories.

Posets as categories

Recall that a poset is a category in which there is at most one morphism between any two objects. Now, since posets are a special case of categories, we might sensibly wonder: if we regard posets as categories, do functors between them correspond to order-preserving functions? Here is a schematic diagram:

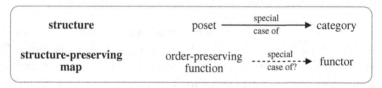

The dotted arrow could mean different things: it could mean that we need to take special cases of functors between posets (expressed as categories) or it could be that we take all the functors, and it's only a special case because we've already restricted the categories. The latter turns out to be true. Later on we'll see that the above "schematic" diagram can be turned into a rigorous piece of theory in its own right.

Things To Think About

T 20.2   Consider posets A and B ordered by $\leq$. Express them as categories, and show that a functor $F\colon A \longrightarrow B$ is precisely an order-preserving function. In what multiple ways have I abused notation here? Does it bother you?

We express a poset as a category by having a morphism $a \longrightarrow a'$ whenever $a \leq a'$. A functor $F\colon A \longrightarrow B$ in that case gives us:

- on objects: a function on objects, that is, a function on the underlying sets $A \rightarrow B$,
- on morphisms: for every $a \rightarrow a'$ we get a morphism $Fa \rightarrow Fa'$, that is, whenever $a \leq a'$ we have $Fa \leq Fa'$, which is exactly telling us we have an order-preserving function.

I have abused notation by using the same notation for the underlying sets, the posets, and also the posets expressed as categories. Personally I think that this is unambiguous and, in this case, clearer than the notational mess that would result if we tried to use different notation for those three structures. Furthermore, I am trying to encourage the idea of thinking of those three things simultaneously, not separately.

Monoids and groups as categories

Recall that a monoid is a category with only one object. Unraveling that, we see it is a set with a binary operation that is associative and unital. A group is then a category with only one object in which every morphism has an inverse, though that's more of a characterization than a direct definition.

──────── **Things To Think About** ────────

T 20.3   What could we then ask ourselves about monoid/group homomorphisms, analogously to what we asked for posets? Can you answer it?

The analogous question is: as monoids are a special case of categories, are monoid homomorphisms precisely the functors between those categories? The answer is yes. Functoriality requires preserving identities and preserving composition. Those precisely correspond to the two conditions for a monoid homomorphism: preserving identities and preserving the binary operation.

20.3 Functors from small drawable categories

Functors *out of* a small drawable category are a way of abstractly finding structure in another category. This is similar to the fact that a function from any 1-element set to a set X simply picks out an element of X.

──────── **Things To Think About** ────────

T 20.4   Here is the "quintessential" category consisting of one arrow, as we mentioned in Section 11.1.

What is a functor from this to **Set**? What about from this to the category of monoids **Mon**? From this to any category $\mathcal{C}$?

Let's call this category $\mathbb{A}$ (for "arrow"), and name its objects and morphism as shown here.

$$a \xrightarrow{\ f\ } b$$

To specify a functor $F: \mathbb{A} \longrightarrow \mathbf{Set}$ we just have to specify sets and functions as shown on the right.

$$Fa \xrightarrow{\ Ff\ } Fb$$

There is nothing left to specify or check as we know that identities must go to identities, and there are no non-trivial composable morphisms anyway. Note that Fa and Fb could be the same set, and Ff could itself be the identity: identities *must* be mapped to identities, but non-identities are allowed to be mapped to identities. So a functor out of the category $\mathbb{A}$ just picks out a function (and by implication, source and target sets as well).

Similarly, a functor $\mathbb{A} \longrightarrow \mathbf{Mon}$ just picks out a monoid homomorphism (and by implication, a pair of monoids for the source and target as well). In general a functor $\mathbb{A} \longrightarrow \mathcal{C}$ just picks out a morphism in $\mathcal{C}$.

In category theory we are liable to say "a functor $\mathbb{A} \longrightarrow \mathcal{C}$ *is* a morphism in $\mathcal{C}$" with a rather categorical[†] use of the word "is". When we look at natural transformations we'll see a way to organize both functors and morphisms into categories themselves, and then say something more precise about "is".

─────────────── Things To Think About ───────────────

T 20.5   Can you show that a functor must map a commutative square to a commutative square? Can you thus make a category $\mathbb{J}$ such that a functor from $\mathbb{J} \longrightarrow \mathcal{C}$ just picks out a commutative square in $\mathcal{C}$?

Suppose we have a commutative square as shown here in some category $\mathcal{E}$, and a functor $F: \mathcal{E} \longrightarrow \mathcal{C}$.

$$\begin{array}{ccc} a & \xrightarrow{\ f\ } & b \\ {\scriptstyle s}\downarrow & & \downarrow{\scriptstyle t} \\ c & \xrightarrow{\ g\ } & d \end{array}$$

Then we certainly get a square in $\mathcal{C}$ as shown here, and the question is whether or not it commutes, that is, whether the composite around the top-and-right equals the composite around the left-and-bottom.

$$\begin{array}{ccc} Fa & \xrightarrow{\ Ff\ } & Fb \\ {\scriptstyle Fs}\downarrow & & \downarrow{\scriptstyle Ft} \\ Fc & \xrightarrow{\ Fg\ } & Fd \end{array}$$

Now, the top-and-right composite is $Ft \circ Ff$ which equals $F(t \circ f)$ by functoriality of F; similarly the left-and-bottom composite equals $F(g \circ s)$. The commutativity of the original square tells us $t \circ f = g \circ s$, so we must also have $F(t \circ f) = F(g \circ s)$. So the second square commutes.

─────────────────────

[†] As always, by "categorical" I mean "pertaining to category theory" rather than "decisive and clear".

This means that we can usefully take a category $\mathbb{J}$ to be a "quintessential commutative square category", that is, it consists of a commutative square as shown on the right, and nothing else.

Then a functor $\mathbb{J} \longrightarrow \mathcal{C}$ will simply pick out a commutative square in $\mathcal{C}$. As with the "quintessential morphism", the commutative square in $\mathcal{C}$ need not have all its corners actually being different objects.

In fact, the functor could perfectly well pick out a commutative square as shown here, which we might call *degenerate*; it's really a single point.

─────────────── **Things To Think About** ───────────────

T 20.6 Can you think of some ways that a functor $\mathbb{J} \longrightarrow \mathcal{C}$ could pick out a single morphism in $\mathcal{C}$ expressed as a degenerate commutative square? That is, some of the edges of the square will be identities.

Here are some ways we could pick out a single morphism $a \xrightarrow{f} b$ as a degenerate square.

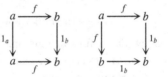

This might seem a bit of a futile exercise but this sort of "degeneracy" can be useful, especially when building up an entire situation out of triangles, which often happens in topology. Allowing degenerate ones gives us important flexibility and generality in the formalism.

We have used a "quintessential" category[†] of a certain shape to pick out diagrams of that shape in another category $\mathcal{C}$. We can now take this idea and use it as an abstract definition of diagrams in a category.

Definition 20.2 Let $\mathbb{J}$ be a finite category and $\mathcal{C}$ be any category. Then a *diagram of shape $\mathbb{J}$ in $\mathcal{C}$* is a functor $\mathbb{J} \longrightarrow \mathcal{C}$.

This is the abstract formalism we use to define general universal properties, as we briefly alluded to in the previous chapter. To define a general limit we start with a diagram of shape $\mathbb{J}$ expressed as a functor. We then define cones over

───────────────

[†] The sorts of categories I'm calling "quintessential" are sometimes called "walking" (as coined by Baez and Dolan), like a man with a mustache that is so big it looks like the man exists only to support the mustache: he's a walking mustache.

that diagram, which we can also do abstractly, but not until we've seen natural transformations.

Another "quintessential" category we could look at is a quintessential isomorphism. In order to use such a category to pick out an isomorphism in another category, we need to know that if we apply a functor to an isomorphism it will still be an isomorphism in the target category.

Things To Think About

T 20.7   See if you can show that the above fact is true about functors and isomorphisms, and construct a "quintessential isomorphism" category such that a functor out of it just picks out two objects and an isomorphism between them.

First it's good practice for us to state this precisely.

Proposition 20.3   *Let $F\colon \mathcal{C} \rightarrow \mathcal{D}$ be a functor, and f an isomorphism in $\mathcal{C}$ with inverse g. Then Ff is an isomorphism in $\mathcal{D}$ with inverse Fg.*

Proof   To prove that Ff and Fg are inverses we compose them and check that the composite is the identity both ways round. Now

$$
\begin{aligned}
Fg \circ Ff \; &= \; F(g \circ f) \qquad && \text{by functoriality} \\
&= \; F1 \qquad && \text{since } f \text{ and } g \text{ are inverses} \\
&= \; 1 \qquad && \text{by functoriality}
\end{aligned}
$$

and similarly (dually) for $Ff \circ Fg$. Thus Fg is an inverse for Ff and Ff is an isomorphism as claimed.   □

Note that I wrote 1 without being very specific what object this is the identity on. It doesn't really matter: it's the identity on the object it needs to be the identity on.

We can make a "quintessential isomorphism" category as shown here, in which f and g are inverses.

A functor from this category to any category $\mathcal{C}$ will then pick out two objects in $\mathcal{C}$ and a pair of inverse morphisms between them. As usual, these two objects could be the same, and the inverse morphisms could both be the identity.

The fact that an isomorphism is still an isomorphism after applying a functor is called *preservation* and we'll come back to that shortly, after we've looked at some more examples of functors.

20.4 Free and forgetful functors

We have secretly been using various functors to **Set** when thinking about sets with structure. For example when we were looking at products of posets we took products of their underlying sets and then constructed a "natural" ordering on that set. Without saying so, we were invoking a functor $\boxed{\textbf{Poset} \longrightarrow \textbf{Set}}$, which takes each poset and "forgets" that it's a poset. That is, the functor sends each poset to its underlying set. On morphisms the functor sends an order-preserving function to its underlying function. We did something similar for products of groups, starting from the product of their underlying sets and constructing a binary operation for it.

Functors that forget structure are called forgetful functors. They are often denoted U (for "underlying", perhaps) and this shows us that when we're talking about a poset A and its underlying set, we should technically call its underlying set UA. Another way we could avoid abusing notation is to call the poset $(A, \leq)$ and then define $U(A, \leq) = A$. However, I think both of these systems are often a little tedious and not terribly enlightening.

Things To Think About

T 20.8   What other forgetful functors have we implicitly thought about?

Here are some other forgetful functors we have met involving sets-with-structure, although we didn't formally say it that way at the time.

$$\begin{aligned} \textbf{Mnd} &\longrightarrow \textbf{Set} \\ \textbf{Grp} &\longrightarrow \textbf{Set} \\ \textbf{Top} &\longrightarrow \textbf{Set} \end{aligned}$$

Forgetful functors can also just forget *some* of the structure.

For example, a group is a monoid with some extra structure (inverses) so we have this forgetful functor:

$$\textbf{Grp} \longrightarrow \textbf{Mnd}$$

We could also take an Abelian group and forget that it's Abelian, giving this forgetful functor:

$$\textbf{Ab} \longrightarrow \textbf{Grp}$$

Similarly we could take a totally ordered set and forget that the ordering was total, giving this forgetful functor:

$$\textbf{Tst} \longrightarrow \textbf{Pst}$$

Sometimes the structure on a set is actually a combination of two types of structure. For example, a *ring* is a set with two binary operations that interact in a particular way. The two binary operations are often thought of as addition and multiplication, and the integers are a key example: we can add and multiply integers, and moreover, addition and multiplication interact nicely via a distributive law $\boxed{a \times (b + c) = (a \times b) + (a \times c).}$

We express this abstractly by saying we have a group with respect to one binary operation and a monoid with respect to the other, and the two binary operations have to interact according to the distributive law as above.[†]

I hope you can guess what ring homomorphisms are — they are simultaneously group homomorphisms and monoid homomorphisms. This makes a category **Rng** of rings and their homomomorphisms, and we have a system of forgetful functors like this.

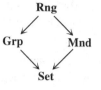

Things To Think About

T 20.9   Do you think this is a pullback square? [‡]

Forgetful functors might seem a bit contentless and it's true that we often use them without thinking about them very explicitly, as indeed we did in some earlier chapters of this book. However, one very powerful aspect of them is the relationship they often have with a functor going the other way which does the "opposite" of forgetting structure: it creates structure freely and is called a *free* functor. By contrast, sometimes forgetful functors do not have a relationship with such a functor, and that is in itself interesting.

Another kind of functor we've implicitly seen that doesn't exactly do much is a "special case" functor. Posets are a special case of categories: we can express any poset as a category with at most one morphism between any two objects. Order-preserving functions between the posets then become functors between those special categories. Thus we get a functor $\boxed{\textbf{Pst} \longrightarrow \textbf{Cat.}}$

Likewise we can express monoids as one-object categories. If we decide in advance what the single object is going to be (say, it's always going to be the symbol $*$) then we get a functor $\boxed{\textbf{Mnd} \longrightarrow \textbf{Cat.}}$

Free functors are much more difficult than "special case" functors or forgetful functors, but we can try them for monoids, where they're not too bad. The idea is to make a functor $\boxed{\textbf{Set} \longrightarrow \textbf{Mnd}}$ that takes a set and turns it into a monoid "freely". This means: without imposing any unnecessary constraints or using any extra information. We have to do exactly what is required to make it a monoid, and do it without imposing any of our own choices. A monoid is a set with some extra structure, so we need to create that structure from somewhere. The extra structure we need is an identity and a binary operation.

We already saw something like this free construction in Section 11.2 when we looked at a category starting from a single object with a single arrow f from the object to itself. We said that if we did not impose any extra conditions

---

[†] We also need to specify the version for $(b + c) \times a$ if multiplication is not commutative.

[‡] It isn't, because a ring isn't just a group and a monoid with the same underlying set: those structures have to satisfy a distributive law. That's not a proof, but it's the main idea.

on the composition in this category then, in addition to f, the category would have an identity 1, and composites f^n for every $n \in \mathbb{N}$, produced by composing the arrow with itself any finite number of times. In fact this is the free monoid generated by a single element f.

─────────────── **Things To Think About** ───────────────

T 20.10   Suppose we start with two elements a, b. You can think of them as arrows in a category with one object, or just elements that we're going to combine using a binary operation. If we keep composing freely without any extra conditions, what composites will we get?

We are starting with a set $\{a, b\}$ and we want to make it into a monoid. What could be the result of doing the binary operation? We need results for doing the binary operation on pairs of elements as shown on the right.

$$\begin{array}{l} a \circ a \\ a \circ b \\ b \circ a \\ b \circ b \end{array}$$

However, the answers can't be a or b because that would impose some unnecessary constraints on the monoid. That is, it would impose some equations that aren't demanded by the definition of a monoid.

The only thing we can do to avoid that is to declare "it is what it is" or something like that: the answer to $a \circ a$ is just that, a new element $a \circ a$ whose entire purpose in life is to be the answer to $a \circ a$. We do something similar for the other answers. Typically we get bored of writing the symbol $\circ$ so we just write these new elements as $aa,\ ab,\ ba,\ bb.$

Now, we can't stop there: we've added in these new elements, so now we need the answers to binary operations involving these new elements. So we need things like $a(ba),\ b(ba),\ b(ab), \ldots$

We get to save a little effort because associativity tells us it doesn't matter where those parentheses go, but we still can't stop because every time we make new elements we will then have to produce answers to binary operations with them. It's just like producing all f^n from a single element f except now we are building up from two elements instead of one. These are called *words* in the elements a and b, because they really are quite a lot like words using an alphabet, it's just that in our example our alphabet only has two letters. A word is a finite string of letters, and the letters are allowed to repeat themselves.

The binary operation is called *concatenation*, and consists of just sticking two words together to make a longer word, as is often done in German to great effect to produce wonderful words like "Schadenfreude" and "Liebestod". (English has some German roots and we do sometimes do this in English too, with words like handbag and toothpaste.)

We haven't yet dealt with the identity: that's the empty word, a word with no letters. It's not a space, because a space is an actual character, and we need something that really has nothing in it so that when we concatenate it with another word nothing happens to that word.

So we can start with any set A and produce the monoid of (finite) words in the elements of A. This is called the *free monoid on A*.

────────────── **Things To Think About** ──────────────

T 20.11   Given sets A and B and a function $f \colon A \longrightarrow B$ can you produce a monoid homomorphism from the free monoid on A to the free monoid on B?

Let us write FA for the free monoid on A. Given a function $\boxed{f \colon A \longrightarrow B}$ we can then make a monoid homomorphism $\boxed{Ff \colon FA \longrightarrow FB}$ as follows. An element of FA is a word $a_1 a_2 \cdots a_n$ where each $a_i \in A$ (and it is possible for the a_i to be equal for different i). We then define

$$Ff(a_1 a_2 \cdots a_n) = fa_1 fa_2 \cdots fa_n$$

which is a word in the elements of B. This is a bit like making a very basic code where you replace every letter of the alphabet by another one, say

A B C D E F G H I J K L M N O P Q R S T U V W X Y Z
Q W E R T Y U I O P A S D F G H J K L Z X C V B N M

and then any normal word can be turned into a coded word such as

$$\text{DELICIOUS} \longmapsto \text{RTSOEOGXL}$$

For codes like this it's important that the function on letters is an isomorphism otherwise it won't be decodable, but for our free monoids it needn't be an isomorphism; for example, if we used the function $\boxed{A \longrightarrow \{*\}}$ then every word in the elements of A would just become a string of $*$'s, like when you type in a password but it stays hidden on the screen. We can check that all this makes F into a functor $\boxed{\textbf{Set} \longrightarrow \textbf{Mnd.}}$ It is called the *free monoid functor*.

The free monoid on a single element is isomorphic to the natural numbers (under addition). This is an abstract encapsulation of the fact that we basically make the natural numbers by starting with the number 1 and adding it together repeatedly. In order for this really to be the free monoid we must include 0 as the additive identity, and this is one reason that I typically prefer including 0 in the natural numbers. Being a "free" structure is a type of universal property, so if we include 0 then the natural numbers have a good universal property among monoids and thus among categories. Whereas if we don't include 0 then the natural numbers only have a good universal property among sets with a binary operation but not necessarily an identity; the lack of identity means that such structures are not an example of a category. (However they are studied in their own right and are called *semi-groups*.)

We can do a similar free construction for groups, but it's more complicated because we have to add in inverses as well. Essentially we throw in new elements to be inverses, and then make words out of elements and their inverses, but we have to be careful to make sure that if a letter finds itself next to its inverse in the word then they cancel out. The free group on a single element happens to be Abelian and in fact is isomorphic to the integers. This encapsulates the fact that the integers are made just from the number 1, as with the natural numbers, except now we want to be able to subtract as well as add.

In earlier chapters we were implicitly using forgetful functors to help us with universal constructions. The idea is that if functors interact well with the structure in their source and target categories, they can help us understand the structure in one via the structure in another. This good interaction can happen both forwards and backwards, and this is the idea of "preserving" and "reflecting" structure.

20.5 Preserving and reflecting structure

When we have a functor, say $\mathcal{C} \xrightarrow{F} \mathcal{D}$, one of the things we do is look at how this relates structure in $\mathcal{C}$ to structure in $\mathcal{D}$, both forwards and backwards. That structure might be any of the things we've seen so far such as isomorphisms, commutative squares, monics and epics, terminal and initial objects, products and coproducts, pullbacks and pushouts, or any limits and colimits.

In the forwards direction we have the question of *preservation*: given a type of structure in $\mathcal{C}$, if we apply F does it still have that structure in $\mathcal{D}$?

In the backwards direction we have the question of *reflection*: if we apply F to something in $\mathcal{C}$ and the result has a particular structure in $\mathcal{D}$, can we conclude that it already had that structure in $\mathcal{C}$?

So far we have seen that isomorphisms and commutative diagrams are preserved by all functors. However, they are not necessarily reflected.

────────────────────── **Things To Think About** ──────────────────────

T 20.12   Can you construct some very small examples of categories and a functor $\mathbb{C} \xrightarrow{F} \mathbb{D}$ in which:

1. There is a morphism $f \in \mathbb{C}$ that is not an isomorphism although Ff is an isomorphism in $\mathbb{D}$.

2. There is a non-commutative diagram in $\mathbb{C}$ such that when we apply F we get a commutative diagram in $\mathbb{D}$.

Hint: for both situations you could take $\mathbb{D}$ to be the category with one object and one (identity) morphism.

In each case we can take $\mathbb{C}$ to be a "quintessential" or "walking" category for the structure in question, and we take $\mathbb{D}$ to be the category $\mathbb{1}$ with one object and one (identity) morphism.

Consider first the "quintessential morphism" category and a functor to $\mathbb{1}$ as shown here.

$$\left(\; \bullet \to \bullet \;\right) \xrightarrow{\; F \;} \left(\; \bullet \;\right)$$

As the target category has only one object and only one morphism, we have no choice about where F can send everything: every object has to go to the single object and every morphism goes to the identity. Functoriality has to hold because the only morphism in the target category is the identity, so every equation of morphisms just reduces to an identity equalling itself.

Now we just check that this functor sends a non-isomorphism to an isomorphism. The single non-trivial morphism in $\mathbb{C}$ is not an isomorphism as it has no inverse, but it is mapped to the identity in $\mathbb{D}$ which is an isomorphism. So we see that F does not reflect isomorphisms.

We can do something similar with a "quintessential non-commutative diagram" category. We could take a square that does not commute.

In fact it could just be a triangle that doesn't commute, or indeed a pair of parallel arrows.

$$\left(\; \bullet \rightrightarrows \bullet \;\right)$$

Technically if we have two parallel arrows that are not equal then the diagram does not commute, as there are two paths with the same endpoints whose composites are not the same. (In fact that is true if we just take a non-trivial loop, that is, a non-identity arrow from an object to itself.) Now if we again take the functor to the category $\mathbb{1}$, the parallel arrows will both be mapped to the identity, so the non-commutative diagram is mapped to a commutative diagram. This is thus a functor that does not reflect commutative diagrams.

--- Things To Think About ---

T 20.13   In Section 15.6 we saw that the monoid morphism $\mathbb{N} \to \mathbb{Z}$ sending everything to itself is epic. However the underlying function is not surjective. What is this a counterexample to, in terms of functors preserving epics? Which functor in particular is this about?

Whenever we're thinking about underlying sets we are implicitly thinking about the forgetful functor of the form $\boxed{\textbf{Sets-with-structure} \longrightarrow \textbf{Set.}}$

In this case we're thinking about the functor U as shown on the right, and its particular action on the inclusion morphism shown.[†]

$$\textbf{Mnd} \xrightarrow{\; U \;} \textbf{Set}$$

$$\left(\; \mathbb{N} \hookrightarrow \mathbb{Z} \;\right) \longmapsto \left(\; \mathbb{N} \hookrightarrow \mathbb{Z} \;\right)$$

[†] Someone more pedantic than I am would write $U\mathbb{N}$ and $U\mathbb{Z}$ for the underlying sets of natural numbers and integers.

The point is that the inclusion $\mathbb{N} \hookrightarrow \mathbb{Z}$ as a map of monoids is epic, but its image under U (that is, where it's mapped to in **Set**) is *not* epic. Abstractly this is saying that U does not preserve epics.

This means that, in general, being an epic is not a very "stable" property, as it might not be preserved by a functor. This is where *split* epics come in.

Recall that a morphism $f: a \rightarrow b$ is a split epic if there is $\quad b \xrightarrow{g} a \xrightarrow{f} b$ a morphism g making the diagram on the right commute. $\qquad\qquad 1_b$

The extra structure of the splitting g is enough to ensure that split epics are always preserved by functors.

———— Things To Think About ————

T 20.14 Can you show that split epics are preserved by all functors?

Suppose we have a split epic f (as in the diagram above) in a category $\mathcal{C}$, and a functor $\mathcal{C} \xrightarrow{F} \mathcal{D}$. We need to show that Ff is a split epic in $\mathcal{D}$.

Now, applying F to the original split epic diagram we $\quad Fb \xrightarrow{Fg} Fa \xrightarrow{Ff} Fb$ certainly get a diagram in $\mathbb{D}$ as shown on the right. $\qquad F1_b = 1_{Fb}$

We just have to check it commutes, which it does by functoriality.

$$
\begin{aligned}
\text{More precisely:} \quad Ff \circ Fg &= F(f \circ g) \quad &&\text{by functoriality} \\
&= F(1_b) \quad &&\text{by hypothesis} \\
&= 1_{Fb} \quad &&\text{by functoriality}
\end{aligned}
$$

One general idea about preservation is that studying whether a functor preserves certain structure might be telling us something about the functor, or it might be telling us something about the type of structure we're thinking about. Structure that is preserved by all functors is generally called *absolute*. We can think of it as particularly "stable" or strong.[†] A general rule of thumb is that anything defined by a commutative diagram will be preserved by functors, but anything involving a "for all" or "there exists" quantifier is trickier because when we move into a different category we will be quantifying over something different. For example, if we move somewhere with fewer morphisms (say because morphisms in the new place are required to preserve some structure), the quantifiers are affected in the following ways, broadly speaking:

- "For all" is now quantifying over fewer morphisms, and so it has more chance of being satisfied.
- "There exists" has fewer morphisms to choose from, so it has less chance of being satisfied.

[†] However if you're British you might be traumatized by the idea of something being "strong and stable". (See 2017 election.)

In the next section we will pin down a bit more precisely some notions of whether a functor takes us to a place with "more" or "fewer" morphisms.

─────────────── **Things To Think About** ───────────────

T 20.15   In Chapter 16 we first looked at terminal and initial objects in **Set** and then in some categories of sets-with-structure. In some cases the terminal and initial set-with-structure had the terminal/initial set as its underlying set, and in others it didn't. Can you recall those examples and see what that is saying about some functors preserving terminal/initial objects, or not preserving them?

In **Set** any one-element set is terminal, and the empty set is initial. We can compare this with various categories of sets-with-structure.

- In the category **Pst** of posets and order-preserving maps, any one-element poset is terminal and the empty poset is initial. So the forgetful functor **Pst** → **Set** preserves terminal and initial objects.

- In the category **Grp** of groups and homomorphisms, the trivial group has one element (the identity) and is both terminal and initial. So the forgetful functor **Grp** → **Set** preserves terminal objects but not initial objects, because if we apply it to the initial object in **Grp** we do not get the initial object in **Set**.

- In the category **Top** of topological spaces and continuous maps any one-point space is terminal and the empty space is initial. So the forgetful functor **Top** → **Set** preserves terminal and initial objects.

─────────────── **Things To Think About** ───────────────

T 20.16   In Chapter 18 we saw that various products of sets-with-structure do have the product of sets as their underlying set. Can you express this in terms of some functors preserving products? Now consider the examples where coproducts and pushouts did not just have the coproduct/pushout of underlying sets as their underlying set. What is this saying abstractly about preservation?

We saw that products of posets, groups and topological spaces did have the product of sets as their underlying set, so the forgetful functor from any of those categories to **Set** does preserve products.

The coproduct of posets did have the disjoint union of its underlying sets as its underlying set, so the forgetful functor **Pst** → **Set** preserves coproducts. However the coproduct of groups was the "free product" which involved taking the disjoint union of underlying sets and then generating a whole lot of new elements for binary products involving elements from both groups. Thus the forgetful functor **Grp** → **Set** does not preserve coproducts; nor does it preserve pushouts, as pushouts in **Grp** are constructed by starting from the free product and then taking a quotient.

20.6 Further topics

Algebraic topology

Functors between large categories of mathematical structures are sometimes how entire branches of mathematics get started. We will look at just one example here, relating algebra and topology.

Topology is the study of topological spaces, and algebraic topology is the study of topological spaces via algebra. The algebra in question is classically group theory, though more recently it has headed towards groupoid theory and higher-dimensional category theory.

The first thing one usually does in an algebraic topology course is the fundamental group. This is a way of taking a space and producing a group from it, by studying loops in the space.

First you have to choose a "basepoint" which is a point in the space where all your loops are going to start and finish. A loop is then a path in the space that starts and finishes at that point, as depicted informally here.

The elements of the fundamental group are the *homotopy classes* of loops. This means, essentially, that we count loops as the same if they're not *really* different, but you can continuously deform one into the other without breaking it or going over a hole in the space.

The binary operation in the group then comes from observing that if you go round one loop and then another, you've got another loop — it's just a loop where you happened to go home to the basepoint in the middle rather than waiting until the end.

This group can detect interesting structure in a space, like holes. For example the fundamental group of a circle is the integers $\mathbb{Z}$, because a loop can go round the circle any finite number of times, forwards or backwards.

The fundamental group of a torus (like the surface of a bagel) starts from two basic loops, one going "through" the hole and the other going around it, as shown on the right. Each of these loops individually generates an entire copy of $\mathbb{Z}$ as we can go round either of these any finite number of times forwards or backwards.

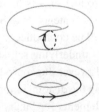

But we can also combine the loops, so we could go round the first one and then the second one and then maybe the first one again, for example. It turns out that the two loops commute up to homotopy, that is, it doesn't matter which order you do them in because one way round can be deformed into the other. So the group we get is the product $\mathbb{Z} \times \mathbb{Z}$. That was, of course, very far from a proof.

In fact the fundamental group construction is a functor $\boxed{\pi_1 : \mathbf{Top}_* \longrightarrow \mathbf{Grp}}$. Here $\mathbf{Top}_*$ is the category of spaces-with-a-chosen-basepoint, and the morphisms are continuous maps preserving the basepoint. In the first instance the fact that π_1 is a functor means that it extends to morphisms: a continuous map between topological spaces (preserving basepoints) produces a group homomorphism between the associated fundamental groups. Functoriality tells us that identities and composition are respected. As it's a functor, we can immediately deduce that isomorphisms are mapped to isomorphisms, that is, if we have an isomorphism between topological spaces we will get an isomorphism between their fundamental groups. This is important because isomorphic topological spaces should count as "the same" and if we're going to study them via associated groups then spaces that count as the same should produce groups that count as the same.

In the previous chapter we mentioned Van Kampen's theorem, which is about when we can take a pushout of spaces and also take a pushout of their fundamental groups, and know that the results will match, enabling us to calculate fundamental groups of complicated spaces by expressing them in terms of simpler ones. If we express this in terms of the fundamental group *functor* (rather than just a "fundamental group construction") the theorem becomes a result about when the fundamental group functor preserves pushouts.

Note that looking at all this is only a starting point. For example we really need π_1 to do more than preserve isomorphisms: the correct notion of sameness for topological spaces is more subtle than isomorphism in the category $\mathbf{Top}$, as we've vaguely seen — it's the notion of homotopy equivalence, where we can continuously deform one space into another. So what we need is for homotopy equivalences to be mapped to group isomorphisms, and this is something much more subtle that basic theory about functors can't directly sort out.

Algebraic topology also uses other functors to groups, and to more complicated categories. Homology and cohomology are different ways of having a functor to groups (in fact, to Abelian groups) and to chain complexes, which are a series of Abelian groups, one for each dimension, and some "chain maps", special group homomorphisms relating the different dimensions.

Another example of a branch of mathematics based on a functor between large categories is linear algebra, which importantly relates the category of finite-dimensional vector spaces to the category of matrices. That example is beyond our scope, but I thought I'd mention it in case it's something you've encountered and might like to look up.

Contravariant functors

There is another type of functor I want to
mention briefly, involving dual categories.
It's called a *contravariant* functor, and the
idea is that it reverses the direction of morphisms. The standard action on morphisms
and the contravariant action are shown on
the right.

standard

F

$$\left(a \xrightarrow{f} b\right) \longmapsto \left(Fa \xrightarrow{Ff} Fb\right)$$

contravariant

F

$$\left(a \xrightarrow{f} b\right) \longmapsto \left(Fb \xrightarrow{Ff} Fa\right)$$

To emphasize a functor that acts in the standard way, the word "covariant"
is used. I'm sorry about that; it's terribly confusing to me because it has the
"co-" prefix so makes it sound like it's reversing something (to me, anyway).

As a result of reversing
the direction of the arrows,
the functoriality condition for
a contravariant functor also
switches the order of composition, as shown on the right.

standard

F

$$\left(a \xrightarrow{f} b \xrightarrow{g} c\right) \longmapsto \left(Fa \xrightarrow{Ff} Fb \xrightarrow{Fg} Fc\right)$$

contravariant

F

$$\left(a \xrightarrow{f} b \xrightarrow{g} c\right) \longmapsto \left(Fc \xrightarrow{Fg} Fb \xrightarrow{Ff} Fa\right)$$

Written out in symbols this says $\boxed{F(g \circ f) = Ff \circ Fg.}$

As all this involves switching the direction of arrows, we can do it formally
by using the dual category $\mathcal{C}^{\mathrm{op}}$.

Definition 20.4 A *contravariant* functor is a functor $\mathcal{C}^{\mathrm{op}} \longrightarrow \mathcal{D}$.

> This is a slightly hazy definition because of course every category is the opposite of
> another category, so every functor is formally a functor from a dual category.[†] But
> the spirit of contravariant functors is that they flip the "obvious" direction of the
> arrows. As with all dual situations I prefer never to draw arrows in $\mathcal{C}^{\mathrm{op}}$, but instead
> to draw the arrows in $\mathcal{C}$ and $\mathcal{D}$ and see them flip when the functor acts.

This might sound like a contrived definition but it turns out that there are
many naturally arising functors that do this rather than keep the morphisms
pointing the same way, especially when we start using functors as tools for
further constructions, not just as ways of comparing structure. We will see a
key example in Chapter 23 when we look at the Yoneda embedding.

[†] Some people would say that a functor $\mathcal{C}^{\mathrm{op}} \longrightarrow \mathcal{D}$ is a contravariant functor $\mathcal{C} \longrightarrow \mathcal{D}$ but for me
that creates more confusion that it resolves.

21

Categories of categories

We now gather categories and functors into a category. We look at structures in that category, much as we have done with other examples of large categories of mathematical structures. We also show how this structure strains at its dimensions and is trying to expand into higher dimensions.

21.1 The category Cat

Categories are mathematical structures, and functors are structure-preserving maps between them. Good categorical instinct tells us that these "should" assemble into a category.

┌─────────────────── **Things To Think About** ───────────────────┐
T 21.1   What do we have to check to make entirely sure that categories and functors form a category? Can you do it?
└──┘

We need to make sure we have an identity functor, and composition of functors, and that the axioms hold. For any category $\mathcal{C}$ we have an identity functor $1_{\mathcal{C}} \colon \mathcal{C} \longrightarrow \mathcal{C}$ which acts as the identity function on objects and on morphisms. Composition also happens by composing the functions on objects and on morphisms. As this is all based on composition of functions it inherits the unit and associativity axioms from the functions.

However, there is one more technical issue with making a "category of categories". Good mathematical instinct will make alarm bells ring here from the self-reference: trying to make a "set of all sets" lands us in a paradox (Russell's paradox) and so there is a similar danger lurking if we try and make a "category of all categories". We avoid this by restricting what size of category we think about at any given moment. We briefly mentioned this in Section 8.5.

The first level is *small categories*. We mentioned that a category is called *small* if its objects form a set (not a large collection) and its morphisms also form a set. We can then make a category **Cat** of small categories and all functors between them. However, this will not include any of the examples of "large categories of mathematical structures" like **Set**, **Pst**, **Top**. It also can't include **Cat** itself, so we avoid a self-referential paradox.

The next level up is *locally small categories*. We said a category is called *locally small* if the morphisms between any two objects form a set; however the objects might form a (large) collection. For example, in the category **Set** the objects are all the possible sets, so these cannot form a set (to avoid Russell's paradox). However, given any sets A and B there is a set of functions $A \to B$. So **Set** is locally small. We can then make a category **CAT** of locally small categories and all functors between them.

Definition 21.1   We write **Cat** for the category of all small categories and functors between them. We write **CAT** for the category of locally small categories and functors between them.

We are going to look at some of the structure inside **Cat**, but first we will warm up by thinking about relationships between **Cat** and some other categories of structures we've seen. As in the previous chapter, it turns out we have been implicitly thinking about some of these functors already. This time the functors in question are between **Cat** and other categories.

─────────────── **Things To Think About** ───────────────

T 21.2   We have seen that posets are special cases of categories. Can you express this as a functor **Pst** $\to$ **Cat**? The analogous situation for monoids is more subtle; can you see why?

A poset is a category in which there is at most one morphism between any pair of objects. This means we have an "inclusion" functor **Pst** $\to$ **Cat** which simply sends each poset to itself as a category.

A monoid "is" a category with only one object, but now the situation is more subtle because there is a direct definition of monoid as a set with a binary operation satisfying some axioms. Unlike in the case of posets this is not *exactly* the same as a one-object category as we have performed a dimension shift and forgotten that the single object was ever there. This means that if we define monoids as sets with a binary operation, then to construct an "inclusion" functor **Mnd** $\to$ **Cat** we have to fabricate a single object from somewhere. We could decide that every monoid is going to become a one-object category with the same single object, say $*$. In any case this is why I put "is" in inverted commas at the start of this paragraph, because sending every monoid to itself expressed as a category is not quite as straightforward as for posets.

─────────────── **Things To Think About** ───────────────

T 21.3   What forgetful functors can you think of from **Cat** to **Set**? How sensible are they?

There are various forgetful functors from **Cat** to **Set** including functors that:

1. send a small category to its set of objects,
2. send a small category to its set of morphisms, or
3. send a small category to its set of *connected components*.

A connected component is a part of the category that is connected by arrows. For example the category on the right has two connected components (it's a diagram of just one category, although it might look like two).

$$a \longrightarrow b$$
$$x \rightrightarrows y \longrightarrow z$$

You could try checking that each of the above three forgetful situations can be made into a functor, at least as far as seeing what happens on morphisms.

As for how sensible these functors are, that of course depends on what you mean by "sensible". The forgetful functors **Grp** $\to$ **Set** and **Mnd** $\to$ **Set** were very sensible in the sense that groups and monoids really are sets with extra structure, so the relationship between the set with the extra structure and the set without it is what I am regarding as "sensible". However, a category is not just a set with extra structure, because its underlying data has objects *and* morphisms. So the forgetful functors to **Set** are in a sense *excessively* forgetful. They don't just forget structure, they also forget data. A more "sensible" forgetful functor would only forget the structure of a category, not the data, and that means that we wouldn't land in **Set** but in the category of graphs.

This notion of "sensible" is not rigorous at all, but it's a feeling about the situation that we can have as humans, and developing those feelings is an important part of developing as a mathematician.

Recall that in Section 17.5 we expressed the underlying data for a category as a graph, that is, a diagram of sets and functions as shown on the right.

$$C_1 \underset{t}{\overset{s}{\rightrightarrows}} C_0$$

Things To Think About

T 21.4   Can you make a category of graphs? What would be a sensible notion of morphism between graphs? If you're stuck, look back to what we said about the underlying data for a functor when categories are expressed in this way.

A morphism of graphs is given by a pair of functions making the diagram on the right serially commute; recall this means the square involving the s morphisms commutes and so does the one involving the t morphisms.

$$\begin{array}{ccc} C_1 & \overset{s}{\underset{t}{\rightrightarrows}} & C_0 \\ {\scriptstyle F_1}\downarrow & & \downarrow{\scriptstyle F_0} \\ D_1 & \overset{s}{\underset{t}{\rightrightarrows}} & D_0 \end{array}$$

Graphs and their morphisms then form a category **Gph**, and we have a sensible forgetful functor $\boxed{\textbf{Cat} \longrightarrow \textbf{Gph.}}$ What about a functor going the other way?

───────── **Things To Think About** ─────────

T 21.5   Can you see how to make a functor going the other way, which makes a "free category" on a graph, just like we made a free monoid on a set? Which part of that construction do we need to modify to allow for the fact that we now have more than one object? Can you make the construction into a functor?

We are going to construct the "free category functor" $\boxed{\textbf{Gph} \xrightarrow{F} \textbf{Cat.}}$

It is a generalization of the free monoid functor $\boxed{\textbf{Set} \longrightarrow \textbf{Mnd.}}$

For the free monoid functor we needed "answers" for the binary operation; for the free category functor we need to make "answers" for composition.

There is nothing to do on objects because a category does not have any extra structure on objects, only on morphisms. So if we start with a graph A, the free category FA has the same set of objects as A. For the morphisms we need to start with the morphisms of A and form composites freely (and an identity). This is just like making the free monoid in which we took strings of letters, except now we're taking strings of morphisms and we only make strings of *composable* morphisms.

So if we just start with morphisms as shown on the right, then we don't have to do anything because those morphisms are not composable. The free category on that data won't need any extra morphisms for composites.

$$a \xrightarrow{f} b \quad x \xrightarrow{g} y$$

However, if we start with morphisms as shown here, then the free category on this data will need a composite $g \circ f$.

$$a \xrightarrow{f} b \quad b \xrightarrow{g} c$$

So in this case we need to add in a new morphism $a \xrightarrow{g \circ f} c$ to be the result of that composition. We will also add an identity at each object.

This is really only a sketch of the construction, but I hope the idea is clear.

Definition 21.2   Given a graph A, the free category on it has

- objects: the same objects as A, and
- morphisms: finite composable strings of morphisms in A.

Note that finite composable strings include strings of length 0 (empty strings), which will be the identities. Composition is then given by concatenation, just like for words in the free monoid, but with the extra composability condition.

—————————————— **Aside on further theory** ——————————————

The relationship between free and forgetful functors is abstractly summed up as shown on the right
and is very important.

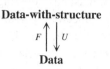

Data-with-structure

$$F \uparrow \quad \downarrow U$$

Data

At the top we have a category of "data equipped with a particular kind of structure", together with structure-preserving maps. In good situations the free and
forgetful functors are in a special relationship called an adjunction, which can
be expressed via universal properties in various ways that are beyond our scope
here. The composite UF is then a functor from the category of underlying data
to itself, and has many excellent properties. It is a prototype example of what is
called a *monad*. Monads are then an abstract way to study "sets with structure",
and more generally the category of data at the bottom could be something other
than **Set**, in which case monads give us a very general way to study algebraic
structure. In fact, at an abstract level this is a *definition* of algebra: something
that can be studied via monads.

Now that we have assembled categories and functors into their own category
we can look for our various types of structure in that category.

21.2 Terminal and initial categories

We are going to look for a few universal properties inside **Cat**, starting with
our most basic universal property, terminal and initial objects.

—————————————— **Things To Think About** ——————————————

T 21.6   What do you think terminal and initial objects in **Cat** are? We have
seen and even used them already.

For terminal and initial things in general, it is often productive to start thinking
about "one-element" and "empty" and see how that goes. It doesn't always
work, as we saw with groups, but it can lead us to find what works.

For categories we need to think about objects and morphisms, so for the case
of "one-element" it's a good idea to think about one object and one morphism,
that is, the category we have already met[†] and called $\mathbb{1}$.

Given any category $\mathbb{C}$ there is a unique functor $\mathbb{C} \longrightarrow \mathbb{1}$ because all objects

————————————————————

[†] We used it when we were finding functors that did not reflect isomorphisms and commutative
diagrams.

of $\mathbb{C}$ must go to the single object of $\mathbb{1}$, and all morphisms must go to the identity. Functoriality holds because everything is the identity, so all the required equations become "$1 = 1$".

For the initial category we can try taking the empty category 0, which has no objects and no morphisms. There is then a unique functor to any category $\mathbb{C}$ as it is vacuously defined, just like the "empty function" from the empty set to any set. Functoriality is then vacuously satisfied.

We have seen that the terminal and initial objects in **Cat** are as we "expected": the category with one object and one morphism is terminal, and the empty category is initial.[†]

21.3 Products and coproducts of categories

Our next universal properties are products and coproducts.

--- **Things To Think About** ---

T 21.7   What do you think products and coproducts in **Cat** are? Can you prove it? Is this coherent with products of posets and monoids expressed as categories?

Given (small) categories $\mathbb{C}$ and $\mathbb{D}$ we define their cartesian product $\mathbb{C} \times \mathbb{D}$ based on the cartesian products of their sets of objects and sets of morphisms.

So the underlying graph of $\mathbb{C} \times \mathbb{D}$ will be this:
$$C_1 \times D_1 \underset{t \times t}{\overset{s \times s}{\rightrightarrows}} C_0 \times D_0$$

That is, objects are ordered pairs (c, d) where $c \in \mathbb{C}$ and $d \in \mathbb{D}$, and a morphism $(c_1, d_1) \rightarrow (c_2, d_2)$ is an ordered pair (f, g) of morphisms as shown on the right.

$$\begin{aligned} c_1 \xrightarrow{f} c_2 &\quad \in \mathbb{C} \\ d_1 \xrightarrow{g} d_2 &\quad \in \mathbb{D} \end{aligned}$$

Composition is "componentwise". This means that to find the composite in $\mathbb{C} \times \mathbb{D}$ we take the composite in $\mathbb{C}$ and the composite in $\mathbb{D}$ individually, and then put them into an ordered pair.

I think it's helpful to have the following sort of picture in mind, where the dashed arrow indicates how the individual arrows in $\mathbb{C}$ and $\mathbb{D}$ combine to make arrows in $\mathbb{C} \times \mathbb{D}$. Personally I find this diagram more compelling than the formula.

$\mathbb{C}$	$\mathbb{D}$		$\mathbb{C} \times \mathbb{D}$
c_1	d_1		(c_1, d_1)
$\downarrow f_1$	$\downarrow g_1$		$\downarrow (f_1, g_1)$
c_2	d_2	⬛⬛⬛➡	(c_2, d_2)
$\downarrow f_2$	$\downarrow g_2$		$\downarrow (f_2, g_2)$
c_3	d_3		(c_3, d_3)

[†] In case you want to look it up: this all follows from general theorems about functors that are part of an adjunction, such as free and forgetful functors.

But anyway here's the formula: $(f_2, g_2) \circ (f_1, g_1) = (f_2 \circ f_1, \; g_2 \circ g_1)$.

Identities are also "componentwise", which means $1_{(c,d)} = (1_c, 1_d)$.

T 21.8   The following formula is a useful result about product categories. Try turning it into a diagram in the spirit of the one above, making it rather more obvious why it's true: $(f, 1) \circ (1, g) = (1, g) \circ (f, 1)$.

	LHS					RHS

As diagrams, the left-hand side and right-hand side of the formula are as shown, and they are both equal to the middle diagram by definition of identities in $\mathbb{C}$ and $\mathbb{D}$.

$$
\begin{array}{cc}
c_1 & d_1 \\
\downarrow 1 & \downarrow g \\
c_1 & d_2 \\
\downarrow f & \downarrow 1 \\
c_2 & d_2
\end{array}
\quad = \quad
\begin{array}{cc}
c_1 & d_1 \\
f \downarrow & \downarrow g \\
c_2 & d_2
\end{array}
\quad = \quad
\begin{array}{cc}
c_1 & d_1 \\
\downarrow f & \downarrow 1 \\
c_2 & d_1 \\
\downarrow 1 & \downarrow g \\
c_2 & d_2
\end{array}
$$

This may look innocuous but it's actually quite profound and is related to many deep structures including what is called "interchange" laws at higher dimensions.

We still need to prove that $\mathbb{C} \times \mathbb{D}$ is a categorical product, by exhibiting its universal property. As usual we consider a diagram as shown on the right, and find a unique factorization K. Here P and Q are projection functors.

The substance of the proof is really no different from the proof in **Set**, it's just that we have to do it at the level of morphisms as well as objects, and make sure that functoriality holds. I'll just sketch it now.

In order to be a factorization the putative functor K must be defined as follows:

- on objects: $Ka = (Fa, Ga)$
- on morphisms: $Kf = (Ff, Gf)$ (which does type-check[†]).

Functoriality follows from functoriality of F and G individually, and the factorization is unique by construction. This proof is abstractly very similar to the proofs we saw for products of posets (expressed as categories with at most one morphism between any two objects) and of monoids (expressed as categories

---

[†] Remember: this means checking that the symbols do at least live in the right "homes". In this case if we start with $a \xrightarrow{f} b$ then Kf is supposed to be a morphism $Ka \longrightarrow Kb$ in $\mathbb{C} \times \mathbb{D}$, and that's what we can type-check.

with only one object). This similarity is a sign that there is some further abstract explanation of things, but it's a little beyond the scope of this book.[†] But more specifically we can observe:

- Products of posets in the category **Pst** correspond to products of posets expressed as categories, in **Cat**.

- Products of monoids in the category **Mnd** correspond to products of monoids expressed as categories, in **Cat**.

This can be expressed abstractly by saying that the functors **Pst** $\longrightarrow$ **Cat** and **Mnd** $\longrightarrow$ **Cat** preserve products.

For coproducts of categories we do something very similar: take the disjoint union of the objects and of the morphisms.

In pictures we get things like the category shown on the right, with two separate disconnected components. This example is a coproduct of the top part and the bottom part; it looks like two separate categories, which is sort of the point.

The underlying graph of a coproduct $\mathbb{C} \sqcup \mathbb{D}$ will then be as shown on the right.

$$C_1 \sqcup D_1 \underset{t \sqcup t}{\overset{s \sqcup s}{\rightrightarrows}} C_0 \sqcup D_0$$

(The notation $s \sqcup s$ means that the C_1 part just maps to the C_0 part by s as before, and the D_1 part just maps to the D_0 part by s as well. Likewise for $t \sqcup t$.) So we have the disjoint union of objects and the disjoint union of morphisms, with no morphisms between objects of $\mathbb{C}$ and objects of $\mathbb{D}$. Composition and identities are then those of $\mathbb{C}$ and $\mathbb{D}$; there is no interaction between the $\mathbb{C}$ part and the $\mathbb{D}$ part so there's nothing subtle about composition.

Coproducts of categories are thus much easier than coproducts of monoids, even though monoids are one-object categories. If you're wondering why coproducts of categories are easier, that's a very good question.

─────────── **Things To Think About** ───────────

T 21.9   See if you can understand why coproducts of monoids are not the same as coproducts of one-object categories. There are two flavors of answer to this question: the idea and the formalism.

The basic idea here is that the coproduct of two one-object categories has two objects, not one object, and this means that no arrows become composable

---

[†] In case you want to look it up, it's to do with these structures all being algebras for monads. This book will cover all the technical background needed to understand that, but will not actually cover that.

that weren't already composable. Whereas in the coproduct of two monoids all the elements can be combined using the binary product; essentially it's like taking the coproduct of one-object categories and then forcing the two single objects to be the same, making everything composable.

21.4 Isomorphisms of categories

Now that we have a category of (small) categories and functors we can try thinking about isomorphisms inside that category. By definition, an isomorphism in **Cat** must be a functor with an inverse functor, but the question is what that *means*: how we might characterize that directly, and what relationship it gives us between isomorphic categories.

──────────── **Things To Think About** ────────────

T 21.10   Can you work out a way to characterize an invertible morphism in **Cat**? Think about the fact that functors are a generalization of functions, operating at the level of objects and morphisms, and that invertible functions are those that are both injective and surjective. It might help to think about the underlying graphs.

Let us think about a pair of functors as shown on the right, where F and G are inverses.

If we think what this does on the underlying graphs of $\mathbb{C}$ and $\mathbb{D}$ we get the diagram of sets and functions shown here. Note that I have drawn the underlying graphs vertically to emphasize the relationship with the diagram above.

Now, for F and G to be inverses they must compose to the identity functor both ways round. We know that the identity functor acts as the identity *function* on both objects and morphisms, and that composition of functors happens by composition of the functions at the level of objects and morphisms. So F_1 and G_1 must compose to the identity function both ways round, and similarly F_0 and G_0. This says we must have a bijection at the level of objects and a bijection at the level of morphisms, as well as the usual functoriality.

In practice this means that the categories $\mathbb{C}$ and $\mathbb{D}$ have the exact same structure, just with different names on the objects and morphisms. This is just like when we said an isomorphism of groups shows that two groups have the same pattern in their multiplication table, just with different names for elements.

For example the categories of factors
shown on the right are isomorphic.

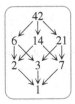

The bijection works on objects as shown on the right. In
this case once we've determined the action on objects
the action on morphisms follows: there is at most one
morphism between any two objects, so no choice about
where it goes.

1	⟷	1
2	⟷	2
3	⟷	3
5	⟷	7
6	⟷	6
10	⟷	14
15	⟷	21
30	⟷	42

--- **Things To Think About** ---

T 21.11 In fact these categories are not *uniquely* isomorphic.

1. What other isomorphisms are there? That is, how
   else could we match up the objects?
2. Why can't we use the bijection on objects shown
   on the right, where the numbers go in size order on
   both sides?

1	⟷	1
2	⟷	2
3	⟷	3
5	⟷	6
6	⟷	7
10	⟷	14
15	⟷	21
30	⟷	42

Note that there are two types of answer to both of
these questions: the technical one, where we simply
give the formal details of the answer, and the "moral"
one, where we give the deep explanation of the answer.

One way of thinking about this is to think about the symmetry in the actual
cube diagrams. We know that the top and bottom vertices play special roles,
but if we keep those cubes standing on one corner we can spin them, and
each set of three vertices that are on the same level will exchange positions.
This symmetry is a manifestation of the fact that the three prime factors play
analogous roles here, so as long as we map the primes to the primes, and then
the products to the analogous products, the structure will be preserved.

A more abstract approach is to observe that this cube is a product of three
copies of the "quintessential arrow" category, one for each distinct prime. If
we call that category I (for interval) then we are looking at $I \times I \times I$. From the
symmetry of products we can deduce that there is an isomorphism permuting
the copies of I in any way we like, which gives an isomorphism between these
categories sending primes to primes.

This shows that saying that two categories "are isomorphic" is different from actually specifying an isomorphism between them. In the first case we are just saying that it is possible to match the structures up, and in the second case we are specifying a particular way of matching them up. That difference, as a general idea, becomes more and more important in higher dimensions.

This more or less answers the second part of the thought: in the bijection matching everything up in size order, the prime 5 on the left is mapped to a non-prime on the right (and the prime 7 on the right is mapped to a non-prime on the left) which is "morally" why this won't work as an isomorphism of categories: it doesn't map primes to primes.

The technical reason it won't work is that we will run into trouble when we try to make the functor act on morphisms: some of the morphisms won't be able to go anywhere. If we define $F(2) = 2$ and $F(6) = 7$ then for example we will have a problem with the morphism $2 \longrightarrow 6$. If F were a functor, this morphism would need to be mapped to a morphism $2 \longrightarrow 7$ on the right, but there isn't one. So this can't be a functor, let alone an isomorphism of categories.

Analogously we have an isomorphism between the category of privilege involving "rich, white, male" and the category of women according to the three types of privilege "rich, white, cisgender"; these categories are depicted on the right (with abbreviations).

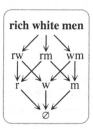 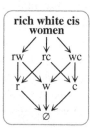

As with the factors of 30 and 42, different versions of this isomorphism are possible with different possible bijections on objects. This tells us which objects play analogous roles and which do not in these contexts. In particular we see that rich white cis women play an analogous role among women to the role that rich white men play in broader society.

As in the situation with factors, these cuboid categories are products of three instances of the single-arrow category shown on the right.

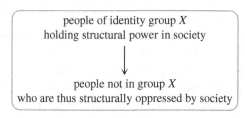

Thus the isomorphism between the cube categories comes down to the fact that we are starting from much simpler isomorphic categories.

In the first case we have these three "single arrow" categories.

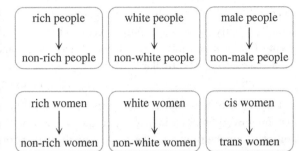

In the second case we have these three categories, which are isomorphic to the previous ones.

As a statement of category theory, the existence of the above isomorphisms of categories is not at all profound or subtle. The difficulty and subtlety about privilege is in how it manifests itself in life, not how the abstract structures are constructed. I just think that understanding abstract structures can help us focus on the subtlety in the right place rather than in an irrelevant place.

One subtlety is that arrows give us a much stronger correspondence between structures than if we're just thinking about sets. I think that when we see some arrow diagrams as the same shape, our eyes are making a bijection visually, and they're making a bijection on objects that respects the arrow structure.

─────────────── **Things To Think About** ───────────────

T 21.12   I like to point out that there is no isomorphism between categories pointing in the directions shown on the right. This is not a rigorous statement. Can you interpret it rigorously and prove it?

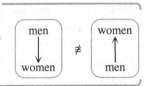

The rigorous statement is that the bijection on objects shown on the right cannot be extended to an isomorphism of categories.

The reason is that, as with when we tried to map 6 to 7 in the lattices of factors, there will be nowhere for the morphism between men and women to go.

This is an abstract explanation of the disagreement that happens when some people complain about the prevalance of men sexually harassing women, and some other people retort "Women do it to men too". Women do do it to men too, but there is a structural difference when it's a group that holds structural power in society harassing a group that does not hold structural power in society, as opposed to the other way round. Those who refuse to acknowledge this difference are looking at an isomorphism of sets as in the bijection above, and those who acknowledge the difference are looking at the impossibility of the isomorphism of categories based on that bijection, even if most people don't express it in quite that way.

We can argue about *how* the structural power difference manifests itself in the experiences of men and women being harassed, and we can argue about whether men really do hold structural power in society (for which there is so much evidence that you have to be quite unreasonable to deny it, or perhaps not understanding what "structural power" means), but we really can't validly argue about isomorphisms between such simple categories, at the abstract level.

One thing that we can and should call into question is whether this is the "right" notion of sameness for categories. We have seen that isomorphism is the "right" notion of sameness for objects *inside* a category, and that we therefore try not to invoke equalities between objects inside a category, rather than isomorphisms. However, now we are thinking about sameness for categories themselves. It turns out that in the definition of an isomorphism of categories we have invoked equalities between objects inside those categories, showing that the definition is "too strict".

Things To Think About

T 21.13   Where in the definition of invertible functor are we invoking an equality between objects?

In order to say that a functor $F \colon \mathcal{C} \longrightarrow \mathcal{D}$ is invertible it needs to have an inverse G, which means $GF = 1_{\mathcal{C}}$ and $FG = 1_{\mathcal{D}}$. Now we unravel what this means: for two functors to be equal they have to have the same action on objects and on morphisms.

So for the first equation, on objects we're saying $\boxed{\forall c \in \mathcal{C},\ GFc = c.}$ We get something similar for the second equation. So we are invoking equalities on objects: uh-oh. This is a hint that the notion of sameness for categories "wants" to be a little weaker than this, where we only invoke isomorphisms between objects in the category and not equalities. It is acceptable to invoke equalities between morphisms in $\mathcal{C}$ and $\mathcal{D}$ because we don't have a notion of isomorphism between morphisms.[†]

So the idea is that a better notion of sameness for categories would be via a functor with the following properties.[‡]

- On morphisms: strictly invertible, so $\boxed{\forall f \in \mathcal{C},\ GFf = f}$ and similarly for the composite FG.
- On objects: only invertible up to isomorphism, so $\boxed{\forall c \in \mathcal{C},\ GFc \cong c}$ and similarly for the composite FG.

We will think about the part on morphisms first, as that is more straightforward. It brings us to the concept of full and faithful functors.

[†] This requires higher-dimensional categories, as we'll see.

[‡] Notation reminders: ∀ means "for all", ∈ means "in", ≅ means "is isomorphic to".

21.5 Full and faithful functors

If we are dealing with a functor that is not strictly a bijection on objects we need to be careful what "bijection on morphisms" means. We're going to warm up by thinking about the "quintessential arrow" category, but adding in some isomorphic copies of the two objects. Categorical instinct says that this new category isn't really different. First we need to make these thoughts precise.

We consider a category with two objects and one morphism between them, as shown on the right.

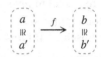

Now we'll try to make a version of this category which is essentially the same, but has two isomorphic versions of a, and two isomorphic versions of b. (The symbol ⫙ is an isomorphism sign $\cong$ on its side.)

─── **Things To Think About** ───

T 21.14   Can you turn this vague idea into something rigorous, by working out what all the arrows should be to make sense of this idea? We need various arrows from a and a' to b and b'.

The idea is that the morphism f has one manifestation from each isomorphic copy of a to each isomorphic copy of b, but they don't really count as "different", just as the isomorphic objects don't really count as "different".

If we continue to represent the isomorphisms by the squiggles then the rest of the arrows are four versions of f as shown on the right (not drawing identities).

This category should count as "the same" as the first one, as we've just included some more isomorphic objects. It just sort of fattens up the category a bit. We'll now try to make that precise. We are going to investigate the following functor which "collapses" the fattened category back to the original lean one.

Isomorphic objects are mapped to the same place, and all the versions of f are mapped to the same place.

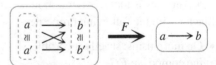

─── **Things To Think About** ───

T 21.15   Can you make precise a sense in which F is a "bijection up to isomorphism" on objects? Also can you see that it is not a bijection on morphisms, but somehow gives a perfect correspondence on morphisms in the context?

We could say F is a "bijection up to isomorphism" on objects, in the sense that objects are only mapped to the same place if they were already isomorphic.

However it is not a bijection on morphisms as four morphisms are sent to the same place, so it is not injective on morphisms.

However, it *is* a bijection on morphisms if we look at individual homsets.[†] By definition, a functor $F: \mathbb{C} \to \mathbb{D}$ gives functions on homsets as follows: given any pair of objects $x, y \in \mathbb{C}$, we have a function $\mathbb{C}(x, y) \to \mathbb{D}(Fx, Fy)$. It is these functions that are all bijections in our example, because once we restrict our attention to a particular pair of objects on the left, we don't see the "extra" copies of the morphism f any more. If you're not sure about this it's worth trying it out for different pairs of objects on the left until you see the point. The overall idea is that this functor is really encapsulating the sense in which these two categories are not *really* different.

The important conclusion here is that we should not look at the set of morphisms overall, but we should look at individual homsets. This comes down to the difference between the two definitions of category we've seen:

- By homsets: for every pair of objects a, b a set of morphisms $\mathbb{C}(a, b)$.
- By total set of morphisms: a set C_1 of morphisms equipped with source and target functions to the set C_0 of objects.

It turns out that the first definition guides us better for thinking about the correct notion of "bijection on morphisms". For a functor $F: \mathbb{C} \to \mathbb{D}$ we need to think about each function $\mathbb{C}(a, b) \to \mathbb{D}(Fa, Fb)$ individually, rather than thinking about the "total" function $C_1 \to D_1$.

Now that we know to think about bijections on homsets, we can unravel this a little further, and break down the concept of bijection into injection and surjection. It turns out that thinking about injectivity and surjectivity separately is fruitful here, so we give those concepts some names.

The following definition piles up notation and terminology. We'll try to remain calm and unravel it afterwards.

Definition 21.3 Let $F: \mathcal{C} \to \mathcal{D}$ be a functor.

- F is called *faithful* if for all $x, y \in \mathcal{C}$ the function $\mathcal{C}(a, b) \to \mathcal{D}(Fa, Fb)$ is injective. We also sometimes call this *locally injective*.

- F is called *full* if for all $x, y \in \mathcal{C}$ the function $\mathcal{C}(a, b) \to \mathcal{D}(Fa, Fb)$ is surjective. We also sometimes call this *locally surjective*.

Note that "locally" in abstract math typically means that we're zooming in to look at a situation close-up, and only paying attention to one small region at a time. In category theory it usually means we're doing something on homsets.

[†] That is, set of morphisms between a fixed pair of objects.

If for all x, y the function $\mathcal{C}(a, b) \longrightarrow \mathcal{D}(Fa, Fb)$ is a bijection then we say F is full and faithful.[†]

Things To Think About

T 21.16   Can you unravel the definitions of injective and surjective in the above context to gain a closer understanding of what full and faithful mean?

In practice we often unravel the definitions as follows.

- Full: $\forall \, Fx \xrightarrow{h} Fy \in \mathcal{D}, \ \exists \, x \xrightarrow{f} y \in \mathcal{C}$ such that $Ff = h$.
- Faithful: $\forall \, f, g : x \longrightarrow y \in \mathcal{C}, \ Ff = Fg \implies f = g$.
- Full and faithful: $\forall \, Fx \xrightarrow{h} Fy \in \mathcal{D}, \ \exists ! \, x \xrightarrow{f} y \in \mathcal{C}$ such that $Ff = h$.

Things To Think About

T 21.17   Are the following functors full and faithful? (The diagrams below are not rigorous definitions, but I invite you to make the "obvious" definitions.)

1.

$$
\begin{array}{c}
\mathcal{C} \\
\left[\begin{array}{c} a \to b \\ a' \to b' \end{array}\right]
\end{array}
\xrightarrow{F}
\begin{array}{c}
\mathbb{D} \\
\left[\, d \rightrightarrows e \,\right]
\end{array}
$$

2.

$$
\begin{array}{c}
\mathcal{C} \\
\left[\begin{array}{c} a \to b \\ a' \to b' \end{array}\right]
\end{array}
\xrightarrow{G}
\begin{array}{c}
\mathbb{D} \\
\left[\, d \longrightarrow e \,\right]
\end{array}
$$

My intention for the first functor is that the two morphisms on the left are sent to the two distinct morphisms on the right. So the functor is bijective on morphisms but not locally: it fails to be locally surjective. Intuitively, the homset $\mathbb{D}(d, e)$ on the right has two morphisms but the homsets on the left only have one morphism each, so local surjectivity is doomed to fail. Formally we could say $Fa = d$ and $Fb = e$ so the function $\boxed{\mathbb{C}(a, b) \longrightarrow \mathbb{D}(Fa, Fb) = \mathbb{D}(d, e)}$ is not surjective.

The second functor is not full and faithful either but the situation is more subtle. This time all the obvious homsets have one morphism each so we need to look more closely. The problem is that some completely unrelated objects on the left end up connected on the right. For example we can consider the function $\boxed{\mathbb{C}(a, b') \longrightarrow \mathbb{D}(Fa, Fb') = \mathbb{D}(d, e).}$ The homset on the left is empty but the one on the right has one element, so this function can't be surjective.

In fact a clue to this is that the functor fails to be "injective on objects", even up to isomorphism: $Fa = Fa'$ but a and a' are not even isomorphic (so

---

[†] Some people call it "fully faithful" but others of us find it odd to make it sound as if the full part is modifying the faithful part. Those really are two separate properties. It would be like calling a bijective function "surjectively injective". Or, as John Baez put it "If I cheat on my wife that's a separate issue from the fact that I ate a big dinner". I can't find a citation for this so I suspect I heard him say it in person.

we could have used the homset $\mathbb{C}(a, a')$ to show that the functor is not locally surjective). We will come back to this idea.

Recall in Section 20.5 we mentioned the idea of reflecting structure, and showed that all functors preserve isomorphisms but functors do not in general reflect isomorphisms.

--------------------- **Things To Think About** ---------------------

T 21.18   Show that a full and faithful functor reflects isomorphisms. As usual the first step is to write down rigorously what this means.

Proposition 21.4   *Let* $F \colon \mathcal{C} \to \mathcal{D}$ *be a full and faithful functor and* $a \xrightarrow{f} b$ *be a morphism in* $\mathcal{C}$. *Then* f *is an isomorphism iff* Ff *is an isomorphism.*

Note that this is "preserves and reflects" in one statement; the "reflects" part is the "if" part, that is: if Ff is an isomorphism then f is an isomorphism. So we'll just prove that part here, as we've done "preserves" before.

Proof   Suppose $Fa \xrightarrow{Ff} Fb$ is an isomorphism in $\mathcal{D}$, with inverse $Fb \xrightarrow{h} Fa$. We aim to find an inverse for f in $\mathcal{C}$.

> The idea is that morphisms $Fb \to Fa$ in $\mathcal{D}$ precisely correspond to morphisms $b \to a$ in $\mathcal{C}$ since F is full and faithful, and so h must have come from an inverse in $\mathcal{C}$ under that correspondence.

We know F is full and faithful, so there is a unique morphism $b \xrightarrow{g} a \in \mathcal{C}$ such that $Fg = h$. We claim that g is an inverse for f.

> We need to compose them and show that the result is the identity in $\mathcal{C}$. If we apply F to the composite we get the identity in $\mathcal{D}$; then we show that F being faithful means that the only thing that can map to the identity is the identity.

First consider the composite $g \circ f$. We know that

$$
\begin{aligned}
F(g \circ f) &= Fg \circ Ff && \text{by functoriality} \\
&= h \circ Ff && \text{since we defined } g \text{ by } Fg = h \\
&= 1_{Fa} && \text{since we defined } h \text{ as the inverse of } Ff \\
&= F1_a && \text{by functoriality.}
\end{aligned}
$$

Since F is faithful, we must have[†] $g \circ f = 1_a$ and similarly we can show $f \circ g = 1_b$. So g is an inverse for f and F reflects isomorphisms as claimed.   □

We will come back to this when we think about "essentially injective", and also when we think about the Yoneda embedding.

[†] This follows from the definition of injectivity.

T 21.19   Earlier on we said that if we express posets as categories then the functors between them are precisely the order-preserving functions. Likewise if we express monoids as one-object categories then the functors between them are precisely the monoid homomorphisms. What is this saying about the "inclusion" functors $\mathbf{Pst} \xrightarrow{K} \mathbf{Cat}$ and $\mathbf{Mnd} \xrightarrow{\Sigma} \mathbf{Cat}$?

The formal version of these statements is that the functors K and Σ are full and faithful. Consider posets A and B. We are saying that the functors $KA \to KB$ are *precisely* the order-preserving functions $A \to B$, which says that this function on homsets is a bijection $\boxed{\mathbf{Pst}(A, B) \to \mathbf{Cat}(KA, KB),}$ that is, K is full and faithful. Likewise for Σ and monoids.

The notion of "full and faithful" has given us a good notion of a functor being an "isomorphism at the level of morphisms". For the level of objects we mentioned above the idea of a "bijection up to isomorphism". The idea is to modify the definition of a bijection. Recall that a function $f \colon A \to B$ is a bijection if it is injective and surjective, that is:

1. Injective: $f(a) = f(a') \implies a = a'$.
2. Surjective: $\forall\, b \in B,\ \exists\, a \in A$ such that $f(a) = b$.

Now if we are in a category these definitions are too strict for the level of objects as they have equalities all over them. We make the following definition instead. You might wonder why we've mysteriously ignored the "injectivity" part; we'll come back to that question shortly.

Definition 21.5   We say that a functor $F \colon \mathcal{C} \to \mathcal{D}$ is *essentially surjective* (on objects) if $\boxed{\forall\, d \in \mathcal{D},\ \exists\, c \in \mathcal{C}\ \text{such that}\ f(c) \cong d.}$

This is an example of a key principle in "categorification", where we take an axiom and replace all the equalities between objects by isomorphisms. We then have this definition of "bijective up to isomorphism" for a functor:

Definition 21.6   A functor $F \colon \mathcal{C} \to \mathcal{D}$ is a *pointwise equivalence* if it is full and faithful, and essentially surjective on objects.

T 21.20   Why don't we need something like "essentially injective on objects"? That is, can you write down what this might mean and show that it is satisfied by any full and faithful functor.

We can make up a definition of "essentially injective on objects" by replacing the equalities by isomorphisms in the definition of injective. This would give something like this: $\boxed{\forall\, c, c' \in \mathcal{C},\ Fc \cong Fc' \implies c \cong c'.}$

But this is just a slightly weak version of "reflects isomorphisms". It is weaker because it just talks about objects being *isomorphic* rather than about specific isomorphisms. In any case it certainly follows from the fact that full and faithful functors reflect isomorphisms.

Proposition 21.7 *Suppose $F: \mathcal{C} \to \mathcal{D}$ is full and faithful. Then F is "essentially injective on objects" as in the above definition.*

Proof Suppose we have an isomorphism $Fc \xrightarrow{h} Fc'$ in $\mathcal{D}$. Since F is full we know that $h = Ff$ for some $c \xrightarrow{f} c'$ in $\mathcal{C}$. But full and faithful functors reflect isomorphisms so f must be an isomorphism, so $c \cong c'$ as required.   □

Note that this result is not very good, because of its vagueness over "isomorphic" objects. The result about F reflecting isomorphisms is the morally correct statement. But in any case we see that "essentially injective on objects" actually follows from "locally bijective on morphisms". This is a very interesting phenomenon meaning that in higher dimensions we don't need to invoke injectivity at every dimension, only (a suitably weak notion of) surjectivity, as injectivity will always follow from the dimension above if there is one.

It is crucial to note that the definition of pointwise equivalence is a weakened version of bijection, not a higher-dimensional version of isomorphism. That is, it uses elements all over the place and so is not categorical, in the sense that we could not immediately place it in other categories. For a categorical version we need some higher-dimensional structure in our category of categories. We have taken equations between objects and weakened them, but what we really need to do to weaken the notion of invertible functor is take the equations in the definition of isomorphism and weaken those.

This would give a pair of functors as shown on the right, satisfying something like $GF \cong 1_{\mathcal{C}}$ and $FG \cong 1_{\mathcal{D}}$.

$$\mathcal{C} \underset{G}{\overset{F}{\rightleftarrows}} \mathcal{D}$$

However, for this we need a notion of (iso)morphism between functors, which is one dimension up from what we have now, in a sense that we are about to investigate. If we could do it, it would give us a *categorical* definition of "weak isomorphism" for functors, that is, one that we can then take into any suitably higher-dimensional category. (We just need one more dimension, so that would be a 2-category.)

There are many situations in math where the full higher-dimensional situation is impractical, unwieldy or just unpalatable, so we try and make do with lower-dimensional structures that somehow encapsulate some features of the higher-dimensional situation. That is what pointwise equivalence does, and it is very useful, but we will look more at the true higher-dimensional structures next.

22

Natural transformations

An appropriate notion of relationship between functors, giving us our first glimpse of two-dimensional structures.

We're going to develop a definition of "sensible relationships between functors", called natural transformations. We will start by a process of gut abstract feeling. This is a key example of how abstraction proceeds once you've developed some abstract intuitions — you can try more or less following your nose. This is when definitions don't need to be memorized, because they *make sense*, just like I generally hope that if I operate on the basis of respecting other humans, not hurting anyone, and being kind and helpful, then I will not break any laws, even though I'm not an expert on the law.

On the other hand, definitions of abstract structures are sometimes made by generalizing specific examples; however if you've never encountered the specific examples then that's not much motivation. Worse, if the definition *only* comes from generalizing a specific example, it has a danger of not making abstract sense, but just being somewhat utilitarian. I will present that approach in the next section, in case it helps.

22.1 Definition by abstract feeling

We start from the principle that the only possible notion of relationship between functions is equality, whereas between functors equality is too strict.

We are going to think about possible relationships between two functions f and g as shown on the right.

Functions are defined on elements, and elements of sets do not (in general) have any notion of relationship between them except equality. Thus the only notion of relationship we can have between functions (in general) is also equality. It is defined element-wise: two functions are called equal if their action on each element produces the same element as a result.

Formally, we say $f = g$ whenever: $\boxed{\forall x \in A, \ f(x) = g(x).}$

Now let us move to considering relationships between two *functors F* and *G* as shown on the right.

$$A \underset{G}{\overset{F}{\rightrightarrows}} B$$

> Note that there is absolutely no technical reason to change the notation at this point; I'm just doing it to remind us that we've gone up a dimension. I sometimes try and write elements in lower case, sets in upper case, and categories in upper case blackboard bold or curly, but it can be tedious keeping those conventions going.

Now that we are working with categories instead of sets, we do have a notion of relationship between elements other than equality: we have morphisms.

So we can replace this form of relationship: $\forall x \in A, \quad f(x) = g(x)$

with this form of relationship: $\forall x \in A, \ F(x) \longrightarrow G(x)$

Note that where previously we were just talking about whether or not two functions were equal, we are now looking at a system of morphisms in B. So we are looking at "senses in which" F and G are related.

The system of morphisms we're looking at has one morphism in B for each object x of A; there remains a question of what we do about morphisms in A. We know they are mapped to morphisms in B, so how do they relate to this system of morphisms in B that we've just come up with?

The action of F on morphisms gives us this: $\forall x \overset{f}{\rightarrow} y \in A, \quad Fx \overset{Ff}{\rightarrow} Fy$

This gives us two paths from Fx to Gy: the two paths around the square shown on the right. (The horizontal morphisms are the ones giving us the relationship between F and G.)

$$\begin{array}{ccc} Fx & \longrightarrow & Gx \\ {\scriptstyle Ff}\downarrow & & \downarrow{\scriptstyle Gf} \\ Fy & \longrightarrow & Gy \end{array}$$

In abstract structures, whenever we have two different ways of doing something, we probably want them to be related. Here we're looking at two (composite) morphisms $Fx \longrightarrow Gy$ in a category, so the only way for them to be related is by an equality. So we ask for the square to commute. This gives us the idea behind how we relate functors, as in the following definition.

Definition 22.1 Given functors $F, G: A \longrightarrow B$, a *natural transformation* α from F to G is depicted as shown on the right, or as $\alpha: F \Longrightarrow G$, and is given by:

$$A \underset{G}{\overset{F}{\Downarrow\alpha}} B$$

- for all objects $x \in A$ a morphism $Fx \overset{\alpha_x}{\rightarrow} Gx \in B$, such that

- for all morphisms $x \overset{f}{\rightarrow} y \in A$ the following square called the *naturality square* commutes:

$$\begin{array}{ccc} Fx & \overset{\alpha_x}{\longrightarrow} & Gx \\ {\scriptstyle Ff}\downarrow & & \downarrow{\scriptstyle Gf} \\ Fy & \underset{\alpha_y}{\longrightarrow} & Gy \end{array}$$

The morphism α_x is called the *component* of α at x.

We think of natural transformations as 2-dimensional, like a "surface" connecting the "paths" made by the functors in the diagram.

0-dimensional   categories   ·

1-dimensional   functors   ·———→·

2-dimensional   natural transformations

─────────── Aside on the word "natural" ───────────

Eilenberg and Mac Lane observed that the point of defining categories was to define functors, and the point of defining functors is to define natural transformations.[†] The word "natural" here is both formal and evocative. It means something completely rigorous: that the naturality squares commute. But there is a certain feeling we can develop, that when some transformation feels "natural" in abstract mathematics it is a sign that there is a natural transformation at work somewhere, and, specifically, the naturality will express the real content.

22.2 Aside on homotopies

If you have seen the formal definition of homotopy then the following comparison may interest you; if not, feel free to skip it. (I am not going to *explain* the formality of homotopy here, as that is beyond our scope.)

We will start with continuous maps $f, g \colon X \to Y$, and consider a homotopy $\alpha \colon f \Rightarrow g$. Recall that we use the unit interval $I = [0, 1]$ and that we can think of this as a unit of "time", during which we're deforming f into g.

Then α is given by a continuous map $\boxed{\alpha \colon I \times X \to Y}$ such that

- $\alpha(0, x) = f(x)$ ("at time 0 we are f"), and
- $\alpha(1, x) = g(x)$ ("at time 1 we are g").

We can try something similar with categories. Instead of a unit interval, we take I to be a "quintessential arrow" category, which we might write as $\boxed{0 \xrightarrow{h} 1}$.

Then, given functors $F, G \colon \mathcal{A} \to \mathcal{B}$ we could try defining a natural transformation $\alpha \colon F \Rightarrow G$ by analogy with the definition of homotopy. This would be a functor $\boxed{\alpha \colon I \times \mathcal{A} \to \mathcal{B}}$ such that $\alpha(0, x) = Fx$ and $\alpha(1, x) = Gx$. In fact this gives us exactly the definition of natural transformation, just expressed rather differently.

─────────────

[†] See Mac Lane, *Categories for the Working Mathematician*, Section I.4.

You might like to check that

- the components α_x are given by $F(h, 1_x) : (0, x) \rightarrow (1, x)$, and

- the naturality squares are the images under F of these commutative diagrams in $I \times \mathcal{A}$:

$$
\begin{array}{ccc}
(0, x) & \xrightarrow{(h, 1_x)} & (1, x) \\
{\scriptstyle (1, f)} \downarrow & & \downarrow {\scriptstyle (1, f)} \\
(0, y) & \xrightarrow[(h, 1_y)]{} & (1, y)
\end{array}
$$

I included this for interest; we won't use it any further.

22.3 Shape

The way we have drawn natural transformations makes them 2-dimensional, with this particular shape.

This shape is a two-sided polygon, coming from the fact that F and G have the same source and target. You might wonder why they have to have the same target. The answer is that it is a choice we make because it's convenient. When we go into higher dimensions we will see that there are various choices we can make about the "shape" of higher-dimensional morphisms, because once we are in more than one dimension, things can have different shapes (whereas in 0 and 1 dimensions there's really no choice). We say that F and G have to be *parallel*, which means they have the same endpoints. The shape of α is then called *globular*.

If instead we wanted to compare *any* two functors we might look at two functors as shown on the right.

$$
\mathcal{A} \xrightarrow{F} \mathcal{B}
$$
$$
\mathcal{C} \xrightarrow{G} \mathcal{D}
$$

This comparison would be a bit futile without some functors going down the sides and so we'd end up needing some "vertical" functors, and then a natural transformation in the middle, as shown here.

$$
\begin{array}{ccc}
\mathcal{A} & \xrightarrow{F} & \mathcal{B} \\
{\scriptstyle P} \downarrow & {\scriptstyle \alpha} \Swarrow & \downarrow {\scriptstyle Q} \\
\mathcal{C} & \xrightarrow[G]{} & \mathcal{D}
\end{array}
$$

These square shapes give a different foundation for higher-dimensional thinking, called "cubical". However, in many situations we define the cubical natural transformations using the definition of globular ones; for example the one above is just a (globular) natural transformation $\alpha : QF \Rightarrow GP$. So usually we might as well stick to the globular definition.

However, in some higher-dimensional situations the cubical approach turns out to be much more naturally arising, and sometimes more productive as well. We will see a glimpse of that in Chapter 24.

22.4 Functor categories

One of the principles of category theory is that if something is interesting we think about other examples of it, and morphisms between them, and assemble all that into a category. We have assembled categories into a category using functors as the morphisms, but we can now assemble functors themselves into a category, using natural transformations as the morphisms between functors.

Definition 22.2 Given categories $\mathcal{C}$, $\mathcal{D}$, the *functor category* $[\mathcal{C}, \mathcal{D}]$ has
- objects: functors $\mathcal{C} \longrightarrow \mathcal{D}$
- morphisms: a morphism $F \longrightarrow G$ is a natural transformation.

Identities and composition for natural transformations are componentwise.

Things To Think About

T 22.1   Can you work out the "componentwise" definition of identities and composition? Remember, "componentwise" means they operate component by component. What do we have to check to make sure this makes sense?

Definition 22.3 Given a functor $F\colon \mathcal{C} \longrightarrow \mathcal{D}$ the *identity natural transformation* $1_F\colon F \Longrightarrow F$ has all identity components.

We should just do a little type-checking: we need a component for each $x \in \mathcal{C}$, and the definition is saying that the components are $\boxed{Fx \xrightarrow{1_{Fx}} Fx.}$

The naturality squares are all trivial: for any $x \xrightarrow{f} y \in \mathcal{C}$ we have the naturality square on the right.

$$\begin{array}{ccc} Fx & \xrightarrow{1_{Fx}} & Fx \\ {\scriptstyle Ff}\downarrow & & \downarrow{\scriptstyle Ff} \\ Fy & \xrightarrow{1_{Fy}} & Fy \end{array}$$

For composition there is a little more to check.

Definition 22.4 Consider functors and natural transformations as shown on the right. The composite natural transformation $\beta \circ \alpha\colon F \Longrightarrow H$ is defined by the following components: for each $x \in \mathcal{C}$, $\boxed{(\beta \circ \alpha)_x = \beta_x \circ \alpha_x.}$

That is: $Fx \xrightarrow{(\beta\circ\alpha)_x} Hx = Fx \xrightarrow{\alpha_x} Gx \xrightarrow{\beta_x} Hx.$

This is also called *vertical composition*, as it looks vertical in the 2-dimensional diagram (and there is a horizontal version which we'll see later).

We still need to check naturality: given any morphism $x \xrightarrow{f} y \in \mathcal{C}$ we need to check the square on the right.

$$\begin{array}{ccc} Fx & \xrightarrow{(\beta\,\circ\,\alpha)_x} & Hx \\ {\scriptstyle Ff}\downarrow & & \downarrow{\scriptstyle Hf} \\ Fy & \xrightarrow{(\beta\,\circ\,\alpha)_y} & Hy \end{array}$$

We can "fill this in" with the squares shown here, which commute by naturality of α and β respectively.

$$\begin{array}{ccc} Fx \xrightarrow{\alpha_x} Gx \xrightarrow{\beta_x} Hx \\ Ff \downarrow \quad \quad Gf \downarrow \quad \quad Hf \downarrow \\ Fy \xrightarrow[\alpha_y]{} Gy \xrightarrow[\beta_y]{} Hy \end{array}$$

Thus the naturality of $\beta \circ \alpha$ follows from the naturality of α and β individually.

Note that all the "action" is going on in the target category $\mathcal{D}$: the components of the natural transformation are in $\mathcal{D}$ and the naturality squares are in $\mathcal{D}$. Since composition is componentwise we inherit associativity and unit axioms from those in $\mathcal{D}$. It is a general principle that functor categories inherit more of their properties from the target category than from the source category.

─────────── **Things To Think About** ───────────

T 22.2   Show that we only get non-trivial natural transformations if the target category has non-trivial morphisms.

A more formal way of saying this is: if $\mathcal{D}$ is discrete (i.e. it only has identity morphisms) then the functor category $[\mathcal{C}, \mathcal{D}]$ is discrete.

This is true because any natural transformation as shown on the right must have components $\alpha_x \colon Fx \to Gx \in \mathcal{D}$.

Thus if $\mathcal{D}$ has only identities then all the components of α must be identities. Then we must have $F = G$, and α is the identity natural transformation. This shows that the existence of this higher dimension of structure (natural transformations) comes from the morphisms in the target category. We will come back to this principle in the last chapter.

The fact that a functor category gets structure from its target category means it can be very useful to study a category $\mathcal{C}$ via functors into a category that we know has a lot of excellent structure, such as **Set**. These particular functors are so useful we give them a name, as follows.

Definition 22.5   Given a category $\mathcal{C}$, a functor $\mathcal{C}^{op} \to \textbf{Set}$ is called a *presheaf* on (or over) $\mathcal{C}$. The *category of presheaves over* $\mathcal{C}$ is the functor category $[\mathcal{C}^{op}, \textbf{Set}]$.

Later we will look at why we often use contravariant functors (that is, from $\mathcal{C}^{op}$ rather than $\mathcal{C}$), and how this helps us study $\mathcal{C}$. For now we will look at a specific use of functors to express some structures we've already seen.

22.5 Diagrams and cones over diagrams

Functor categories give us a way to define cones over general diagrams, in order to define limits over general diagrams. Recall that given a small "shape" category $\mathbb{I}$, a diagram of shape $\mathbb{I}$ in a category $\mathcal{C}$ is a functor $D: \mathbb{I} \longrightarrow \mathcal{C}$. You can imagine $\mathbb{I}$ to be a quintessential commuting square category, or a parallel pair of arrows, if you want to visualize something. The functor D then picks out a specific diagram of that shape in the category $\mathcal{C}$.

Now, a cone over that diagram needs a vertex v and then morphisms from v to each object in the diagram, making everything in sight commute. To pick out a vertex we use a little technical trick: a constant functor. This is like a constant *function*, that sends every object to the same place. The only difference is that it's a functor so it acts on objects and morphisms: it sends every object to the same object, say v, and every morphism to the identity on v. That is the content of the following definition.

Definition 22.6   Let $\mathbb{I}$ and $\mathcal{C}$ be any categories, and v an object of $\mathcal{C}$. We write $\Delta_v: \mathbb{I} \longrightarrow \mathcal{C}$ for the *constant functor* at v, defined as follows.

$$\mathbb{I} \xrightarrow{\quad \Delta_v \quad} \mathcal{C}$$

$$\text{on objects:} \qquad i \quad \longmapsto \quad v$$

$$\text{on morphisms:} \qquad i \xrightarrow{f} i' \quad \longmapsto \quad v \xrightarrow{1_v} v$$

The idea is that no matter what the category $\mathbb{I}$ is, this functor collapses the whole thing to a single point in $\mathcal{C}$.

We now have everything we need to define a cone. This formalism might appear plucked out of thin air; we'll work out what it means immediately.

Definition 22.7   Given a diagram $D: \mathbb{I} \longrightarrow \mathcal{C}$, a *cone* over it with vertex v is a natural transformation $\Delta_v \Longrightarrow D$.

Things To Think About

T 22.3   Try working through this definition to understand why this gives a cone. If you're confused, you could try with some simple examples of $\mathbb{I}$, for example the quintessential arrow category.

You could also try the shapes we've used for our examples of limits: two individual objects (for products), and the shape for pullbacks as shown on the right.

Let's try this for the quintessential arrow category. It will help us to have names of things so let us set $\mathbb{I}$ to be this category: $\boxed{0 \xrightarrow{\ f\ } 1}$

Then a functor $D: \mathbb{I} \longrightarrow \mathcal{C}$ gives us this diagram in $\mathcal{C}$: $\boxed{D(0) \xrightarrow{D(f)} D(1)}$

A natural transformation $\alpha: \Delta_v \Longrightarrow D$ has a component $\Delta_v(i) \xrightarrow{\alpha_i} D(i)$ for each object $i \in \mathbb{I}$.

But $\Delta_v(i) = v$ for all objects, so we have components shown on the right, giving us projections from the vertex v to each object of the diagram.

$$\Delta_v(i) = v \xrightarrow{\alpha_0} D(0)$$
$$\Delta_v(i) = v \xrightarrow{\alpha_1} D(1)$$

For naturality in this case we only have one non-trivial morphism $f \in \mathbb{I}$ to consider, and its naturality square is shown on the right.

$$\begin{array}{ccc} \Delta_v(0) & \xrightarrow{\alpha_0} & D(0) \\ \Delta_v(f) \downarrow & & \downarrow D(f) \\ \Delta_v(1) & \xrightarrow{\alpha_1} & D(1) \end{array}$$

However, we again use the fact that Δ_v produces v for all objects and the identity for all morphisms, so the square evaluates as shown on the right.

$$\begin{array}{ccc} v & \xrightarrow{\alpha_0} & D(0) \\ 1_v \downarrow & & \downarrow D(f) \\ v & \xrightarrow{\alpha_1} & D(1) \end{array}$$

Furthermore the identity on its left means it has effectively collapsed to a triangle as shown here. This is exactly a cone over our diagram, with vertex v.

$$\begin{array}{cc} & D(0) \\ v \nearrow^{\alpha_0} & \downarrow D(f) \\ \searrow_{\alpha_1} & D(1) \end{array}$$

   In general, naturality corresponds to the commutativity of every triangle necessary in a cone over a diagram D, with each naturality square collapsing to one triangle. This way of expressing general cones is the basis of the general definition of "limit over a diagram of shape $\mathbb{I}$". However, the full definition is slightly beyond our scope.

22.6 Natural isomorphisms

One of the first things to do when we've made any category is to look at isomorphisms in it and see what they're like, so we'll now look at that inside functor categories. An isomorphism in a functor category is a natural transformation with an inverse; as usual the question is what that means when we unravel it.

──────── **Things To Think About** ────────
T 22.4   Can you show that isomorphisms in functor categories are componentwise isomorphisms? The first thing to do is interpret that statement.

This is saying that a natural transformation is an isomorphism iff all its components are isomorphisms. Let's state that formally.

Proposition 22.8 *A natural transformation as shown on the right is an isomorphism in the category* $[\mathcal{C}, \mathcal{D}]$ *if and only if for all* $x \in \mathcal{C}$ *the component* $\alpha_x \colon Fx \to Gx$ *is an isomorphism in* $\mathcal{D}$.

$$\mathcal{C} \underset{G}{\overset{F}{\rightrightarrows}} \Downarrow \alpha \; \mathcal{D}$$

Proof   First suppose α has an inverse β in $[\mathcal{C}, \mathcal{D}]$, that is, a natural transformation as shown on the right such that $\beta \circ \alpha = 1_F$ and $\alpha \circ \beta = 1_G$. We aim to show that it has componentwise inverses.

> We just unravel this on components and we get an inverse for each component.

We claim that for each $x \in \mathcal{C}$ the component β_x is an inverse for α_x. Now by the definition of vertical composition, $(\beta \circ \alpha)_x = \beta_x \circ \alpha_x$.

So $\beta \circ \alpha = 1_F$ means: $\forall x \in \mathcal{C}$, $\beta_x \circ \alpha_x = 1_{Fx}$, as shown in the diagram on the right.

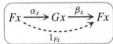

Similarly $\alpha \circ \beta = 1_G$ means: $\forall x \in \mathcal{C}$, $\alpha_x \circ \beta_x = 1_{Gx}$, as shown on the right.

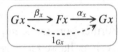

So each β_x is an inverse for α_x, giving componentwise inverses as claimed.

> For the converse we need to take individual inverses for each component of α and show that they compile into a valid natural transformation, that is, that they satisfy naturality. This part has more content.

Conversely suppose that each component α_x has an inverse β_x. We claim that the β_x are the components of a natural transformation.

We need to show that for any morphism $x \xrightarrow{f} y \in \mathcal{C}$, the square on the right commutes.

$$
\begin{array}{ccc}
Gx & \xrightarrow{\beta_x} & Fx \\
{\scriptstyle Gf}\downarrow & & \downarrow{\scriptstyle Ff} \\
Gy & \xrightarrow[\beta_y]{} & Fy
\end{array}
$$

> The idea here, morally, is that if the top and bottom morphisms of that square are isomorphisms then the sides are "more or less the same" whichever way we travel along the isomorphisms. We need to make that rigorous though.

Consider (carefully) the diagram on the right. We are trying to show that the right-hand square commutes, and we know that all the other small regions commute, along with the outside.

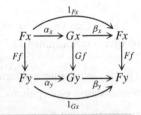

> Generally with commutative diagrams we know the outside commutes if all the inside parts commute. However if we are trying to show that one of the *inside* parts commutes it does not generally follow from knowing that the outside and all the other inside parts commute: care is needed.

We have the string of equalities shown below, which we read dynamically from the diagram above: at each stage we are just looking at the black arrows and ignoring the gray ones until the next step. See if you can see what commutativity is enabling us to move across each equals sign, and then see if you can read the sequence of equalities directly from the single diagram, dynamically.

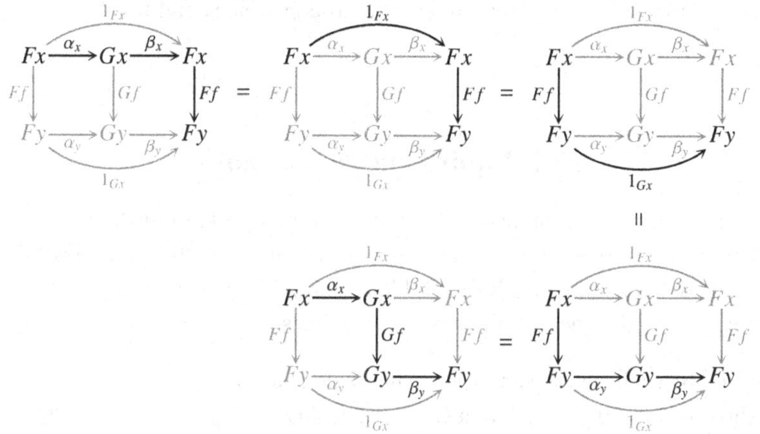

This tells us that the right-hand square commutes upon pre-composition with α_x. But α_x is an isomorphism so we can cancel it out by its inverse and conclude that the right hand square commutes.

> For this last part we actually only need to know that α_x is epic or has a one-sided inverse. In algebra we have this: $Ff \circ \beta_x \circ \alpha_x = \beta_y \circ Gf \circ \alpha_x$ and so just being able to cancel α_x on one side gives us the square we want.

Thus the morphisms β_x are the components of a natural transformation that is inverse to α, as claimed. □

Definition 22.9 An invertible natural transformation is called a *natural isomorphism*. If there is a natural isomorphism $F \Rightarrow G$ we often write $F \overset{\sim}{\Rightarrow} G$ or $F \cong G$ and say F and G are naturally isomorphic.

We now have two equivalent characterizations of natural isomorphisms:

- abstract/categorical: isomorphisms in a functor category
- concrete/pointwise: each component is an isomorphism.

As is often the case, the abstract version is better for theory and use, where the concrete one is better for checking. That is, when we check that something is a natural isomorphism it usually comes down to checking the components, but then when we use the fact that it is a natural isomorphism we often use the fact that it is an isomorphism in a functor category, rather than the fact that its components are isomorphisms. This is often the point of having two equivalent characterizations in abstract math. One is a concrete set of conditions to check to ensure that some abstract principle holds, and the abstract principle is then the thing that enables us to do things. The key is to know how the two are related. This is often generally referred to as coherence conditions.

22.7 Equivalence of categories

The first thing we can immediately look at is how we use these "morphisms of functors" to make our more nuanced version of sameness for categories, which is called equivalence. Recall the problem with isomorphism of categories was that we invoked some equalities between objects.

This happened because isomorphism involves a pair of functors as shown on the right, such that $GF = 1$ and $FG = 1$.

Now that we have the concept of isomorphism between functors we can do something more appropriately nuanced, where "appropriate" means making full use of every level of morphism available to us: the above equalities between functors are too strict, when we can use isomorphisms between functors instead. This gives us the following definition of a more nuanced notion of sameness for categories.

Definition 22.10   A functor $F: \mathcal{C} \longrightarrow \mathcal{D}$ is an *equivalence* if there is a functor $G: \mathcal{D} \longrightarrow \mathcal{C}$ and natural isomorphisms $GF \cong 1_{\mathcal{C}}$ and $FG \cong 1_{\mathcal{D}}$. Then we say the categories $\mathcal{C}$ and $\mathcal{D}$ are *equivalent*, and that F and G are *pseudo-inverses*.

Note that the prefix "pseudo" is often used in category theory when something holds up to isomorphism. Here F and G are like "inverses up to isomorphism".

There is a lot of "existence" in this definition at the moment. We characterize F by saying "there exists a pseudo-inverse" and we characterize a pseudo-inverse by saying "there exist some isomorphisms". If we wanted to be more specific we would actually ask for G and the isomorphisms to be specified. In that case it is also good to ask for a relationship between the isomorphisms. This is beyond our scope, but that structure is called an *adjoint equivalence*.

T 22.5 Recall this functor F from
Section 21.5, which "collapses" a
fattened category to a lean one.

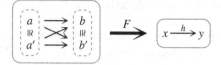

Isomorphic objects are mapped to the same place, and all the non-isomorphism
arrows are mapped to the same place. Show that this functor is an equivalence,
by constructing a functor G going the other way, and isomorphisms $GF \cong 1_{\mathbb{C}}$
and $FG \cong 1_{\mathbb{D}}$. Is the pseudo-inverse unique?

To construct a pseudo-inverse G we need to start by deciding where x and y are
going to go. This is supposed to be like an inverse, so they should be mapped
to things that were in turn mapped to them; it would be silly for G to take
x to b for example. However, we still have some choices: x could go to a or
a', and y could go to b or b'. In either case, once we've made that decision
there is no choice about where the morphism $x \rightarrow y$ goes as there is only one
non-invertible morphism available once we've fixed where x and y go; this
comes down to the fact that F is full and faithful.

So we can define G as shown on the right. We then
need to check the composites FG and GF.

$$\begin{aligned} x &\longmapsto a \\ y &\longmapsto b \end{aligned}$$

The composite FG is actually *equal* to the identity
functor since on objects the equations on the right
hold, and on morphisms f is mapped back to itself.

$$\begin{aligned} FG(x) &= F(a) = x \\ FG(y) &= F(b) = y \end{aligned}$$

On the other hand GF is not *equal* to the identity functor since a' and b' are not mapped back to
themselves, as shown on the right.

$$\begin{aligned} GF(a') &= G(x) = a \\ GF(b') &= G(y) = b \end{aligned}$$

However GF takes everything back to somewhere *isomorphic* to where it
started, which enables us to define a natural isomorphism $GF \overset{\sim}{\Rightarrow} 1$.

In order to do this fully it will
help us to have some more
names for things, so let's name
everything, as shown here.

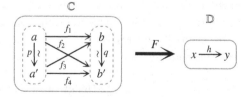

We'll now take a moment to elucidate the structure of $\mathbb{C}$. First note that $\mathbb{C}$ also
has inverses to p and q which I have not drawn (and also identities, as usual).
Every diagram in $\mathbb{C}$ commutes, by definition: the whole point of the morphisms

f_1, f_2, f_3, f_4 is that they are really "the same" morphism, just with the wiggle of the isomorphism built in.

So in fact in the spirit of not drawing composites that we can deduce, the category $\mathbb{C}$ could be drawn as shown here.

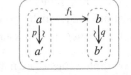

However, although it's usually clearer to omit composites, on this occasion I don't like it as it makes it look like f_1 is somehow more fundamental than, or different from, the other morphisms, when really they all play the same role.

Anyway, we now define G as shown on the right. It then remains to define a natural isomorphism $\alpha\colon GF \Rightarrow 1_{\mathbb{C}}$.

	$\mathbb{D}$	$\xrightarrow{G}$	$\mathbb{C}$
on objects:	x	$\longmapsto$	a
	y	$\longmapsto$	b
on morphisms:	$x \xrightarrow{h} y$	$\longmapsto$	$a \xrightarrow{f_1} b$

> At this point if you haven't already done so you might stop and see if you can just work out where the components need to "live" (i.e. their source and target); once you've done that there's not much choice of what they actually are as morphisms.

We need one component for each object of $\mathbb{C}$, and we define them as shown in the table below.

We could think through it like this: we see what the source and target of the component are by definition, then we evaluate those objects in $\mathbb{D}$, and then we decide what morphism in $\mathbb{D}$ the component will be.

component	defined as
$\alpha_a : GFa \longrightarrow a$	$a \xrightarrow{1_a} a$
$\alpha_b : GFb \longrightarrow b$	$b \xrightarrow{1_b} b$
$\alpha_{a'} : GFa' \longrightarrow a'$	$a \xrightarrow{p} a'$
$\alpha_{b'} : GFb' \longrightarrow b'$	$b \xrightarrow{q} b'$

Each component is certainly an isomorphism, so provided this is a natural transformation it will be a natural isomorphism. Thus it just remains to check all the naturality squares. But these all live in $\mathbb{C}$, and everything in $\mathbb{C}$ commutes, so all the naturality squares must commute.

───────── **Things To Think About** ─────────

T 22.6   It's still a good exercise to see if you can write down some naturality squares. There is one for each morphism of $\mathbb{C}$.

Here's the naturality square for $a \xrightarrow{f_2} b'$. First I've drawn what it is *a priori* (by definition, before we evaluate everything) and then I've drawn what it actually comes out to be in $\mathbb{C}$.

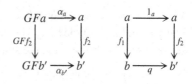

The definition we have just been using is a "categorical" definition of equivalence, in that we have not mentioned elements of sets anywhere. In fact it's a 2-dimensional version of categorical definition: a "2-categorical" definition, as we'll see when we do higher dimensions. For now the main thing is to relate it back to the elementary definition of "pointwise equivalence" that we saw in the previous chapter. You might have felt that the above proof was a bit tedious; the elementary definition is indeed much easier to check.

Proposition 22.11 *A functor is an equivalence if and only if it is a pointwise equivalence, that is, full and faithful and essentially surjective on objects.*

This proof is a little beyond our scope. It's not hard, but it involves quite a bit of "diagram chasing" and a few little tricks, together with holding quite a lot of notation in your head at once.

> See Riehl[†] for example; we have done enough of the technicalities for you to follow that proof at least step by step.

The general idea is that equivalent categories are "the same" in an appropriate sense. Categories are isomorphic if they have *exactly* the same object and arrow structure, so the only difference is that everything is renamed. Categories are equivalent if they have the same arrow structure but the objects can be "fattened up" a bit by having many isomorphic copies, or "thinned down" by identifying isomorphic objects (making them the same). In the process of fattening or thinning we just have to be careful to adjust the morphisms appropriately so that they *effectively* stay the same, as in the example above.

─────────── **Things To Think About** ───────────

T 22.7 Construct a two-object category $\mathbb{C}$ that is equivalent to the terminal category $\mathbb{1}$ (one object and one identity morphism). [Hint: the two objects must be uniquely isomorphic in order for $\mathbb{C}$ to be "more or less the same as" $\mathbb{1}$.] Then try doing one with three objects, and n objects.

For the example with two objects we need to take the "quintessential isomorphism" category as shown on the right; in this category $gf = 1_a$ and $fg = 1_b$.

[†] Emily Riehl, *Category Theory in Context*, Dover, 2017.

We might draw this as shown here, to emphasize the fact that there
really is no loop or hole here. The double-headed arrow indicates a $a \longleftrightarrow b$
sense of "reversibility", and the symbol ~ signifies isomorphism.

Now, the terminal category $\mathbb{1}$ has one object, say $*$, so a functor $\mathbb{C} \xrightarrow{F} \mathbb{1}$ must
map every object to $*$. Then F being full and faithful means for every $x, y \in \mathbb{C}$
the following function on homsets is an isomorphism: $\mathbb{C}(x, y) \xrightarrow{F} \mathbb{1}(*, *)$. But
the homset $\mathbb{1}(*, *)$ has only one element (the identity) so each $\mathbb{C}(x, y)$ must also
have exactly one element. We'll now unravel what that means.

First, for any $x \in \mathbb{C}$, the homset $\mathbb{C}(x, x)$ can only contain the identity 1_x.
Next, for any distinct $x, y \in \mathbb{C}$ there is exactly one morphism $x \rightarrow y$ and one
morphism $y \rightarrow x$. Furthermore these must be inverses: the composite gf is a
morphism $x \rightarrow x$ but the only morphism $x \rightarrow x$ is the identity 1_x. Similarly
the composite $fg = 1_y$. Thus all objects of $\mathbb{C}$ must be uniquely isomorphic.

This generalizes to any category equivalent to $\mathbb{1}$, with any number of objects.

For 3 objects we could draw it like this:	For 4 objects we could draw it like this:	which we could interpret as a tetrahedron:

Categories equivalent to $\mathbb{1}$ are surprisingly useful, so we give them a name.

Definition 22.12   A category is called *indiscrete* (or *chaotic*) if it is equivalent
to the terminal category.

Recall that we call a category *discrete* if it has only identity morphisms. Discrete and indiscrete categories are two ways to create a category canonically
from a set of objects. Indiscrete categories might sound a bit silly as we've
essentially made a bunch of objects the same from the point of view of the
category, but in fact that's the whole point: it's a way to make objects the same
without actually identifying them. Identifying them consists of making objects
equal, which is not the right level of nuance in a category.

Indiscrete categories are used in the theory of *cliques*; I'm saying that in case you
want to look them up.

Discrete and indiscrete categories have opposite universal properties. They
are analogous (ideologically and formally) to discrete and indiscrete spaces in
topology: in a discrete space all points are considered to be far apart, and in an
indiscrete space all points are considered to be close together. The analogy is

made formal by expressing the universal properties, for example by something called *adjunctions*, but this is beyond our scope; I mention it in case you would like to look it up.

22.8 Examples of equivalences of large categories

We have secretly seen various examples of equivalences of categories throughout the earlier chapters, without being able to say so. Every time we say something "is" something else in category theory, it is quite likely there is some nuance going on and that the nuance is an equivalence of categories (unless it's higher-dimensional and thus even more nuanced).

One key example is every time we have said that a monoid "is" a one-object category. In some sense this is not strictly true, as a monoid (directly defined) has as underlying data a set of elements, whereas a one-object category has an object and morphisms. But we can make sense of "is" via an equivalence of (large) categories of those two types of structure.

─────────────── **Things To Think About** ───────────────

T 22.8   See if you can construct an equivalence between the category **Mnd** of monoids and homomorphisms, and the category **ooCat** of small one-object categories and functors between them.

First note that when we take all one-object categories, we have many copies of the same monoid expressed as a one-object category in rather trivially different ways — just by having the one object being called different things.

This means that if we try and define a pair of pseudo-inverse functors as shown on the right, one direction is easier than the other.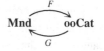

Defining G is "easy" as we just send a one-object category to the monoid of its morphisms. But defining F is "hard" because we have to decide *which* one-object version of the monoid to choose. However, it's not *excessively* hard: we could just decide that the single object is always going to be $*$, say. In that case GF really is the identity on **Mnd**, but FG is only isomorphic to the identity on **ooCat**. This is because if we start with a one-object category whose single object is not $*$, when we come back again via FG we will land on a category with the same monoid of morphisms, just a differently named single object. This is not *really* different, just isomorphic.

Of course, that's not a proof, but gives the idea. Incidentally the pointwise way of looking at it is that G is surjective but F is only essentially surjective.

Both are full and faithful because a functor between one-object categories is precisely a monoid homomorphism.

This is why category theorists are happy saying "a monoid is a one-object category" where others might get upset about the level shift. Those who get upset might accuse category theorists of not being rigorous, but I reckon we are being rigorous, we're just comfortable using "is" to refer to an equivalence of categories, not just an isomorphism.

22.9   Horizontal composition

I said in passing that the more abstract (not pointwise) definition of equivalence is actually 2-categorical, without saying what that means. Recall that a "categorical" definition is one that only uses objects and morphisms inside an ambient category, which means we can try taking that definition to other ambient categories to see what it gives. Our definition of equivalence didn't just use the objects and morphisms of **Cat**, as we saw that would only give us the concept of isomorphism, which is too strict for categories. Instead we escalated and used objects (categories), morphisms (functors) and "2-dimensional morphisms" (natural transformations).

The idea is that although categories and functors do form a category, there is a better, richer, more nuanced 2-dimensional environment if we take categories, functors *and* natural transformations between them. When we first defined categories, we had a prototype category given by sets and functions. Now we get a prototype 2-dimensional category given by categories, functors and natural transformations. We say "2-category" instead of "2-dimensional category" so that it's less of a mouthful.

There is one more structure that we need to consider in order to go fully 2-dimensional, and that is horizontal composition.

When we did vertical composition our natural transformations all had to be between functors going to and from the same pair of categories, as shown here.

However, if we're going to organize categories, functors and natural transformations into a good structure for having them all interact, I hope you feel that this is rather restrictive. After all, we can compose functors going across the page, so what about natural transformations?

So-called horizontal composition deals with categories, functors, and natural transformations in the configuration shown here.

$$\mathcal{A} \underset{G}{\overset{F}{\Downarrow \alpha}} \mathcal{B} \underset{K}{\overset{H}{\Downarrow \beta}} \mathcal{C}$$

Looking at the geometry, we might expect this to produce a natural transformation as shown here.

$$\mathcal{A} \underset{KG}{\overset{HF}{\Downarrow}} \mathcal{C}$$

This would have components for all objects $a \in \mathcal{A}$, $\boxed{HFa \longrightarrow KGa.}$

Things To Think About

T 22.9 See if you can produce such a morphism from the components of α and β together with some functor action. In fact, there are two ways to do it; can you produce both and show that they're equal?

The components for α and β give us the morphisms shown on the right.

$$\boxed{\begin{array}{ll} \forall a \in \mathcal{A} & Fa \xrightarrow{\alpha_a} Ga \\ \forall b \in \mathcal{B} & Hb \xrightarrow{\beta_b} Kb \end{array}}$$

Now it's a case of following our nose. Starting at HFa we can make this composite.

$$\boxed{HFa \xrightarrow{H\alpha_a} HGa \xrightarrow{\beta_{Ga}} KGa.}$$

Note that for the first morphism we applied the functor H to the component α_a, but for the second one we did the component of β at the object $Ga \in \mathcal{B}$.

Here is another way to make a candidate component for the composite natural transformation we're looking for.[†]

$$\boxed{HFa \xrightarrow{\beta_{Fa}} KFa \xrightarrow{K\alpha_a} KGa.}$$

This time we are doing a component of β first, and a component of α afterwards. These two ways of producing a potential component $HFa \longrightarrow KGa$ look different, but are equal.

They are two ways around a naturality square for β at the morphism α_a as shown here.

$$\begin{array}{ccc} HFa & \xrightarrow{\beta_{Fa}} & KFa \\ {\scriptstyle H\alpha_a}\downarrow & & \downarrow{\scriptstyle K\alpha_a} \\ HGa & \xrightarrow{\beta_{Ga}} & KGa \end{array}$$

Now, we would like to use these morphisms $HFa \longrightarrow KFa$ (expressed either way round) to define the components of a natural transformation $HF \Rightarrow KG$, so we need to check naturality.

Things To Think About

T 22.10 Can you check naturality for these putative components?

[†] That is, another way to make a morphism which goes from HFa to KGa and so has the potential to be a component of the composite natural transformation we're trying to define, which goes from HF to KG.

This is another case of filling in a diagram with smaller diagrams. Consider a morphism $x \xrightarrow{f} y \in \mathcal{A}$. I will use the first expression of the component above, that is $HFx \xrightarrow{H\alpha_x} HGx \xrightarrow{\beta_{Gx}} KGx$.

The square we need is the outside of the diagram on the right, and we fill it in as shown. We will observe that the two smaller squares are individual naturality squares.

$$
\begin{array}{ccc}
HFx & \xrightarrow{H\alpha_x} HGx \xrightarrow{\beta_{Gx}} & KGx \\
{\scriptstyle HFf}\downarrow & \quad \downarrow{\scriptstyle HGf} & \downarrow{\scriptstyle KGf} \\
HFy & \xrightarrow[H\alpha_y]{} HGy \xrightarrow[\beta_{Gy}]{} & KGy
\end{array}
$$

> If you weren't able to check naturality yourself, you might take a moment now to see if you can see what naturality squares these two small ones are.

The left-hand square is the naturality square for α at the morphism f, with H applied to the whole square; recall that all functors preserve the commutativity of diagrams so after applying H the square still commutes. The right-hand square is the naturality square for β, at the morphism Gf. They both commute, so the outside commutes.

Thus we have indeed defined a natural transformation $HF \Rightarrow KG$, and this is a new form of composition. Here is the whole definition.

Definition 22.13 Let α and β be natural transformations as shown.

We define the *horizontal composite*

to have components $HFa \xrightarrow{H\alpha_a} HGa \xrightarrow{\beta_{Ga}} KGa$

or equivalently: $HFa \xrightarrow{\beta_{Fa}} KFa \xrightarrow{K\alpha_a} KGa$.

Note the standard notation which is to use a star $*$ for horizontal composition and a circle $\circ$ for vertical composition. Of course "horizontal" and "vertical" depend on which way up we draw things on the page; what really defines the type of composition is the dimension of the boundary along which we've stuck the natural transformations together. For vertical composition the boundary is a functor (1-dimensional) and for horizontal composition the boundary is a category (0-dimensional). When we look at higher dimensions we will see that this characterization allows us to generalize into higher dimensions.

22.10 Interchange

You might be wondering what the *meaning* is of the two different ways of defining the horizontal composite. Following our nose by type-checking can be a fun game to play, but I do like understanding the meaning of things. We are going to examine the meaning of the two ways of expressing components of the horizontal composite $\beta * \alpha$. This leads us to thinking about how horizontal composition interacts with vertical composition, which is called *interchange*.

Let's think about this composite first: $HFa \xrightarrow{H\alpha_a} HGa \xrightarrow{\beta_{Ga}} KGa$.

The first morphism $H\alpha_a$ is a component of α, with the functor H applied. We might draw it as on the right, since the functor H is applied after doing α.

The second morphism β_{Ga} involves applying the functor G *first* and then taking the component of β at the resulting object Ga. We might draw that as shown here.

In fact those are perfectly valid ways of composing a natural transformation with a functor to produce a new natural transformation.[†] The first one is written $H\alpha$ and the second βG.

We can then compose $H\alpha$ and βG vertically, which we might draw like this.

This is informally referred to as *whiskering* as the functors sticking out look a bit like whiskers.

───────────── **Things To Think About** ─────────────

T 22.11   Can you draw the diagram for whiskering the other way round, corresponding to the other way of defining horizontal composition?

The other expression for a component of $\beta * \alpha$ is: $HFa \xrightarrow{\beta_{Fa}} KFa \xrightarrow{K\alpha_a} KGa$.

Here the first morphism says do F first and then take the component of β, so it's from the natural transformation βF. The second morphism is a component of α with K applied afterwards, so it's from the natural transformation $K\alpha$. The diagram is on the right.

We've now seen that the two expressions for a component of $\beta * \alpha$ come from two different ways of "whiskering" α and β.

The equality of these two situations follows from the definition of naturality, and can be succinctly summed up in this diagram.

───────────────────────

[†] Note that we are composing a 1-dimensional morphism (functor) with a 2-dimensional morphism (natural transformation) but it works.

For natural transformations this equality follows from the definitions, but we're going to take this as inspiration for an axiom in higher dimensions, something we demand is true in order to give a suitably coherent structure.

In fact we typically use a situation with a 2×2 grid of natural transformations as shown here.

There are two different ways of building this composite from horizontal and vertical composites:

- compose pairs horizontally first, and then compose the result vertically, or
- compose pairs vertically first, and then compose the result horizontally.

Things To Think About

T 22.12   See if you can draw diagrams to show these two different schemes, and then prove they give the same answer for natural transformations. The first thing to do is to give everything in the diagram a name.

Let us name everything in sight as follows:

$$A \xrightarrow[H]{\overset{F}{\underset{G}{\quad}}} B \xrightarrow[R]{\overset{P}{\underset{Q}{\quad}}} C$$

with α, β on the left (F, G, H) and γ, δ on the right (P, Q, R).

The equality we're looking for is summed up in this diagram, with the equation written in algebra below it.[†]

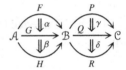

$$(\delta * \beta) \circ (\gamma * \alpha) \;=\; (\delta \circ \gamma) * (\beta \circ \alpha)$$

I definitely find the diagrammatic form more intelligible. However, in the algebra note the way that the types of composition swap places, and the inside terms also swap places, whereas the outside terms stay put. This is called *interchange*.

Definition 22.14   The above equation is called the *interchange law*.

Proposition 22.15   *Natural transformations satisfy the interchange law.*

The proof is just some fiddling around with components.

[†] Remember that the algebraic notation for composition is "backwards" from how we draw the arrows.

I think the hardest part is writing down what we have to show is equal, which is the outside part of the following diagram; we can then see that the square in the middle is a naturality square for γ at the morphism β_a, so the whole thing commutes.

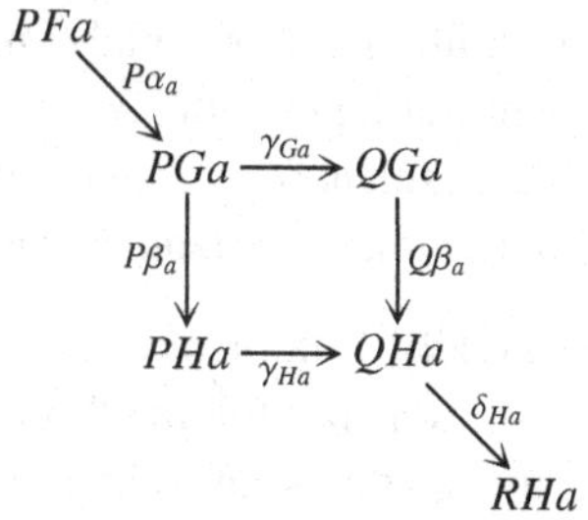

$$
\begin{array}{ccc}
PFa & & \\
\quad\searrow^{P\alpha_a} & & \\
PGa & \xrightarrow{\gamma_{Ga}} & QGa \\
\downarrow^{P\beta_a} & & \downarrow^{Q\beta_a} \\
PHa & \xrightarrow[\gamma_{Ha}]{} & QHa \\
& & \quad\searrow^{\delta_{Ha}} \\
& & RHa
\end{array}
$$

> I do think this is the kind of thing where if you've understood what's going on it's easier to construct the diagram yourself than decipher somebody else's.

I hope that you are yearning for a more illuminating explanation; if so the following might help. Recall that we defined horizontal composition by whiskering, and started off by showing that whiskering either way round gave the same answer. We can break the interchange law down into instances of that whiskering principle.

We just need this one small thing:

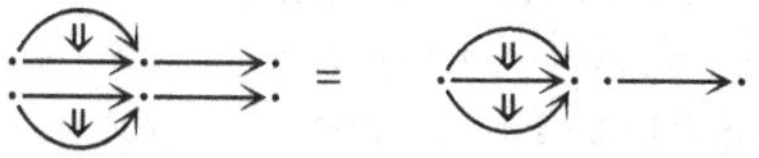

If you write out what this means you will see that it follows from functoriality; we also need the version with the functor sticking out on the other side but that one follows by definition.[†]

  We then have the following string of equalities.

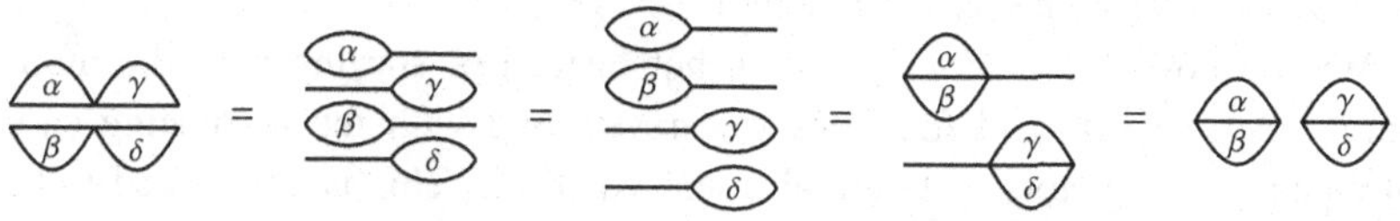

> I have omitted much of the notation here to simplify the diagrams, on the understanding that there are certain conventions: all the 1-dimensional arrows point right, all the 2-dimensional arrows point down, everything at each dimension is distinct, and so on.

─────────── **Things To Think About** ───────────

T 22.13   It's a good idea to verify that you know where each of these equalities is coming from.

Here is how the string of equalities arises, from left to right:

[†] This asymmetry is important in higher dimensions when we're doing things weakly, as it means things are weaker on one side than the other

1. Definition of horizontal composition.
2. Whiskering the other way round with the middle two "fish".
3. The result above about combining the tails of two "fish".
4. Definition of horizontal composition.

Yes, I really do think of them as fish sometimes; also the interchange law is like those little paper folding "fortune telling" things where you fold up a square of paper so that it has four corner compartments, and you stick your fingers in and make it open in opposite directions.[†]

Now that we have the interchange law we have everything we need to think about the totality of categories, functors, and natural transformations.

22.11 Totality

One way of approaching the definition of a category is to look at the motivating example of sets and functions and see what sorts of things they satisfy. At a basic level functions can be composed, and this composition is unital and associative. This is inherent to the definition of function, but when we define categories we then *demand* all this in axioms, and go round looking for things that satisfy it. Essentially what we're doing is saying: "Look at this handy behavior exhibited by sets and functions. Let's see what other places exhibit such behavior, because then we can go there and work somewhat analogously to how we work with sets and functions."

We are now doing the same thing but instead of starting with the motivating case of sets and functions we are starting with the motivating case of categories, functors and natural transformations. This is considered to be a 2-dimensional structure, in which categories are the 0-dimensional objects, functors are 1-dimensional, and natural transformations are 2-dimensional.

There are some things we know we can do inside this world. We know that functors can be composed, and that this composition is unital and associative. We also know that natural transformations can be composed in two ways, and that those satisfy the interchange law. This is essentially the definition of a *2-category*, and we'll do it more formally in Chapter 24 on higher dimensions. But first we will spend one chapter taking in the view from where we are.

[†] If you don't know what I'm talking about you can do a search for "origami fortune teller" and it should come up.

23

Yoneda

Bringing together all that we've done in one of the pinnacles of abstraction in category theory.

We have now climbed to a certain height of abstraction, and it's time to pause to take in a breathtaking view of all of the country laid out before our eyes. In a way this chapter is an aside; you could move smoothly from the previous chapter to the following one about higher dimensions. For me, the abstractions in this chapter are one of the great joys of category theory. However, I will warn you that it is very formal and abstract, and that if it is the applications and examples that interest you this might just seem like too much formalism to you. But if you can get anything out of it at all, you have definitely started to appreciate the joy of abstraction.

23.1 The joy of Yoneda

We're going to look at one of the most important and beautiful uses of natural transformations. It is very abstract and formal but encapsulates some of the most deep fundamental truths about algebraic structures, while somehow at the same time being almost trivial to the point of being barely any more than type-checking. This apparent triviality is something that leads some people to dismiss category theory as contentless, but leads some other people to adore its ability to cut through to the core so cleanly that all mess seems to fall away, leaving just a fundamental nugget of simple but profound truth. As a friend of mine[†] put it: anyone can complicate things, but it takes real intelligence to simplify them. And I will note that simplifying something is not the same as what I'll call "simplisticating" them; that is, there is a difference between making something simple and making it simplistic.

The term "Yoneda" is named after Japanese mathematician Nobuo Yoneda. It may refer to a functor called the Yoneda embedding, and also a result called the Yoneda Lemma. It is so fundamental and ubiquitous that sometimes I joke

---

[†] Thank you Tyen-Nin Tay.

with category theory friends that the whole point of our research is to find the sense in which any particular thing is an instance of Yoneda.[†]

We have been hinting at Yoneda at various times in earlier chapters. We also mentioned that Yoneda would be an important use of contravariant functors, and we are now going to see that it's an important use of functor categories.

23.2 Revisiting sameness

In Section 14.4 we looked at the sense in which a category sees isomorphic objects as "the same". We're now going to revisit it at a higher level of abstraction which explains and encapsulates more of what is going on there.

In that section, we drew diagrams like the one on the right, with an isomorphism between a and b exhibited by inverses f and g. We considered an object x "looking" at a and b, and saw that x has the same relationships with a as it does with b.

The correspondence between how x interacts with a and with b is shown informally on the right.

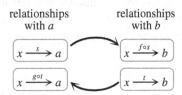

I have drawn this suggestively to look like yet another diagram exhibiting an isomorphism, with inverse arrows going in each direction. And in fact it's not just a schematic diagram: it's something we now have the abstract technology to make precise.

First assume that we're in a category $\mathcal{C}$ that is locally small, so that we really do have sets of morphisms. Now what I informally called the "relationships of x with a" is really the set $\mathcal{C}(x, a)$, and likewise the "relationships with b" is really the set $\mathcal{C}(x, b)$.

The above schematic diagram then becomes a diagram of sets and functions as shown on the right. The notation needs some explaining.

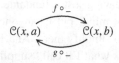

The little horizontal line ("underscore") is a deliberate gap left for us to place a variable. The function $f \circ \_$ takes an element of $\mathcal{C}(x, a)$ as an input, that is, a morphism $x \xrightarrow{s} a$. The function tells us to put s where the underscore is, so the

---

[†] In fact the Australian school of category theory seems to be so good at this that we sometimes go further and joke that "Yoneda" is the Australian category theory version of Mornington Crescent, but that's a rather esoteric joke for listeners of BBC Radio 4 of a certain generation.

output is $f \circ s$. The function going backwards takes as input a morphism $x \xrightarrow{t} b$ and gives the output $g \circ t$. Thus these functions with underscores correspond to the large arrows shown informally in the previous diagram of "relationships". Note that we've gone up a level of abstraction here, and are now considering morphisms themselves as elements of sets, with functions acting on them.

> This has the potential to be quite confusing because of the different levels involved. I remember being very mind-blown about it when I first encountered it. The way we annotate composition like $f \circ s$ "backwards" adds further potential confusion, but if in doubt I draw out all the morphisms including source and target and the type-checking sorts everything out.

Now, so far this just *looks* like a diagram showing an isomorphism of sets. We can check that the functions in question are really inverses by following their action on elements around. Here's the composite from $\mathcal{C}(x, a)$ to itself:

$$\mathcal{C}(x, a) \xrightarrow{f \circ \_} \mathcal{C}(x, b) \xrightarrow{g \circ \_} \mathcal{C}(x, a)$$
$$s \longmapsto f \circ s \longmapsto g \circ (f \circ s) = s$$

For the last step, note that $g \circ (f \circ s) = (g \circ f) \circ s$ by associativity, and $g \circ f$ is the identity because f and g are inverses (the whole point of this story). So the composite from $\mathcal{C}(x, a)$ to itself sends s to s, so is the identity function.[†]

An analogous argument shows that the composite the other way round is the identity function $\mathcal{C}(x, b) \to \mathcal{C}(x, b)$, so we have an isomorphism of sets as required. Note how this was a direct result of the morphisms f and g being inverses; in fact each half of the inverse definition for f and g produced half of the inverse definition at the level of function on the homsets.

Dually, we can think about morphisms *to* x rather than *from* x.

───── **Things To Think About** ─────

T 23.1 Try the dual version: see if you can draw the diagram, write down the functions with underscores correctly, and check the action on elements.

Here is the diagram for the dual version.

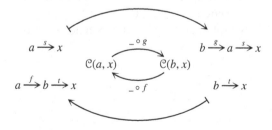

───────────────

[†] You might prefer this to say "so it is the identity function". I'm using language in the spare way mathematics is often written, in case you want to go on to read more formal texts.

Note that the directions have switched:

When acting on morphisms *out* of x, we use *post-composition* with f. The direction of the resulting function is shown on the right.

$$
\begin{array}{ccc}
\mathcal{C}(x,a) & \xrightarrow{\ f\circ\_\ } & \mathcal{C}(x,b) \\
x \xrightarrow{s} a & \longmapsto & x \xrightarrow{s} a \xrightarrow{f} b
\end{array}
$$

When acting on morphisms *into* x, we use *pre-composition* with f. The direction of the resulting function is reversed.

$$
\begin{array}{ccc}
\mathcal{C}(b,x) & \xrightarrow{\ \_\circ f\ } & \mathcal{C}(a,x) \\
b \xrightarrow{s} x & \longmapsto & a \xrightarrow{f} b \xrightarrow{s} x
\end{array}
$$

We are going to see that these situations are each part of a functor, and the switching of directions means the second (dual) one is *contravariant*.[†]

Expressing all this as functors will also make precise the sense in which an isomorphism between objects of $\mathcal{C}$ "induces" an isomorphism of homsets. The functors in question are called representable functors.

23.3 Representable functors

So far we have established that, given an isomorphism $a \xrightarrow{f} b$ in $\mathcal{C}$, we can produce an isomorphism on sets of morphisms as shown below.

An isomorphism in $\mathcal{C}$ produces an isomorphism in **Set**.

$$
\begin{array}{ccc}
\in \mathcal{C} & & \in \textbf{Set} \\
a \underset{g}{\overset{f}{\rightleftarrows}} b & \longmapsto & \mathcal{C}(x,a) \underset{g\circ\_}{\overset{f\circ\_}{\rightleftarrows}} \mathcal{C}(x,b)
\end{array}
$$

I hope that this suggestive diagram has caused you to wonder whether this is actually down to some sort of functor $\mathcal{C} \to \textbf{Set}$. This is the kind of situation in category theory where I think "everything you want to be true is true". (Of course, this depends on wanting the right sort of things.)

I am claiming there is a functor with an action as follows; note that here x is fixed, but now a and b can be any objects, and f any morphism (not necessarily an isomorphism). We will call this functor H^x to remind us that x is fixed.

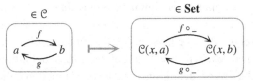

$\mathcal{C}$	$\xrightarrow{\ H^x\ }$	**Set**
on objects:	a $\longmapsto$	$\mathcal{C}(x,a)$
on morphisms:	$a \xrightarrow{f} b$ $\longmapsto$	$\mathcal{C}(x,a) \xrightarrow{f\circ\_} \mathcal{C}(x,b)$

[†] The type of functor that switches the direction of morphisms.

Things get a little hairy here as there are so many levels at play, but if we proceed slowly we can show that this does indeed define a functor.

─────────────────── **Things To Think About** ───────────────────

T 23.2   So far all we have established is that for each morphism $f \in \mathcal{C}$, we have a function $f \circ \_$. See if you can work out what we need to check to show that this construction makes H^x a functor. (And then check it: however, I personally think working out what needs to be checked is the hard part.)

To check functoriality we need to keep our wits about us a bit but it's basically just type-checking. We just need to keep in mind that when we're looking at a function on homsets we'll probably want to examine it on elements, but the elements in that case are elements of homsets, that is, morphisms of $\mathcal{C}$.

First we need to check that H^x sends identities to identities.

The action of H^x on identities is shown on the right, so we need to check the function $1_a \circ \_$ is the identity function.

$$\mathcal{C} \xrightarrow{\quad H^x \quad} \textbf{Set}$$
$$\left(a \xrightarrow{1_a} a \right) \longmapsto \left(\mathcal{C}(x,a) \xrightarrow{1_a \circ} \mathcal{C}(x,a) \right)$$

We check the action of $1_a \circ \_$ on an element $f \in \mathcal{C}(x,a)$ as shown on the right: f is sent to f so $1_a \circ \_$ is indeed the identity function.

$$\mathcal{C}(x,a) \xrightarrow{1_a \circ} \mathcal{C}(x,a)$$
$$f \longmapsto 1_a \circ f = f$$

For composition we consider a pair of composable morphisms $a \xrightarrow{f} b \xrightarrow{g} c$ in $\mathcal{C}$ and compare two things:

If we apply the functor H^x first and then compose the results in **Set** we get:

$$\mathcal{C}(x,a) \xrightarrow{f \circ} \mathcal{C}(x,b) \xrightarrow{g \circ} \mathcal{C}(x,c)$$
$$s \longmapsto f \circ s \longmapsto g \circ (f \circ s)$$

If instead we compose in $\mathcal{C}$ first and then apply the functor H^x we get:

$$\mathcal{C}(x,a) \xrightarrow{(g \circ f) \circ} \mathcal{C}(x,c)$$
$$s \longmapsto (g \circ f) \circ s$$

The action of these two functions on s is the same, by associativity. So in fact, functoriality in this case comes down to associativity of composition in $\mathcal{C}$.

It is good practice to observe this relationship, rather than just ignoring the parentheses because we know that associativity holds. Ideologically this is because category theory is a discipline of seeing how and why structure arises; in practice it's important because if we move into situations in which associativity doesn't hold strictly, then it's important to know what the consequences are.

So far we have made a functor by fixing an object x and looking at sets $\mathcal{C}(x, a)$. The dual version looks at sets $\mathcal{C}(a, x)$ instead but, as we mentioned at the end of the last section, the functor is now contravariant, that is, it reverses the direction of morphisms.

─────────────── **Things To Think About** ───────────────

T 23.3   Try and follow through aaall of the above construction, making sure to take into account the functor now being contravariant.

As before we fix an object $x \in \mathcal{C}$, but this time we are making a functor that sends an object x to the set $\mathcal{C}(a, x)$. We are making a contravariant functor, that is, a functor $\mathcal{C}^{\mathrm{op}} \longrightarrow \mathbf{Set}$. We will call this one H_x with the subscript (rather than the previous superscript H^x) indicating that we are now looking at homsets of morphisms *into* x. The functor H_x acts as follows (but note that as always I will only draw morphisms in $\mathcal{C}$, not in $\mathcal{C}^{\mathrm{op}}$):

	$\mathcal{C}^{\mathrm{op}}$	$\xrightarrow{\;H_x\;}$	$\mathbf{Set}$
on objects:	a	$\longmapsto$	$\mathcal{C}(a, x)$
on morphisms:	$a \xrightarrow{f} b \in \mathcal{C}$	$\longmapsto$	$\mathcal{C}(b, x) \xrightarrow{\;-\circ f\;} \mathcal{C}(a, x)$

In fact it's the same construction as before, but starting with $\mathcal{C}^{\mathrm{op}}$ instead of $\mathcal{C}$. We've just unraveled it using the definition $\mathcal{C}^{\mathrm{op}}(x, a) = \mathcal{C}(a, x)$. Thus functoriality follows by waving the BOGOF[†] wand and saying "dually"; we don't have to check anything. It is perhaps just worth noting that identities are preserved because they act as identities on the other side this time.

These functors are sometimes written as $\mathcal{C}(x, \_)$ and $\mathcal{C}(\_, x)$ with the underscore or "blank" indicating to us where the variable goes. Otherwise we use H or h, perhaps standing for "hom".[‡]

In summary we have the following functors for a fixed object x.

$$\mathcal{C}(x, \_) = H^x \text{ or } h^x : \mathcal{C} \longrightarrow \mathbf{Set}$$
$$\mathcal{C}(\_, x) = H_x \text{ or } h_x : \mathcal{C}^{\mathrm{op}} \longrightarrow \mathbf{Set}$$

─────────────── **Terminology** ───────────────

These things are so widespread and formally useful that we give them names. We mentioned before that functors $\mathcal{C}^{\mathrm{op}} \longrightarrow \mathbf{Set}$ are called *presheaves* on $\mathcal{C}$. When we use the prefix "pre-" in abstract math it's usually to indicate that the structure is a starting point for something more complicated that will need more conditions. In this case a presheaf is indeed the starting data for a sheaf.

───────────────

[†] Buy One Get One Free, that is, prove one and get the dual for free.
[‡] Some people write the homset $\mathcal{C}(a, b)$ as $\mathrm{hom}(a, b)$ or $\mathrm{hom}_{\mathcal{C}}(a, b)$.

Functors of the form H^x and H_x are called *representable*; more accurately any functor isomorphic to one of these (via a natural isomorphism) is called representable. We'll come back to this. Representable functors are particularly well-behaved, so if a functor can be expressed in this way it's very handy.

Here is one thing we can immediately do using our newly coined functors H_x and H^x. We began all this by studying the fact that if f is an isomorphism then it induces an isomorphism on the homsets. That is, if f is an isomorphism then $H^x(f)$ is also an isomorphism, and dually so is $H_x(f)$. This is saying that the functors H^x and H_x *preserve* isomorphisms. But in fact all functors do that, as we saw in Chapter 20.

You might be feeling your head spinning at the different levels operating here, and I think that's quite a reasonable response. And you will either be excited or horrified that there is yet another level. You might have wondered why we kept having to fix an object x to start. That's a good instinct: this is itself part of a functor that takes an object x and sends it to the functor H^x (or dually to H_x). This is a functor sending objects of $\mathcal{C}$ to functors themselves, which means we need to invoke the categories of functors that we just met in the previous chapter. The functor sending x to H_x is the *[drumroll]* Yoneda embedding.

23.4 The Yoneda embedding

We will now make use of the functor categories we saw in the previous chapter: we saw that given categories $\mathcal{C}$ and $\mathcal{D}$ we have a category $[\mathcal{C}, \mathcal{D}]$ whose objects are functors $\mathcal{C} \longrightarrow \mathcal{D}$ and morphisms are natural transformations.

We now want to send an object $x \in \mathcal{C}$ to a functor $H_x \colon \mathcal{C}^{op} \longrightarrow \mathbf{Set}$. But H_x is itself an object of the functor category $[\mathcal{C}^{op}, \mathbf{Set}]$, so we can build a functor as follows. We are going to call it $H_\bullet$ where the "blob" $\bullet$ is like an empty spot for us to place the variable x, or f.

	$\mathcal{C}$	$\xrightarrow{\ H_\bullet\ }$	$[\mathcal{C}^{op}, \mathbf{Set}]$
on objects:	x	$\longmapsto$	H_x
on morphisms:	$\boxed{x \xrightarrow{f} y}$	$\longmapsto$	$\boxed{H_x \overset{H_f}{\Longrightarrow} H_y}$

To define this functor we need to define H_f. At this point it is important to remain calm[†] and trust that a little type-checking will see us through this.

─────────────── **Things To Think About** ───────────────

T 23.4   See if you can work out what the natural transformation H_f must do. Just note that like all natural transformations it must have one component for each object $a \in \mathcal{C}$. Remain calm about the fact that if you're defining H_f then x and y have been fixed. Write down what the endpoints of the component must be (by type-checking) and then see what is the only way you can construct a component in that place using the given information.

To define H_f, let's first see where its components must "live". Remember through all of this $x \xrightarrow{f} y$ is fixed.

We are defining a natural transformation $H_f \colon H_x \Rightarrow H_y$, so by definition we must have components as shown on the right.

$$\forall a \in \mathcal{C}, \quad H_x(a) \xrightarrow{(H_f)_a} H_y(a)$$
$$\text{that is,} \quad \mathcal{C}(a, x) \longrightarrow \mathcal{C}(a, y)$$
$$a \xrightarrow{s} x \longmapsto a \xrightarrow{?} y$$

It remains to decide where an element $a \xrightarrow{s} x$ should be mapped to. Now it's like a type-checking jigsaw puzzle: we're starting with a morphism $a \xrightarrow{s} x$ and a morphism $x \xrightarrow{f} y$, and we're trying to produce a morphism $a \to y$, so there's really only one thing we can do: compose s and f.

This gives us the following definition of the component of H_f at a.

$$\mathcal{C}(a, x) \xrightarrow{(H_f)_a} \mathcal{C}(a, y)$$
$$a \xrightarrow{s} x \longmapsto a \xrightarrow{s} x \xrightarrow{f} y$$

> This might remind you of what we did when we were defining the action of the functor H_x on morphisms in the first place. This can get confusing because so many things get defined as post- or pre-composition, but if you remain calm and keep type-checking everything will stay in place.

─────────────── **Things To Think About** ───────────────

T 23.5   Can you check that this definition of H_f satisfies naturality?

Naturality is then largely another case of remaining calm and keeping the notation in place.

We are trying to define a natural transformation as shown on the right.

$$\mathcal{C}^{\mathrm{op}} \underset{H_y}{\overset{H_x}{\rightrightarrows}} \mathbf{Set} \quad \Downarrow H_f$$

─────────────────────

[†] My wonderful PhD supervisor Martin Hyland often reminds us to remain calm, and it's very sound advice in abstract math and also in life.

We need to check naturality squares, but we also need to remember that H_x and H_y are *contravariant* functors so they flip the direction of arrows. If you forget this then the square won't type-check, so you'll get a reminder.

> Technically we have a naturality square for each morphism of $\mathcal{C}^{op}$, but as usual I prefer never drawing morphisms in $\mathcal{C}^{op}$, but keeping them drawn in $\mathcal{C}$ and flipping directions when relevant.

Recall that throughout this whole scenario we have fixed $x \xrightarrow{f} y$ in $\mathcal{C}$. Now we need a naturality square for every $a \xrightarrow{p} b$ in $\mathcal{C}$.

Writing out the definition of naturality square gives the first square here. (Remember that H_x and H_y flip the direction of p.) It evaluates to the second square shown.

$$
\begin{array}{ccc}
H_x(b) & \xrightarrow{(H_f)_b} & H_y(b) \\
{\scriptstyle H_x(p)}\downarrow & & \downarrow{\scriptstyle H_y(p)} \\
H_x(a) & \xrightarrow{(H_f)_a} & H_y(a)
\end{array}
\qquad
\begin{array}{ccc}
\mathcal{C}(b,x) & \xrightarrow{f\,\circ\,\_} & \mathcal{C}(b,y) \\
{\scriptstyle \_\,\circ\,p}\downarrow & & \downarrow{\scriptstyle \_\,\circ\,p} \\
\mathcal{C}(a,x) & \xrightarrow{f\,\circ\,\_} & \mathcal{C}(a,y)
\end{array}
$$

> Once you've evaluated the corners of the square there's really no choice of what the morphisms can be, just by type-checking.

One way to check this commutes is to sort of blindly insert a variable at the top left corner and follow it around according to the formulae, as shown here. We see that going round in either direction gives the same answer by associativity of composition in $\mathcal{C}$.

$$
\begin{array}{ccc}
s & \xmapsto{f\,\circ\,\_} & f \circ s \\
{\scriptstyle \_\,\circ\,p}\downarrow & & \downarrow{\scriptstyle \_\,\circ\,p} \\
 & & (f \circ s) \circ p \\
 & & \| \\
s \circ p & \xmapsto{f\,\circ\,\_} & f \circ (s \circ p)
\end{array}
$$

However, if you want to type-check it properly (to feel better that we did everything right) you can draw all the morphisms out as shown here.

$$
\begin{array}{ccc}
\boxed{b \xrightarrow{s} x} & \xmapsto{f\,\circ\,\_} & \boxed{b \xrightarrow{s} x \xrightarrow{f} y} \\
{\scriptstyle \_\,\circ\,p}\downarrow & & \downarrow{\scriptstyle \_\,\circ\,p} \\
\boxed{a \xrightarrow{p} b \xrightarrow{s} x} & \xmapsto{f\,\circ\,\_} & \boxed{a \xrightarrow{p} b \xrightarrow{s} x \xrightarrow{f} y}
\end{array}
$$

> The first method is fine, but I think the second is safer and more illuminating: it helps us see that there was no choice as to which morphisms were pre-composed and which were post-composed. Drawing out the sources and targets forces us to compose things on the correct side, whereas if you just write strings like $f \circ s$ you could quite merrily write $s \circ f$ and never know that anything was wrong.

We have now constructed the key functor for Yoneda.

Definition 23.1 The functor $\mathcal{C} \xrightarrow{H_{\bullet}} [\mathbb{C}^{\mathrm{op}}, \mathbf{Set}]$ is called the *Yoneda embedding*.

We have slightly skipped ahead of ourselves here because we've called it an "embedding" before proving that it is an embedding; in fact we haven't even said what an embedding is.

—————————————————— Aside on embeddings ——————————————————

The idea of an embedding is that it's a bit like a subset inclusion, but one dimension up. For example, if we observe that the natural numbers are a subset of the integers, there's a function which just performs that inclusion sending every natural number to itself regarded as an integer. We often denote it like this: $\mathbb{N} \hookrightarrow \mathbb{Z}$, with a little "hook" at the beginning to remind us that it's like a subset inclusion $\subset$.

Philosophers may worry about whether the natural number n is actually the same as the integer n or not, but we can get around this by observing that the key is this function is injective, and then it doesn't really matter whether the elements of $\mathbb{N}$ are actually the same ones as in $\mathbb{Z}$ or not.

If we now go up to the level of categories we might want to encapsulate the idea of a *subcategory* inclusion, but now things get a bit hazy. Do we want strict subcategory inclusions or weak ones? That is, do we want to be strictly or weakly injective on objects? Do we want only full subcategory inclusions?

Regardless of quite what definition you take of an embedding, the "Yoneda embedding" is generally considered to be worthy of the name because it is full and faithful, as we'll now show.

———————————————————————————————————————

Theorem 23.2 *The Yoneda embedding is full and faithful.*

This proof essentially falls into place by type-checking, if you remain calm. I find it enormously satisfying to re-prove it rather than look up a proof. It's like doing an abstract jigsaw puzzle again even though you've done it before: it's still satisfying to me no matter how many times I do it. (On the other hand if it's not satisfying to you the first time it might never be satisfying.) It does require one abstract "trick", which I prefer to think of as a general principle.

—————————————————— General Yoneda-y principle ——————————————————

When you're dealing with a functor $H_x = \mathcal{C}(\_, x)$ one thing you can always do in the absence of any special information is evaluate it at the object x, giving $H_x(x) = \mathcal{C}(x, x)$. And then after that, the one thing that you know exists in that homset is the identity 1_x. This is the principle behind all Yoneda-y[†] things: that

———————————————————————————————————————
[†] "Yoneda-y", like "chocolatey".

the one thing we know we have in any world of homsets is the identity, and all the Yoneda functors and natural transformations are acting by composition on one side or the other. This encapsulates all the structure of a category, and thus everything else that follows is deeply inherent to the structure of a category. Typically we don't hold all the proofs in our brain, just this principle, and then everything else flows from it.

─────────── **Things To Think About** ───────────

T 23.6 See if you can prove Theorem 23.2 for yourself using my Yoneda-y principle. The whole thing is basically just unraveling definitions and type-checking. You might well get stuck, in which case I suggest reading through my proof but stopping once I've shown how to use that principle and then trying to proceed yourself. This is in general a good way to read math proofs: try and do it yourself, get stuck, read someone else's proof just to the point where they reveal a step you didn't get, then try and continue by yourself, get stuck again, and iterate.

Proof First let us show that $H_\bullet$ is faithful, which means "locally injective".

So we fix $x, y \in \mathcal{C}$, consider morphisms $f, g \colon x \longrightarrow y$ and check the
implications shown on the right.

$$H_\bullet(f) = H_\bullet(g) \implies f = g$$
$$\text{that is} \quad H_f = H_g \quad \implies f = g$$

Now, H_f and H_g are natural transformations, so being equal means all their components are equal, as functions.

Their respective components for each $a \in \mathcal{C}$
are shown on the right, and we are suppos-
ing that these are equal *for all* $a \in \mathcal{C}$.

$$(H_f)_a \colon \mathcal{C}(a, x) \xrightarrow{f \circ \_} \mathcal{C}(a, y)$$
$$(H_g)_a \colon \mathcal{C}(a, x) \xrightarrow{g \circ \_} \mathcal{C}(a, y)$$

Now we invoke the Yoneda-y principle: we can put $a = x$ and consider $1_x \in \mathcal{C}(x, x)$. We know $(H_f)_x$ has to be the same function as $(H_g)_x$, so they need to send everything to the same place, and in particular have to send 1_x to the same place. But one of them sends 1_x to f and the other to g.

Set $a = x$. The component $(H_f)_x$ is a
function acting on 1_x as shown here.

$$(H_f)_x \colon \mathcal{C}(x, x) \xrightarrow{f \circ \_} \mathcal{C}(x, y)$$
$$1_x \longmapsto f \circ 1_x = f$$

Similarly the component $(H_g)_x$ is a
function acting on 1_x as shown here.

$$(H_g)_x \colon \mathcal{C}(x, x) \xrightarrow{g \circ \_} \mathcal{C}(x, y)$$
$$1_x \longmapsto g \circ 1_x = g$$

By hypothesis[†] the components $(H_f)_x$ and $(H_g)_x$ are equal as functions; this means they have to produce the same result on every element of $\mathcal{C}(x, x)$, and in particular on 1_x. So $f = g$ as required.

This proof went in steps in a process of gradually homing in on the one crucial piece of information at 1_x, like this:

1. $H_f = H_g$ by hypothesis.
2. So they're equal at every component $(H_f)_a = (H_g)_a$; in particular $(H_f)_x = (H_g)_x$.
3. These are equal as functions, so in particular $(H_f)_x(1_x) = (H_g)_x(1_x)$.
4. But if we evaluate these we find that the left is f and the right is g so we can conclude $f = g$ as required.

If you read a proof in a standard textbook you are quite likely to find that the proof of faithfulness consists of one line, essentially the line (3) above. Once you are able to type-check and unravel proficiently, that one line is enough!

We will now show that $H_\bullet$ is full. By definition this means: any natural transformation $\alpha: H_x \Rightarrow H_y$ must be of the form H_f for some $x \xrightarrow{f} y$. So first we need to produce a morphism f from the natural transformation α.

There are two ways to proceed here. One way is to follow your nose and find the only morphism $x \longrightarrow y$ that falls out of the information we have so far. Another way is to think about what we've already done: we discovered that $(H_f)_x(1_x) = f$, which is saying $\alpha_x(1_x) = f$ for the case $\alpha = H_f$; perhaps that works more generally? Both of these approaches produce the same result.

Now α has components for all $a \in \mathcal{C}$, $\boxed{\alpha_a: \mathcal{C}(a, x) \longrightarrow \mathcal{C}(a, y).}$ These α_a are functions, and we can use the same "Yoneda-y" principle:

The one thing we know we have here is the identity $1_x \in \mathcal{C}(x, x)$. So we can look at the action of α_x on 1_x as shown on the right.

$$\mathcal{C}(x, x) \xrightarrow{\ \alpha_x\ } \mathcal{C}(x, y)$$

$$\boxed{x \xrightarrow{\ 1_x\ } x} \longmapsto \boxed{x \xrightarrow{\ \alpha_x(1_x)\ } y}$$

We have produced a "canonical" morphism $x \longrightarrow y$ from α, so we can just try it: set $f = \alpha_x(1_x)$ and claim that $\alpha = H_f$.

At this point if you did the "follow your nose" method, you might think there's no earthly reason to suspect that α should equal H_f; this is where going back to $(H_f)_x(1_x) = f$ helps. That is, it worked in the case $\alpha = H_f$ so we are in with a fighting chance here.

[†] This means "using what we assumed at the start of this proof", which in this case is $H_f = H_g$.

To show that $\alpha = H_f$ we need to show that all their components are equal, that is, for all $a \in \mathcal{C}$, $\alpha_a = (H_f)_a$.

But $(H_f)_a$ acts as shown here.

$$
\begin{array}{ccc}
\mathcal{C}(a, x) & \xrightarrow{(H_f)_a \; = \; f\circ\_} & \mathcal{C}(a, y) \\[4pt]
p & \longmapsto & f \circ p
\end{array}
$$

So we need to show that for any morphism $a \xrightarrow{p} x$, we have $\alpha_a(p) = f \circ p$.

The information we haven't yet used about α is the naturality, so it's just as well to write down what a naturality square looks like: for any morphism $a \xrightarrow{p} b$ the square on the right commutes.

$$
\begin{array}{ccc}
\mathcal{C}(b, x) & \xrightarrow{\alpha_b} & \mathcal{C}(b, y) \\
\downarrow{\scriptstyle \_\circ p} & & \downarrow{\scriptstyle \_\circ p} \\
\mathcal{C}(a, x) & \xrightarrow{\alpha_a} & \mathcal{C}(a, y)
\end{array}
$$

Now we use the Yoneda-y principle again: the one thing we know exists is the element $1_x \in \mathcal{C}(x, x)$. We also know we are trying to prove something for all morphisms $a \xrightarrow{p} x$. This is all pointing towards setting $b = x$ in this diagram and seeing what happens.

Set $b = x$ in the above naturality square. Then for any morphism $a \xrightarrow{p} x$ we have the naturality square shown here.

$$
\begin{array}{ccc}
\mathcal{C}(x, x) & \xrightarrow{\alpha_x} & \mathcal{C}(x, y) \\
\downarrow{\scriptstyle \_\circ p} & & \downarrow{\scriptstyle \_\circ p} \\
\mathcal{C}(a, x) & \xrightarrow{\alpha_a} & \mathcal{C}(a, y)
\end{array}
$$

These are sets and functions, so we can see what happens to a particular element. If we start from the identity $1_x \in \mathcal{C}(x, x)$ and follow it around the diagram in two ways we get the result shown here.

$$
\begin{array}{ccc}
\boxed{1_x} & \xrightarrow{\alpha_x} & \boxed{\alpha_x(1_x) = f} \\
\downarrow{\scriptstyle \_\circ p} & & \downarrow{\scriptstyle \_\circ p} \\
& & \boxed{f \circ p} \\
& & \| \\
\boxed{1_x \circ p = p} & \xrightarrow{\alpha_a} & \boxed{\alpha_a(p)}
\end{array}
$$

Naturality tells us both ways round the square are equal, so we can deduce that $\alpha_a(p) = f \circ p$ and thus that $\alpha = H_f$. So $H_\bullet$ is full as claimed. $\square$

───────────── **Aside on research and proofs** ─────────────

One of the reasons proving something new is harder than re-proving something you already know is true is that when it's new you don't actually know it's true until you've proved it. The best you can do is have a very strong hunch, either from seeing a lot of examples or (my preferred way) by feeling something structural going on in the situation. Whereas if you know that something is true we can do what we did above and say "Well the only way to construct this

f is like this, so it must be that". When you're doing research (or at least, when I'm doing research) I get a very strong hunch that something is true but if I run into trouble proving it then I can just end up in a big mess of doubt, and flip over to trying to prove it's not true. One can spend a lot of time flipping backwards and forwards like that. But the doubt is important because if you trust your hunch too much then you might just miss some subtle point in your conviction that the thing is true. Plenty of erroneous "proofs" have arisen in research like that. Sometimes it just means the proof has a hole but the result turns out to be still correct, but sometimes the result is actually wrong.

The proof that the Yoneda embedding is full and faithful is something I find supremely, joyfully satisfying in its own abstract right, but it is also of deep importance. The idea is that we can start with any category $\mathcal{C}$ and embed it in its presheaf category† via the Yoneda embedding $\mathcal{C} \xrightarrow{H_\bullet} [\mathcal{C}^{op}, \mathbf{Set}]$.

Now there is a general principle we mentioned before which is that the data for a natural transformation lives in the *target* category, and so the structure of a functor category is largely inherited from its target category. The target category for presheaves is the category **Set**, which is a particularly well-behaved category. This means that the category $[\mathcal{C}^{op}, \mathbf{Set}]$ of presheaves inherits excellent structure from **Set**, even if $\mathcal{C}$ does not have that structure, for example limits and colimits.

The fact that $H_\bullet$ is an embedding means we can then sort of regard $\mathcal{C}$ as a subcategory of the presheaf category. Or at least, a copy of it appears as a subcategory of the presheaf category: we know that the embedding takes an object $x \in \mathcal{C}$ to the functor H_x, so if we take all the functors of this form and all the natural transformations between them, we get a copy (or model) of $\mathcal{C}$ as presheaves. Note that taking *all* the natural transformations between them is right because the embedding is full; if it weren't full then taking all the natural transformations would give us extra morphisms that are not in $\mathcal{C}$.

This is one of the reasons that presheaves of the form H_x are so important, and why they get a name: they are called *representable functors*. However, as these are objects in a category, it's better to define them only up to isomorphism, so functors are called representable more generally if they are isomorphic to one of these.

One of the brilliant things about the Yoneda embedding is that $\mathcal{C}$ is not just any old subcategory of $[\mathcal{C}^{op}, \mathbf{Set}]$. Something highly canonical has happened in that $[\mathcal{C}^{op}, \mathbf{Set}]$ is the *closure of $\mathcal{C}$ under colimits*. That is, we throw in all colimits for things in $\mathcal{C}$ and nothing more. So the parts of the presheaf category

---

† Recall: functors $\mathcal{C}^{op} \longrightarrow \mathbf{Set}$ are called presheaves on $\mathcal{C}$.

that are not in the (model of) the subcategory $\mathcal{C}$ are all colimits of the things in $\mathcal{C}$ in a canonical way. The other side of this coin is then to say that *every presheaf is canonically a colimit of representables*. These are all very profound results in category theory, illustrated by the schematic diagram below.

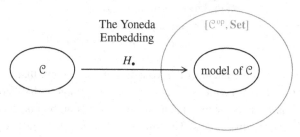

In fact, the principles we used in proving that the Yoneda embedding is full and faithful are part of a more general theorem which we will now describe.

23.5 The Yoneda Lemma

The Yoneda Lemma is called a "lemma" although it's really very profound, important, and ubiquitous. Usually something is called a lemma if it's just something slightly trivial that helps us prove something else. I like the fact that this makes the Yoneda Lemma sound self-deprecating. In a way it really is very trivial and the proof, though complex, is essentially just many layers of type-checking. However the ramifications are very widespread.

It essentially captures the "Yoneda-y principle" we used above, which was that our natural transformations kept being entirely determined by looking at the identity. We were looking at natural transformations $H_x \Rightarrow H_y$ but the principle only really depended on the *source* functor being of the form H_x. The target functor could have been any functor $F \colon \mathcal{C}^{\mathrm{op}} \to \mathbf{Set}$.

So we consider a natural transformation as shown on the right. By definition, it must have a component α_a for every object $a \in \mathcal{C}$, living here: $H_x(a) = \mathcal{C}(a, x) \xrightarrow{\alpha_a} F(a)$.

$$\mathcal{C}^{\mathrm{op}} \Downarrow \alpha \; \mathbf{Set}$$

We can now use the Yoneda-y principle. We set $a = x$ and look at the action of α_x on $1_x \in \mathcal{C}(x, x)$, as shown on the right.

$$\mathcal{C}(x, x) \xrightarrow{\alpha_x} F(x)$$
$$\left(x \xrightarrow{1_x} x \right) \longmapsto \alpha_x(1_x)$$

The image[†] of 1_x, that is $\alpha_x(1_x)$, is now not a morphism, because $F(x)$ is a set, not necessarily a homset. But the idea is that in order to define a natural

---

[†] This means: where 1_x lands when we apply the function in question.

transformation α we have a free choice of where 1_x can go, and moreover this completely determines the rest of the components of α, by following a naturality square around as we did in the proof of $H_\bullet$ being full. Thus:

> natural transformations $H_x \Rightarrow F$
> correspond precisely to
> elements of $F(x)$

This is essentially the content of the Yoneda Lemma, except that because we now understand the principles of category theory we know that "correspond precisely to" is a rather vague statement and sounds ambiguously a bit like it's just a bijection between sets. The things in this case don't just correspond, they correspond in a way that uses the identity and composition structure of the categories all over the place. That is in turn encapsulated by some naturality. Here is what the formal statement often looks like:

Theorem 23.3 (Yoneda Lemma) *Let $\mathcal{C}$ be a locally small category and F a functor $\mathcal{C}^{op} \to$ Set. Then there is an isomorphism $F(x) \cong [\mathcal{C}^{op}, \text{Set}](H_x, F)$ natural in x and F.*

If you have a feeling that there are too many levels of naturality going on here I sympathize. In fact I very vividly remember that feeling from when I was going over my Category Theory lecture notes the very first time I saw the Yoneda Lemma, and feeling like I had missed something very serious. The point here is that where we've said "natural in..." in the statement of the theorem, we have cut a corner by invoking naturality of a natural transformation that we never even defined. We didn't even define the functors it lived between. But whenever we say "natural in X" what we mean is that X lives in some category, and if we use a different X and a morphism in between the two different X's, we will get a square and it needs to commute.

In this case x lives in the category $\mathcal{C}$ so that's not too surprising; naturality in x is thus based on a morphism $x \xrightarrow{f} y$ and a square like this (remembering that F is contravariant so flips the direction of f):

$$\begin{array}{ccc} Fy & \xrightarrow{\sim} & [\mathcal{C}^{op}, \text{Set}](H_y, F) \\ {\scriptstyle Ff}\big\downarrow & & \big\downarrow{\scriptstyle \_ \circ H_f} \\ Fx & \xrightarrow{\sim} & [\mathcal{C}^{op}, \text{Set}](H_x, F) \end{array}$$

Now F is a functor, but as an object it lives in the category $[\mathcal{C}^{op}, \text{Set}]$ so "naturality in F" is based on a natural transformation $F \xRightarrow{\alpha} G$ and a square like this:

$$\begin{array}{ccc} Fx & \xrightarrow{\sim} & [\mathcal{C}^{op}, \text{Set}](H_x, F) \\ {\scriptstyle \alpha_x}\big\downarrow & & \big\downarrow{\scriptstyle \alpha \circ \_} \\ Gx & \xrightarrow{\sim} & [\mathcal{C}^{op}, \text{Set}](H_x, G) \end{array}$$

The proof is then basically a lot of type-checking, similar to what we've done already in this chapter. I'm sorry not to include it, but I hope you are now well-placed to try it yourself, and, if you get stuck, to read it elsewhere.[†] It looks complicated when you write everything out but I hope to have conveyed a sense that all it really means is that *everything you could possibly hope to be true is true* — as long as you hope for the right things. This is one of the reasons I love category theory: it's a place where all my dreams come true.

23.6 Further topics

Mac Lane said that all concepts are Kan extensions. In fact Kan extensions are examples of adjunctions, which are examples of universal properties, which can all be expressed via representability of certain very souped-up functors. That is in turn all based on Yoneda, so one could also say that *all concepts are Yoneda*.

This doesn't mean everything else is redundant: sometimes you have to soup things up rather a lot to see how it's really Yoneda, and it's important to understand the unraveled version as well. But the skill of souping things up in order to see how everything is related is, I think, a key skill in abstract math and particularly category theory. It can seem like complicating things unnecessarily, but the point is that although we're complicating a framework, we're doing it in order to simplify a thought process.

My personal favorite way to do that process of "complicating in order to simplify" is to go into higher dimensions. We've already been doing that somewhat: we had to go up a dimension from sets to categories in order to understand structure better. We then found that we had to go up a dimension to 2-categories to deal with natural transformations, and although we haven't done it, this is also what enables us to express general limits and colimits formally, as well as opening up the way to adjunctions and monads.

In the next and last chapter we will briefly explore the ideas and principles of higher-dimensional category theory.

[†] For example see *Category Theory in Context* (Riehl).

24

Higher dimensions

We continue applying the principle of looking at relationships between things, because we also need to look at relationships between those relationships, and so on. This gives us more dimensions, possibly infinitely many.

In this final chapter I'll give a broad sweeping overview of my particular field of research: higher-dimensional category theory. This will not be detailed or rigorous; the aim is to give an idea and flavor of how the field develops into higher dimensions. Rigorous details are far beyond our scope, but I hope the ideas feel like a natural development of all we've done so far.

24.1 Why higher dimensions?

We began our journey into category theory with the idea of relationships. We said that we wanted to study things in the context of their relationships with other things, rather than in isolation. This gave us the idea of morphisms between objects; taking objects and morphisms together is what gave us the notion of a category. This gave us, among other things, a more nuanced notion of sameness, and a way of studying various kinds of abstract structure arising from patterns in relationships.

But now, if we believe in studying relationships between things, what about relationships between morphisms? What about a more nuanced notion of sameness between morphisms? What about a way of studying various kinds of abstract structure on morphisms? If we're interested in categories rather than sets, how about categories of morphisms instead of sets of them?

This is a way in which abstract principles nudge us into higher dimensions just by us using our imagination. A way in which something more like practice nudges us that way is if we think about totalities of structures. We have seen that sets are 0-dimensional structures, as they just consist of points; but sets and functions together make a category, which is 1-dimensional.

Then, categories come with functors and also natural transformations. The natural transformations are "morphisms between functors" so they are one dimension higher, and they arise from the existence of morphisms *in* categories. So the extra dimension *inside* categories produced an extra dimension in the *totality* of categories.

When we have objects, morphisms, and morphisms between morphisms, this is called a 2-category. For more succinct terminology we sometimes call these 0-cells, 1-cells and 2-cells. This terminology also generalizes well into more dimensions (rather than saying morphisms between morphisms between morphisms between...).

We have a growing system of structures and their totalities, as shown on the right. This definitely suggests that things are not going to stop at 2-categories. As you might have guessed, the totality of 2-categories is a 3-category, and this continues so that we have a notion of *n*-category for every finite *n*, and *n*-categories form an *n* + 1-category.

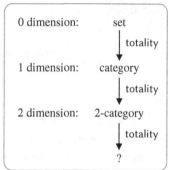

More specifically, if we look at a totality of 2-categories it seems quite likely that we will gain an extra level of structure beyond functors and natural transformations. Natural transformation arose from the existence of morphisms in categories, but now we also have 2-cells. The 2-cells *inside* a 2-category go on to produce morphisms between natural transformations.

We are also nudged towards higher-dimensional thinking just by dreaming of morphisms between morphisms between morphisms, and so on, for dimension after dimension. In fact, why do we ever need to stop? The answer is we don't. In an *n*-category we have morphisms at every dimension up to *n*, but we could have a higher-dimensional structure with morphisms at every dimension "forever", with no upper limit. This is called an *infinity category*.[†]

This is all very well as an idea, but just like most dreams it is much harder to realize in practice than just to dream about. In this chapter we're going to sketch out how to do it, why it's hard, and why therefore there are reasons to try and keep things finite-dimensional if we can.

This is the last chapter in the book and is about a huge open research field, so much of this will be a sketch and a glimpse rather than anything rigorous. I will include more Things to Think About that I will not fully write out afterwards.

[†] They are sometimes also called ω-categories, where ω is the Greek letter omega which is often used to denote the "first" infinity in set theory, that is, the size of the natural numbers. We use this because really we're saying there is a dimension of morphism for every natural number.

24.2 Defining 2-categories directly

As a starting point we can look at the full definition of 2-category, which we have hinted at. Many of the issues with higher dimensions can be illustrated in 2 dimensions, including the fact that there are many different possible approaches and they diverge more and more as dimensions increase.

For the definition of a 2-category we basically take the structure of categories, functors and natural transformations and encapsulate it as data, structure and properties. The most elegant ways to do it require some more theory, but here is a direct definition first.

——————————— 2-categories: elementary definition ———————————

A 2-category $\mathcal{A}$ consists of:

DATA

- A collection of 0-cells.
- For any 0-cells a, b a collection $\mathcal{A}(a, b)$ of 1-cells from a to b, drawn as shown.
- For any parallel 1-cells $f, g \colon a \to b$ a collection $\mathcal{A}(f, g)$ of 2-cells from f to g, drawn as shown.

STRUCTURE

- Identity 1-cells: ∀ 0-cells a a 1-cell $a \xrightarrow{1_a} a$

- Composition of 1-cells: ∀ 1-cells $a \xrightarrow{f} b \xrightarrow{g} c$ a 1-cell $a \xrightarrow{g \circ f} c$

- Identity 2-cells: ∀ 1-cells $a \xrightarrow{f} b$ a 2-cell $a \Downarrow 1_f \ b$

- Composition of 2-cells

    vertical:[†]   ∀ 2-cells $a \ \Downarrow\alpha \ \Downarrow\beta \ b$ a 2-cell $a \Downarrow \beta \circ \alpha \ b$

    horizontal: ∀ 2-cells $a \ \Downarrow\alpha \ b \ \Downarrow\beta \ c$ a 2-cell $a \Downarrow \beta * \alpha \ c$

PROPERTIES

- Unit laws: the identity 1-cells act as identities with respect to 1-cell composition, and identity 2-cells act as identities with respect to vertical composition of 2-cells.

———————————

† In vertical composition, the 1-cell g isn't there in the composite because it's like the middle object — the composite goes all the way from end to end.

- Associativity: composition of 1-cells, and both horizontal and vertical composition of 2-cells are associative.
- The interchange law (Section 22.10), summed up in this diagram:

Perhaps you feel that this definition has a "nuts and bolts" feel to it, and that it doesn't really have a satisfying logic behind it; if so then you are thinking like a category theorist. I will now sketch two more abstractly elegant ways of approaching this definition, both of which come to the same thing when unraveled. When we are making abstract definitions there is no real concept of right and wrong, as long as we don't actually cause a contradiction. There are instead guiding principles:

- External: are there many examples of this structure?
- Internal: is there some compelling internal logic behind the definition?

The nuts and bolts definition came from looking at a motivating example: the totality of categories, functors and natural transformations. We'll now try two ways of looking at the internal logic, coming from the two approaches to defining categories in the first place: by homset, or by underlying graph.

24.3 Revisiting homsets

In this approach we are going to take seriously the idea of thinking about relationships between morphisms. Instead of having *sets* of morphisms, we are going to have *categories* of morphisms. Then any part of the definition that was previously a function between sets of morphisms is now going to have to be a *functor* between categories of morphisms. This is a general abstract process called *enrichment*: we "enrich" a category by allowing its morphisms to have some structure on them. As we'll see, it is a formal definition, not just an idea.

First we need to revisit the definition of category in an abstract way that best suits this generalization. The following expression is the same definition we're used to, just expressed slightly more "categorically", that is, using sets and functions rather than sets and elements.

Definition of category: revisited

A (locally small) category $\mathcal{C}$ is given by

DATA

- A collection of objects.
- For any objects a, b a set $\mathcal{C}(a, b)$ of morphisms from a to b.

STRUCTURE

- Identities: for any object a, a function as shown,[†] picking out an identity morphism $a \xrightarrow{1_a} a$.

$$1 \xrightarrow{\text{id}} \mathcal{C}(a,a)$$
$$* \mapsto 1_a$$

- Composition: for any objects a, b, c, a function as shown, sending composable pairs to composites.

$$\mathcal{C}(b,c) \times \mathcal{C}(a,b) \xrightarrow{\text{comp}} \mathcal{C}(a,c)$$
$$b \xrightarrow{g} c, \ a \xrightarrow{f} b \ \mapsto \ a \xrightarrow{g \circ f} c$$

For the last part we need to recall that the product of two sets contains ordered pairs with one element from each set. So the product $\mathcal{C}(b,c) \times \mathcal{C}(a,b)$ is a set containing pairs of morphisms, one with target b and one with source b, so they are composable at b. The reason we've put it in the "backwards" order is to match up with the "backwards" order in which we write composition.

PROPERTIES

The axioms for units and associativity can now be expressed entirely according to these *functions* on homsets, without reference to *elements* of homsets.

The two triangles are the unit axioms saying that identity morphisms act as identities on both the left and the right; the square underneath is associativity.

I've put the diagrams here for you to see, but won't go into much explanation; you might like to chase elements around them and see if you can understand why they correspond to the axioms in question.

$$\mathcal{C}(a,b) \cong 1 \times \mathcal{C}(a,b) \xrightarrow{\text{id} \times 1} \mathcal{C}(a,a) \times \mathcal{C}(a,b)$$

with 1 to $\mathcal{C}(a,b)$ and comp down to $\mathcal{C}(a,b)$

$$\mathcal{C}(a,b) \cong \mathcal{C}(a,b) \times 1 \xrightarrow{1 \times \text{id}} \mathcal{C}(a,b) \times \mathcal{C}(b,b)$$

with 1 to $\mathcal{C}(a,b)$ and comp down to $\mathcal{C}(a,b)$

$$\mathcal{C}(c,d) \times \mathcal{C}(b,c) \times \mathcal{C}(a,b) \xrightarrow{1 \times \text{comp}} \mathcal{C}(c,d) \times \mathcal{C}(a,c)$$

$$\text{comp} \times 1 \downarrow \qquad\qquad \downarrow \text{comp}$$

$$\mathcal{C}(b,d) \times \mathcal{C}(a,b) \xrightarrow{\text{comp}} \mathcal{C}(a,d)$$

Now that we are only referring to homsets and functions between them (not elements), we have made the definition more "categorical". All the diagrams above are just diagrams in the category **Set**, so we can take this definition and try to put it inside any other category $\mathcal{V}$. This means that everywhere we

---

[†] Here 1 denotes the terminal set; remember a function from the terminal set just picks out an element of the target set. I've written $*$ for the single object of the terminal set.

referred to a set of morphisms we now have an object of $\mathcal{V}$ instead, and everywhere we have a function between homsets we now have a morphism in $\mathcal{V}$. If you look at the definition carefully you will see that at one point we refer to products of homsets, so actually it looks like we need products in $\mathcal{V}$. It turns out that something slightly less stringent than a categorical product will do, called a monoidal structure. But we'll come back to that.

If we move all our homsets from **Set** to some category $\mathcal{V}$, this is the notion of a *category enriched in* $\mathcal{V}$. We can then define 2-categories as categories enriched in **Cat**. This means that instead of homsets we have hom-categories, and structure is now given by functors between hom-categories. We will now sketch out that definition.

Definition of 2-category by enrichment

A (locally small) 2-category $\mathcal{C}$ is a category enriched in **Cat**. That is:

DATA

- A collection of objects.
- For any objects a, b a *category* $\mathcal{C}(a, b)$ of morphisms from a to b.

> Note that the objects and morphisms of this hom-category are then the 1-cells and 2-cells of the 2-category. Composition in this category is vertical composition.

STRUCTURE

- Identities: for any object a a *functor* $\mathbb{1} \to \mathcal{C}(a, a)$ picking out an identity 1-cell $a \xrightarrow{1_a} a$.

> Note that $\mathbb{1}$ is the terminal category, with one object and just the identity morphism. The identity morphism in $\mathbb{1}$ has to be sent to the identity morphism in $\mathcal{C}(a, a)$ so we get no further information from looking at the action on morphisms.

- Composition: for any objects a, b, c a functor

$$\mathcal{C}(b, c) \times \mathcal{C}(a, b) \longrightarrow \mathcal{C}(a, c)$$

on objects (1-cells): $\quad b \xrightarrow{g} c \,,\ a \xrightarrow{f} b \ \longmapsto\ a \xrightarrow{g \circ f} c$

on morphisms (2-cells): $\quad b \underset{g_2}{\overset{g_1}{\Rightarrow\beta}} c \,,\ a \underset{f_2}{\overset{f_1}{\Rightarrow\alpha}} b \ \longmapsto\ a \underset{g_2 \circ f_2}{\overset{g_1 \circ f_1}{\Rightarrow\beta*\alpha}} c$

So the composition functor gives us both composition of 1-cells and horizontal composition of 2-cells.

PROPERTIES

The axioms for units and associativity are now the same commutative diagrams we drew for the categorical definition of a category, they just happen to live in **Cat** now rather than **Set**.

─────────────────── Things To Think About ───────────────────

T 24.1   For me the most interesting thing about this definition is the fact that functoriality of the composition functor corresponds to the interchange law. See if you can unravel that.

Note that once we start going into higher dimensions we may feel a compulsion to keep asking about the next dimension. With the definition of category enriched in $\mathcal{V}$ we can iterate the process as long as we can always form products. Then 2-categories will form a category with products, and if we enrich in that we'll get 3-categories, and so on.

You might notice that 2-categories are really supposed to form a 3-category. That's a good thing to notice, and is one of the reasons things get more subtle. You might also have noticed that we blithely said we would ask for some commutative diagrams of functors. This is not really the right level of sameness as it involves equality of functors, and we now know that we could (and perhaps should) ask for them to be naturally isomorphic instead. This is the question of strictness and weakness, which we'll come back to shortly.

24.4 From underlying graphs to underlying 2-graphs

A different way to go up a dimension starts from the definition of a (small) category via an underlying graph, as shown here.
$$C_1 \underset{t}{\overset{s}{\rightrightarrows}} C_0$$

As this is a diagram in **Set**, we could try putting the entire diagram inside **Cat** instead of **Set**. This is called *internalization* but it gives us something slightly different: we'll get a category of morphisms and a category of objects as well. We will come back to that.

For now we'll directly extend the graph by one dimension, to include a set of 2-cells like this:
$$C_2 \underset{t}{\overset{s}{\rightrightarrows}} C_1 \underset{t}{\overset{s}{\rightrightarrows}} C_0$$

However we now need a condition on the source and target maps, to ensure that our 2-cells are globular shaped, as shown here.

This shape occurs because the source and target 1-cells of α themselves have the same source and target. That is $ss = st$ and $ts = tt$. This is called the *globularity condition*. A diagram of sets and functions satisfying this condition is called a (2-dimensional) *globular set* or 2-globular set[†] and generalizes into n dimensions. Our original graphs are 1-globular sets.

Here are the three most interesting 1-globular sets that live inside the above 2-globular set. The first two are somewhat obvious, involving just the right-hand part or the left-hand part; the third comes from composing all the way from end to end.

$$C_1 \underset{t}{\overset{s}{\rightrightarrows}} C_0$$

$$C_2 \underset{t}{\overset{s}{\rightrightarrows}} C_1$$

$$C_2 \xrightarrow[ts = tt]{ss = st} C_0$$

In the 2-category of categories, functors and natural transformations, the first sub-globular set (collection) above has categories as its objects, and functors as its morphisms. We know this has the structure of a category: it's the basic category **Cat** from before we even thought about natural transformations.

The second sub-globular set (involving C_2 and C_1) has functors as objects and natural transformations as morphisms. This also forms a category. Previously we have defined functor categories $[\mathcal{C}, \mathcal{D}]$ only after fixing the source and target categories, but we can throw all functors together into a large category of functors and natural transformations with any endpoints.

The third sub-globular set is a little curious. It considers natural transformations with respect to their source and target *categories* rather than the functors.

So the natural transformation shown on the right is considered to have source $\mathcal{A}$ and target $\mathcal{B}$.

We will then consider composition when the target category of one natural transformation matches the source category of another, that is, horizontal composition.

So asserting that each of these three sub-globular sets is a category gives us, respectively, composition of 1-cells, vertical composition of 2-cells, and horizontal composition of 2-cells. In this framework we have to assert interchange separately, but we essentially have the following definition.

[†] We could also call it a 2-graph but technically that is defined slightly differently.
[‡] That is not actually a *small* 2-category but the ideas still hold, just with collections instead of sets.

2-categories: globular set definition

A (small) 2-category is a 2-globular set in which

- every sub-1-globular set has the structure of a category, and
- interchange holds.

This definition gives the same structure as the one defined by enrichment (aside from size issues) but it has a different feel to it, and leads to different types of generalization. We saw that even for ordinary categories these two different characterizations (by homset, or by underlying graph) give us slightly different insights. For example, from the graph approach we saw how duality naturally arises from the symmetry between s and t. In a 2-globular set we have *two* pairs of source/target functions, each with analogous symmetry. We can flip either pair independently. Flipping the left-hand pair consists of flipping the direction of the 2-cells; flipping the right-hand pair consists of flipping the direction of the 1-cells; or we can flip both.

- Flipping the direction of the 1-cells is like what we do with ordinary categories, and this gets called "op" just like for categories.
- Flipping the direction of the 2-cells means we're taking the duals of the hom-categories; this gets called "co".
- Flipping both means we flip everything, and is inevitably called "co-op".

It recently occurred to me to try and think of the opposite of "FOMO" (the Fear Of Missing Out). During the COVID-19 era I was very grateful not to have to join in with anything, being lucky to be able to work from home. I thought of calling it "FOJI" for Fear Of Joining In, but then I saw that others had called it "JOMO" for Joy Of Missing Out. Those are not quite the same thing, just as the op and the co of a 2-category are not in general the same thing. I decided that if this were a higher-dimensional structure then the emotion (fear/joy) would be a higher dimension than the thing itself (missing out/joining in), so

- $(\text{FOMO})^{\text{op}} = \text{FOJI}$
- $(\text{FOMO})^{\text{co}} = \text{JOMO}$
- $(\text{FOMO})^{\text{coop}} = \text{JOJI}$ — Joy Of Joining In.

I think JOJI is really not the same as FOMO; the latter seems a little sad, whereas JOJI is a much more active and positive idea.

That is all very far from rigorous but is the sort of way one's brain might start working when one spends a lot of time thinking about higher dimensions. (Well, mine does.)

Once we know what 2-categories are, there are various natural things to do to develop category theory. We could develop "2-category theory": take everything we've done in one dimension and try to do it in two, by replacing equalities with isomorphisms and working out what axioms those isomorphisms should satisfy one dimension up.

Another key way to progress is to express everything 2-categorically. Recall that one of the principles of 1-category theory was to take a "familiar" structure involving sets, and express it just using sets and *functions* rather than sets and elements. We then go round looking for the structure in categories other than **Set**. Once we know about 2-categories we can take a familiar structure involving categories, and express it just using categories, functors and natural transformations (rather than objects and morphisms inside any categories). We can then go round looking for it in 2-categories other than **Cat**.

────────────── **Things To Think About** ──────────────

T 24.4   Think about the two definitions of equivalence of categories. Which one is 2-categorical? Can you see how this gives us a notion of sameness for 0-cells in any 2-category, slightly weaker than isomorphism?

The first definition we saw was "pointwise" equivalence of categories. The definition was expressed in terms of a functor being full, faithful and essentially surjective on objects. The phrase "on objects" is a clue to this involving elements, as is the term "pointwise"; note that the definitions of full and faithful involve thinking about morphisms *inside* categories, rather than just functors *between* them.

The second definition of equivalence of categories was expressed in terms of pseudo-inverses: functors F and G going back and forth, and then natural isomorphisms $GF \cong 1$ and $FG \cong 1$. As this definition only involves categories, functors and natural transformations between them, it is something we can now do in any 2-category instead of **Cat**. We would just express it using 0-cells, 1-cells and 2-cells, instead of categories, functors and natural transformations.

This gives us the "correct" notion of sameness for 0-cells in a 2-category. Isomorphism is too strong as it invokes equalities between 1-cells. However, for 1-cells isomorphism is the "correct" notion of sameness as it only invokes equalities between 2-cells; 2-cells are the top dimension, so equality is the only relationship we have for them.

In summary, here are the "correct" notions of sameness for cells at each dimension in a 2-category.

dimension	sameness
0-cells	equivalence
1-cells	isomorphism
2-cells	equality

We see that the more dimensions that a cell has *above* it, the more subtle a notion of sameness we can get. The number of dimensions that a cell has above it is called its *codimension*.

So far the only example of a 2-category we have seen is the 2-category of categories, functors and natural transformaions. We will now look at an important special case of 2-categories: those with only one 0-cell. These are called monoidal categories.

24.5 Monoidal categories

Monoidal categories are one of the most widespread examples of a *slightly* higher-dimensional structure. They are a "categorification" of the concept of monoids. Categorification[†] is the process of going up a dimension by turning sets into categories. It is not exactly a rigorously defined term, because it is not a straightforwardly definable process. However, it is a good guiding principle.

A monoid is a set equipped with a binary operation; the categorification will be a category equipped with a binary operation, and is called a monoidal category. So far this is an idea, not a definition. The abstract framework of 2-categories helps us to make the definition in a way that we know will have good internal logic. We can proceed by analogy with monoids. Monoids arise as one-object categories, and when we unravel this we find that the non-trivial information amounts to a set equipped with a binary product. We can proceed analogously, but this time starting from 2-categories, as follows.[‡]

Definition 24.1 A *monoidal category* is a 2-category with only one 0-cell.

If we unravel this we can perform a dimension shift just like we did for monoids:

- The single 0-cell gives us no information so we can ignore it.
- We have non-trivial 1-cells and 2-cells; we regard these as objects and morphisms in the new structure.
- *All* 1-cells are now composable, so ∘ on 1-cells becomes a binary operation on our new objects, which we often write as ⊗.
- *All* 2-cells are now horizontally composable so ∗ on 2-cells becomes a binary operation, that is, ⊗ extends to morphisms in our new structure.

[†] This term was coined by Louis Crane.

[‡] The original definition was made directly as a category with a binary operation, just as the original definition of monoid was made directly as a set with a binary operation.

Here is a diagram of the dimension shift:

2-category		monoidal category
unique 0-cell	$----\blacktriangleright$	
1-cells	$----\blacktriangleright$	objects
2-cells	$----\blacktriangleright$	morphisms
composition of 1-cells	$----\blacktriangleright$	$\otimes$ on objects
horizontal composition of 2-cells	$----\blacktriangleright$	$\otimes$ on morphisms
vertical composition of 2-cells	$----\blacktriangleright$	composition of morphisms

We can think about this more abstractly using the enriched definition of a 2-category, in which we have hom-categories $\mathcal{C}(a, b)$ for any objects a, b.

For any objects a, b, c we have a composition *functor* as shown here, giving composition of 1-cells and horizontal composition of 2-cells.

$$\mathcal{C}(b, c) \times \mathcal{C}(a, b) \longrightarrow \mathcal{C}(a, c)$$

Now if our 2-category has only one 0-cell, say $*$, then there is just one hom-category $\mathcal{C}(*, *)$ and a single composition functor:

$$\mathcal{C}(*, *) \times \mathcal{C}(*, *) \longrightarrow \mathcal{C}(*, *)$$

The "dimension shift" then consists of not bothering to think about the single object $*$, and writing the single hom-category $\mathcal{C}(*, *)$ simply as $\mathcal{C}$. This is the underlying category of our monoidal category.

The composition functor then becomes a functor as shown here. We often write it as $\otimes$, pronounced "tensor", and it's not to be confused with composition $\circ$.

$$\mathcal{C} \times \mathcal{C} \xrightarrow{\ \otimes\ } \mathcal{C}$$

on objects: $\quad a, b \longmapsto a \otimes b$

on morphisms:
$$\begin{array}{ccc} a & b & \\ f\downarrow & \downarrow g & \\ a' & b' \end{array} \longmapsto \begin{array}{c} a \otimes b \\ \downarrow f \otimes g \\ a' \otimes b' \end{array}$$

Things To Think About

T 24.5 Earlier on I said I found it interesting that functoriality of the composition functor in a 2-category gives interchange; if you like, try and see what the functoriality of this $\otimes$ functor does.

There are further categorified analogies we can make with monoids. Recall that we can think of the binary operation in a monoid as being like a form of multiplication, but it need not actually come from multiplication.

Likewise for monoidal categories we can think of the $\otimes$ operation as being like a multiplication, and while it can come from categorical products (with some caveats about weakness which we'll come to) it need not. We mentioned this when we gave the definition of enriched categories: we can enrich in a category with products, but we can also do it in a monoidal category where the $\otimes$ operation isn't a categorical product.

One motivating example for abstract mathematicians is the tensor product of vector spaces; tensor products in general are subtle binary operations on abstract structures, which are not the same as products but can be studied a little like products. The framework of monoidal categories gives us a way to do that. As tensor products arise in a vast range of mathematical fields, monoidal categories are one of the most widespread examples of categorical structures.

The main subtlety comes from the issue of what the axioms should be. This is the question of weakness.

24.6 Strictness vs weakness

We have not been terribly rigorous in our exploration of 2-categories, but you might have noticed that we did not do things as subtly as we might: we asked for some equalities at levels where something more nuanced was possible. This is the issue of "weakness" for higher-dimensional categories. If we ask for strict equalities even though a more nuanced type of sameness is available, we are doing things *strictly*. So far we have really been talking about *strict* 2-categories. To do things *weakly*, we must use the most subtle version of sameness available at whatever dimension of cell we're looking at.

In the definition of 2-category the issue arises where we asked for diagrams to commute in **Cat**. Those diagrams gave us the unit and associativity laws, but were at the "wrong" level of weakness: the more nuanced notion of sameness for functors is natural isomorphism. So we did something unnecessarily strict.

It is a little easier to see with monoidal categories.

─────────────────── **Things To Think About** ───────────────────

T 24.6   Suppose we have a category $\mathcal{C}$ and a functor $\otimes$ giving us a binary operation on objects and on morphisms. What would associativity look like? What's wrong with that?

───

Associativity on objects would look like this: $(a \otimes b) \otimes c = a \otimes (b \otimes c)$.

This is an equality between objects in a category, so it's the "wrong" level of sameness. The correct level is an isomorphism. And the best thing to do is not just to say that these things are isomorphic, but to specify an isomorphism showing it, like this: $(a \otimes b) \otimes c \xrightarrow{\alpha_{abc}} a \otimes (b \otimes c)$.

T 24.7   What would the isomorphisms look like for the unit laws?

For the unit laws the same thought process leads us to realize that we don't want the identity to act as a strict identity but as a weak one. As a result we tend not to write it as 1 any more, but perhaps I. The unit isomorphisms look like this: $\boxed{a \otimes I \xrightarrow{r_a} a}$ and $\boxed{I \otimes a \xrightarrow{l_a} a.}$ The letters r and l stand for "right" and "left" as they're dealing with a unit on the right or a unit on the left.

The "special" isomorphisms that we specify to replace the equalities for associativity and unit laws are generally called "structure isomorphisms" or "coherence isomorphisms". There are now several more complications:

- What happens at the level of morphisms?
- How are the structure isomorphisms for different objects related?
- Do the structure isomorphisms satisfy some axioms?

The general idea now is that "everything that could commute does commute". That is, whenever there are two ways to get from one place to another using the coherence isomorphisms, they should give the same answer so that we don't get abstract anarchy.

T 24.8   Try working out what happens for associativity of $\otimes$ on morphisms. You might think we still have to ask for strict associativity on morphisms, because that is the top level of dimension. However, if you draw out the endpoints you should find that that would not typecheck. What structure isomorphisms could you insert to fix that? If you can do that, see if you can "follow your abstract nose" to write down some other things you might like to be true.

Here is a case where writing down strings of symbols could lead us astray. We could merrily write down an axiom for associativity of morphisms, like what we have in the strict case: $\boxed{(f \otimes g) \otimes h = f \otimes (g \otimes h).}$ However this doesn't typecheck any more, as we'll now see.

Suppose the original morphisms are f, g, and h as shown here; I've drawn them vertically as it might help when we start doing $\otimes$ on them.

$$\begin{array}{ccc} a & b & c \\ \downarrow f & \downarrow g & \downarrow h \\ a' & b' & c' \end{array}$$

Then the two sides of the supposed associativity equation on morphisms are the morphisms shown here.

$$\begin{array}{c|c} (a \otimes b) \otimes c & a \otimes (b \otimes c) \\ \downarrow {\scriptstyle (f \otimes g) \otimes h} & \downarrow {\scriptstyle f \otimes (g \otimes h)} \\ (a' \otimes b') \otimes c' & a' \otimes (b' \otimes c') \end{array}$$

The issue is that in a weakly associative situation those endpoints don't match, so we can't ask for the morphisms to be equal.

However, we do have structure isomorphisms mediating between the ends, so
we can ask instead for this diagram to
commute.

$$(a \otimes b) \otimes c \xrightarrow{\alpha_{abc}} a \otimes (b \otimes c)$$

$$\downarrow (f \otimes g) \otimes h \qquad f \otimes (g \otimes h) \downarrow$$

$$(a' \otimes b') \otimes c' \xrightarrow{\alpha_{a'b'c'}} a' \otimes (b' \otimes c')$$

You might notice that this looks a bit like a naturality square. In fact, it looks
so much like a naturality square, it *is* a naturality square. The structure isomorphisms α_{abc} compile into a natural transformation, as do the ones for units.

We still need some axioms for the structure isomorphisms. We want them to
behave well enough that we can manipulate things somewhat as if they were
strict equalities, even though they're not. That means that if we move parentheses around using the isomorphisms, it shouldn't matter in what steps we do
it — one of the key things about equalities is that we can pile them up on each
other, and keep substituting things for other things safe in the knowledge that
the equalities will still hold. This is what we want here.

We can look for possible issues by thinking about situations in which different paths might have been possible, and then ask for them to be equal. For
example, if we're faced with the expression $(a \otimes b) \otimes c$, there is only one way
to use a structure isomorphism to move the parentheses to the right. However
with the expression $((a \otimes b) \otimes c) \otimes d$, there are two things we could do.

─────── **Things To Think About** ───────

T 24.9   Can you find the two ways to move parentheses to the right in the
expression $((a \otimes b) \otimes c) \otimes d$, using structure isomorphisms? Once you've done
the first step (in two ways), you can keep moving parentheses to the right without any further branching choices. This makes two paths built from structure
isomorphisms. Follow them around until they meet up again; those will be two
paths that we want to ensure compose to the same thing.

We get this pentagon called
the *associativity pentagon*;
it is one of the axioms for
a weak monoidal category.
The other axiom involves
the unit object I.

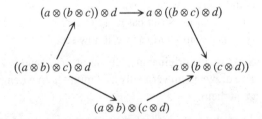

─────── **Things To Think About** ───────

T 24.10   Can you find two ways to use structure isomorphisms to "get rid of
the I" in the expression $(a \otimes I) \otimes b$? Again, when the arrows meet again we
get a diagram that we will ask to commute.

We get this triangle, which is called the *unit triangle*, and is the other axiom for a weak monoidal category.

$$(a \otimes I) \otimes b \xrightarrow{\alpha_{aIb}} a \otimes (I \otimes b)$$
$$r_a \otimes 1_b \searrow \quad \swarrow 1_a \otimes l_b$$
$$a \otimes b$$

The pentagon and the triangle are the axioms we need for a weak monoidal category, to ensure that we can manipulate structure isomorphisms as if they were equalities. In fact, as this is a one-object 2-category, those are the axioms we need for a weak 2-category, also called a *bicategory*. The structure isomorphisms are also called *coherence constraints*, and the question of how constraints in general interact is one of the big questions of higher-dimensional category theory. It is the question of coherence.

24.7 Coherence

Coherence questions in category theory are essentially about which diagrams commute in a particular type of category, as a consequence of the axioms. At least, that's what they're about on the surface. Deep down they're about the interaction between different ways of presenting the structure in question. We briefly mentioned in Chapter 8 the two different ways of presenting basic associativity (of multiplication or composition, say):

1. Local: for any a, b, c, $(ab)c = a(bc)$.
2. Global: any finite string $a_1 a_2 \cdots a_n$ has a unique well-defined product.

For monoidal categories we have something analogous:

1. Local: for any objects a, b, c, d the above pentagon and triangle commute.
2. Global: all diagrams of structure isomorphisms commute.

The point of the "global" description is that if all diagrams of structure isomorphisms commute, then we can move around them without worrying exactly which path we took, which is somewhat how we manipulate equalities.

The fact that those two presentations of monoidal categories (or bicategories) are equivalent is a deep theorem of Mac Lane called a *coherence theorem*, specifically, *coherence for monoidal categories*. Coherence typically involves a question of which presentation is the definition and which one is given as a theorem. This is a question for ordinary associativity as well as for higher-dimensional structures. Monoidal categories are typically defined according to the two axioms, and then we prove a theorem saying that "all diagrams commute" (this has to be made precise of course, so this is more of a slogan than a theorem). The point is to find a small set of axioms that generates all the commuting diagrams. However, another point of view is to define monoidal categories by saying "all diagrams commute" and then prove a coherence theorem saying that it suffices to know that the pentagon and triangle commute.

The two presentations have different uses in practice. When we are checking that something is a monoidal category in the first place it is necessary to have a small set of axioms to check because we can't physically check that all diagrams commute. But when we are working with (or in) a monoidal category we usually use the fact that all diagrams commute, as the diagram we're looking at is typically not precisely a pentagon or triangle. The important thing is that we have both possibilities and we know they're equivalent.

The small set of generating axioms can seem a little arbitrary sometimes, and I personally think this is one of the reasons abstract math can seem baffling. By contrast, the global presentation makes sense in some fundamental way. As a case in point, when Mac Lane was first making the definition of monoidal category his list of axioms was longer; it was only later that Max Kelly proved that some were redundant. However, arguably Mac Lane had a point, because more axioms turn out to be needed to understand the next dimension up properly.

Another way of interpreting coherence for monoidal categories is that weakness doesn't matter *that* much, and we can more or less behave as if it's a strict monoidal category without anything going too wrong. This is why, for example, we can take cartesian products of three sets and not worry too much whether we're really taking ordered triples (a, b, c) or $((a, b), c)$ or $(a, (b, c))$.

There are many ways to state that coherence theorem and they're all beyond our scope, but here's one for the record anyway.

Theorem 24.2 (Coherence for monoidal categories)   *Every weak monoidal category is monoidal equivalent to a strict one.*

As I hope you can guess by now, "monoidal equivalent" is the correct notion of sameness for a monoidal category, and is essentially an equivalence of categories where the functors in question respect the monoidal structure. This coherence theorem is a one-object version of the one for weak 2-categories, which are also called bicategories:

Theorem 24.3 (Coherence for bicategories)   *Every bicategory is biequivalent to a (strict) 2-category.*

Sometimes it's easier to think about monoidal categories just because we don't have to draw so many dimensions. Studying higher-dimensional structures in which the bottom dimensions are trivial (containing only one cell) can help us understand and deal with higher dimensions at many levels. A structure with trivial lower dimensions is called *degenerate*.

24.8 Degeneracy

A degenerate category is one where the lowest dimension is trivial, that is, there is only one object. Once we're in a 2-category there is one further level of degeneracy we could consider. We have seen that a 2-category with just one 0-cell "is" a monoidal category, but what about a 2-category with just one 0-cell and also one 1-cell? This is called "doubly degenerate".

If there is only one 0-cell and only one 1-cell our non-trivial data is now just a set of 2-cells. We now have two types of composition, horizontal and vertical, so we get two binary operations, one coming from vertical composition and the other from horizontal composition. We also know one thing about the interaction between those binary operations, coming from the interchange law.

At this point something rather amazing happens, called the *Eckmann–Hilton argument*. This was originally an argument in algebraic topology used to show that higher homotopy groups are always commutative, but it comes down to a general algebraic principle.

For our doubly degenerate 2-category we consider the interchange law in the following special case, with only two of the 2-cells being non-trivial.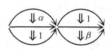

Remember that as the structure is doubly degenerate all the 0-cells are the same, and all 1-cells are identities. We then have the following which I like to call the "Eckmann–Hilton Clock":

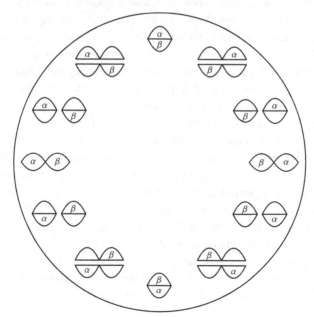

The "clock" depicts an argument showing this amazing result: horizontal and vertical composition must be the same binary operation, and moreover, it is commutative. The clock is not a formal proof, but it's the beginning of one: the idea is that every time you move to the next "hour" on the clock face you are performing a coherence isomorphism. (All the unmarked 2-cells are also identities.)

Things To Think About

T 24.11   See if you can work out what each isomorphism is as you move around this clock. There is some subtlety around 12 o'clock and 6 o'clock, as we have to use a vertical unit horizontally, but everything is strict so it works. (However it becomes critical in weaker situations.) It's worth also understanding why this argument only works when all the 1-cells are identities.

In summary, we start with a set with two monoid structures on it that are *a priori* different, but satisfy the interchange law. We then show this means that the second monoid structure isn't really extra structure, but has the effect of forcing the first one to commute. So a doubly degenerate 2-category "is" a commutative monoid.

This is a sign of something rather subtle going on in higher dimensions. Higher-dimensional categories are difficult to study for many reasons, and studying various degenerate versions can be a good way to get a handle on subtle coherence questions. As we go up dimensions, the gap between strict and weak versions gets wider and wider. We have seen that for 2-categories the strict and weak versions aren't *extremely* different. However for 3-categories the gap is wider. Weak 3-categories are called tricategories, and coherence for tricategories is much harder. In particular, it is crucially *not* the case that every tricategory is triequivalent to a strict 3-category. As part of the same phenomenon, it is not the case that every diagram of constraints commutes.

It is much easier to see this effect for *doubly degenerate* tricategories, that is, tricategories with only one 0-cell and only one 1-cell. This is like a categorified version of what we just did for doubly degenerate bicategories. The Eckmann–Hilton argument showed us that doubly degenerate bicategories produce commutative monoids. When we go up a dimension we have a sort of weak Eckmann–Hilton argument showing that doubly degenerate tricategories produce "weakly commutative monoidal categories".

This is very far from a rigorous definition, but the idea is that if a monoidal category is weakly commutative it is "commutative up to isomorphism", which means $A \otimes B$ is isomorphic to $B \otimes A$ for any objects A and B. However, as good category theorists, we prefer to specify an actual isomorphism $A \otimes B \xrightarrow{\gamma} B \otimes A$. Furthermore, that isomorphism then also needs to satisfy some axioms ensuring we can manipulate it as if it were an equality.

This isomorphism γ is called a *braiding*, because it is quite a lot like braiding hair. This idea starts from the idea that commutativity can be shown using physical objects by moving them around each other physically. For example the diagram below depicts 4 + 2 on the left, and a way of moving the objects past one another to produce 2 + 4 on the right.

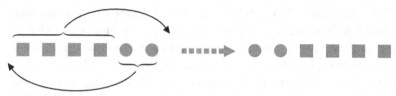

Note that we have to be in at least a 2-dimensional world to do this; if we only have one dimension that means the objects are stuck in a line and won't be able to move sideways. But in a 2-dimensional world we can use the second dimension to get around the first. This is essentially what the Eckmann–Hilton argument was doing in a doubly degenerate bicategory: using the second dimension to get around the first.

Now imagine that those physical objects are sitting on a table but attached to pieces of string hanging from the ceiling. (In the above diagram we are looking at the objects from directly above.) Now when you move the objects around each other the string will get crossed over, or braided. This is how Maypole dancing works.†

This is essentially what happens in a tricategory — we can move things past each other but a trace remains of how we did it; we can also think of this as recording the passage of time in a third dimension.

Crucially this also means that if we move the objects past each other the other way, the strings duly cross over the other way, and we get a *different* braid; for example in the above example we could slide the squares round the circles instead in the anti-clockwise direction instead.

We often represent these two possibilities as the two different string crossings shown on the right.

---

† This is a traditional English folk dance where everyone is in a circle and holds onto a long ribbon attached to the top of a tall pole in the center. The dancers weave around each other while going round the pole, making intricate braids in the ribbons. When I was little I was enthralled by how that worked. (I still am.)

Anyone accustomed to braiding hair knows that the under-
and over-crossing methods produce different braids, especially if you're doing a French
braid, as I've done in the pictures on the right.

If we were braiding hair in 4-dimensional space we could hop into the fourth
dimension to bring one of those strands "through" the other (somewhat like
when we do the commutativity in the first place we hop into the second dimension to bring the objects past each other). In 4-dimensional space the braid
wouldn't hold, as one strand of the braid could just move through any other
strand. In category theory this question arises when we start thinking about
degenerate 4-categories.

The *periodic table of n-categories* was proposed by Baez and Dolan as a table showing the sorts of structures that arise with different levels of degeneracy
in weak *n*-categories. It essentially encapsulates increasingly subtle versions of
commutativity. At the level of monoids, things are simply commutative or not.
With every added dimension there are more senses in which things could be
commutative (or not) rather than just "yes" and "no".

I like to think that studying these lower-dimensional remnants of higher-
dimensional structures is like footprints in the snow. I love the book *Miss
Smilla's feeling for snow* by Peter Høeg. Miss Smilla is from Greenland and
grew up playing a game of deciphering footprints in the snow. One person
would close their eyes while another jumped around in the snow making footprints. The first person would then study the footprints and be challenged to
reconstruct the actions of the first person. Miss Smilla's understanding of snow
is so intimate that she can read the footprints perfectly, even if someone has
spun around in the air and landed right back in the same place.

Studying degenerate higher-dimensional categories is like studying the footprints left in the snow by some mysterious higher-dimensional creatures, and
helps us to understand how much more subtle they are than 2-categories.

24.9 *n* and infinity

Higher-dimensional categories are difficult, and get more and more difficult as
the dimensions increase. They are already difficult when $n = 3$ and are not very
well understood beyond that. In fact, above $n = 3$ they are so difficult that we

typically revert to general n at that point, and don't try to study any particular dimension.

There have been many approaches to defining n-categories, and describing any of them here is far beyond our scope. However, I would like to describe some of the different ideologies behind the approaches.

Here are the general features that a definition needs to have.

Data A way of producing more dimensions. Higher-dimensional morphisms are sometimes called "cells".

Structure A way of dealing with all the different levels of composition and all the structure/constraint cells that need to exist (and these should all be equivalences at the right level for the codimension).

Properties A way of deciding what axioms the constraint cells should satisfy in order to produce a coherent enough structure. The general principle is that at each dimension there are constraint cells whose interaction is mediated by constraint cells at the dimension above; this continues to the top dimension where there is no dimension above, so we need some axioms in the form of equalities. Of course, if we are doing infinity-categories there never is a top dimension so we never have any axioms invoking equalities.

Notation One of the most fundamental problems that makes this all hard is the problem of expressing higher-dimensional cells in the first place. We can't easily draw them any more as we're stuck with our 2-dimensional pieces of paper. Even once we've succeeded in making a definition, doing any sort of calculations or proofs in higher dimensions remains a difficult problem.

Here are some of the ways in which definitions of higher category can take on different flavors.

Iteration vs all-at-once

Some definitions iterate a process of adding a dimension, whereas others start with data giving all dimensions at once. We saw this difference in our two ways of defining 2-categories: one way was by enrichment, where we took the definition of category and added another dimension by using hom-categories instead of hom-sets. The other way involved starting with a 2-globular set as our data, straight off.

Iteration always has the appealing feature that you do one simple thing and then iterate it to produce something very complex. However, for higher categories this idea comes with a complicated feature: as we go up in dimensions, enrichment needs to happen more and more weakly. If we iterate *strict* enrichment we'll only get *strict* n-categories. We can work out how to do weak enrichment to produce bicategories at the first step, but iterating that will not produce something as weak as tricategories at the next stage (although it will be slightly weaker than a strict 3-category).

Instead of adding in one dimension at a time by iteration we could throw in all dimensions at once. In that case we have to come up with all new ways of dealing with those dimensions. We can't really rely on the existing low-dimensional theory to extend to the higher dimensions, which is perhaps why that branch of research is progressing somewhat slowly. We'll come back to that in "Algebraic vs non-algebraic".

Enrichment vs internalization

Even if we've decided to do iteration, there are (at least) two different flavors of it. One is by enrichment as we've seen, where we replace the hom-sets by hom-categories. The other is called "internalization" and involves the "categorical" definition of category, starting from graphs.

We briefly saw that we could define a category starting from an underlying graph, which is a diagram in **Set** as shown here.
$$C_1 \underset{t}{\overset{s}{\rightrightarrows}} C_0$$

In this form of definition, composition and identities are also defined as morphisms in **Set**, and the unit and associativity axioms are expressed as commutative diagrams in **Set**. As this is categorical, we can pick this definition up and place it inside other categories.[†]

This gives us the definition of a category *internal* to another category, and is slightly different from a category *enriched* in another category. For an internal category, the whole underlying graph is now a diagram inside another category, which means something has happened to the objects as well as the morphisms. This is different from enrichment, where the objects are still just a set.

For example, if we take a category enriched in topological spaces, we have a set of objects, and for every pair of objects a space of morphisms between them. If we take a category internal to topological spaces, we have a space of objects and a space of morphisms.

We can also compare categories enriched in **Cat** and categories internal to **Cat**. Categories enriched in **Cat** are 2-categories. When we defined them, we mentioned that the objects did not become a category as they already had morphisms between them. However if we instead take categories internal to categories then the objects are themselves a category before we even think about the morphisms.

[†] The definition of composable pair involves a pullback, so in order to do it we would need to be able to take the relevant pullback in the new category.

The underlying graph as shown on the right is now a diagram in **Cat**, that is, a diagram of categories and functors.

$$C_1 \underset{t}{\overset{s}{\Rightarrow}} C_0$$

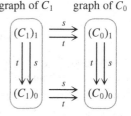

graph of C_1 graph of C_0

This means that C_1 and C_0 themselves have underlying graphs. If we unravel that, we have the diagram in **Set** shown on the right. You might find yourself getting a bit dizzy trying to understand all the parts of this; if so that's understandable.

In this structure the 0-cells $(C_0)_0$ have two types of 1-cell between them:

- $(C_0)_1$: those coming from the original category of objects C_0, and
- $(C_1)_0$: the objects of the category of morphisms C_1.

These are usually depicted as horizontal and vertical arrows. We then have 2-cells given by $(C_1)_1$ and these have a source and target in $(C_1)_0$ and also a source and target in $(C_0)_1$, with some commuting conditions.

The end result is that 2-cells have this shape; remember that the horizontal and vertical 1-cells come from different parts of the abstract structure and can't be composed: they're different.

$$
\begin{array}{ccc}
a & \xrightarrow{f} & b \\
p\downarrow & \Downarrow\alpha & \downarrow q \\
c & \xrightarrow[g]{} & d
\end{array}
$$

Here f and g are the source and target of α in $(C_1)_0$ and p and q are the source and target of α in $(C_0)_1$. We could have picked the horizontal and vertical conventions the other way round, but I like it this way as it shows the generalization from globular cells vividly: globular 2-cells can be regarded as square ones in which all the vertical 1-cells are identities (collapsing the sides to a point). This comes from the category C_0 being discrete, that is, $(C_0)_1$ being trivial; the underlying data then collapses to a 2-globular set.

This structure with square-shaped 2-cells is called a *double category*. It still has horizontal and vertical composition and interchange. The square-shaped cells complicate the situation but turn out to be quite efficacious for addressing some questions in algebraic topology. When this process is iterated we get *n*-dimensional cubes at each dimension, and indeed these types of *n*-categories are called cubical rather than globular.

──────────── **Things To Think About** ────────────

T 24.12   It's quite interesting to try fully unraveling all the definitions to see where the structure of a double category comes from. You could start by checking that my diagram of a 2-cell makes sense — how do we know the endpoints of the source and target 1-cells need to match up like that? The key is to think about the fact that the source and target maps in the underlying graph are functors. So on underlying data they are morphisms of graphs, thus some commuting conditions arise in the unraveled diagram.

Algebraic vs non-algebraic

Instead of building dimensions by iteration we can start with all dimensions of data at once, with an underlying infinite-dimensional globular set or something similar. We've seen from the previous discussion that there are different ways of expressing the underlying data: it could be a globular set, or we could have cells of different shapes such as cubes. There are in fact many other different underlying shapes that are possible, and they are usually described by using a different adjective on the word "set". Thus so far we have had globular sets and cubical sets, but there are also simplicial sets (which have been well studied by topologists for some time) and then shapes called things like "opetopic", "multitopic", "dendroidal", "multisimplicial".

The shapes are not just arbitrary choices; they're related to the way in which we then go on to express the structure on the underlying data. In some cases the definition involves specifying things like binary operations at every dimension. This is called *algebraic*, because algebra is about operations combining multiple inputs into one output. In higher-dimensional algebra we often need to combine more than two inputs, so we might generalize to "k-ary operations" with k inputs.

These generalized operations are often expressed by means of monads. We have not really talked about monads but they are special functors which are particularly good for generating algebraic structures "freely", as we did for the free monoid construction. Once we have generated the structure freely we then say "'right, now we want this structure actually to have a value in our underlying data". So for a monoid we would start with an underlying set A, generate the free monoid on it (all the words) and then say: given any word abc, say, I want this actually to have a value in my original underlying set. For example in the natural numbers we would say "$2 \times 4 \times 3$ is a valid possible multiplication, now what does it actually equal?"

Abstractly this amounts to giving a function $\boxed{\text{free monoid on } A \longrightarrow A.}$

This is something that we can do using *algebras for monads*. (I'm giving the name just so that you can look it up elsewhere if you are interested.) What we do for n-categories is something like this: start with an underlying globular set A, generate the free n-category on it, and then evaluate all those operations by means of a morphism (of globular sets) $\boxed{\text{free } n\text{-category on } A \longrightarrow A.}$

A key feature of expressing structure algebraically is that we have definite answers to things. We know that $2 \times 4 \times 3$ definitely equals 24 in the natural numbers. However, there are situations in which answers are less definite. For example, suppose that instead of doing multiplication of natural numbers we are doing products of sets. We know that a product of three sets *exists* but there are many possible candidates for it. So instead of having one definite answer we have many possible answers, and they are all uniquely isomorphic.

This idea is expressed by *non-algebraic* approaches to *n*-categories. To express structure non-algebraically we don't actually specify what the results of operations are, we just verify that there is enough structure around making valid possible results exist. For example, composition might be possible in various ways; even in a weak 2-category we have two different composites of this diagram $\boxed{a \longrightarrow b \longrightarrow c \longrightarrow d}$ because we don't have strict associativity.

Non-algebraic approaches typically use shapes of structure more complicated than globular cells, in order to be able to build in enough scope for expressing what the candidates for composites are. Arguably, the approaches based on simplicial sets have made by far the most progress. Simplicial sets are based on "simplices", which are higher-dimensional generalizations of triangles. In three dimensions the shape in question is a tetrahedron, which we saw in Section 8.6 is a shape that gives the geometry of associativity for three composable arrows. In *n* dimensions the shape gives the geometry of associativity for *n* composable arrows.

The simplicial approaches are greatly helped by a huge body of existing work on simplicial sets from the fields of topology and homotopy theory. I think this accounts for why non-algebraic approaches have made more progress than algebraic approaches: homotopy theorists have long been used to asking for structure to exist rather than worrying about exactly what it is, and have developed many techniques for dealing with it. However, algebraic approaches are not made redundant by this; in fact I think it's particularly important to remember the algebraic questions that are left unanswered by the non-algebraic approaches.

Finite dimensions vs isomorphisms forever

We have seen that a set "is" a category with no non-trivial morphisms, that is, all morphisms are identities. If we relax that condition slightly we get a category in which all the morphisms are *isomorphisms*. This is the definition of a groupoid.

We could do this one dimension up, to express categories as 2-categories: we could say that a category "is" a 2-category in which all 2-cells are identities, and then we could relax that and say that all 2-cells are isomorphisms.

But as is typical in higher-dimensional category theory we could now say "Why stop there?" Whenever we cut off at n dimensions, we are essentially saying that every morphism at a dimension above n is the identity. Instead, we could say that every morphism above dimension n is an isomorphism, or rather, an equivalence at the appropriate level. If we decide we're never going to cut ourselves off, we will keep doing this forever into infinite dimensions. We will get an infinity category in which every morphism above dimension n is an equivalence, for an appropriate infinite-codimensional[†] notion of equivalence.

┌─────────────────── **Things To Think About** ───────────────────┐

T 24.13   You might like to see if you can create that definition of equivalence for yourself. Start by recalling how we went from inverses to pseudo-inverses. Now find the place in the definition of pseudo-inverse that involves inverses, and replace those by pseudo-inverses. And then keep going because there is no top dimension, so there is no point where we should talk about equalities between k-cells.

└──┘

These are sometimes called (∞, n)-categories. They arise naturally in topology because topological spaces effectively have no top dimension — there is always another level of homotopy that can be defined. But sometimes one wants to focus on a few dimensions at the bottom and sort of let the others take care of themselves, hence we look at n dimensions closely but without having to chop off and "seal up" the top dimensions.

One of the advantages of this approach is that you can do everything maximally weakly right from the start, which might help with iteration. You don't start with equalities that you then have to weaken to isomorphisms when you go up a dimension (and so on), because you had infinite dimensions all along. However, only a finite number of the dimensions are really non-trivial, so you don't end up with the full complications of infinite dimensions. This approach has been particularly fruitful in non-algebraic definitions, as a certain amount of non-specified-ness is helpful for allowing all those equivalences to whiz around all the way up to infinity. The theory of $(\infty, 1)$-categories is particularly well developed, so much so that they are sometimes called "infinity-categories".

Having infinite dimensions might sound much more complicated than strictly having n, but I proved a theorem[‡] which shows that there are some advantages:

[†] Remember "codimension" is the number of dimensions *above* a cell, so if we're in an infinity-category then all cells have infinite codimension.

[‡] An ω-category with all duals is an omega groupoid. In *Applied Categorical Structures*, 15(4):439–453, 2007.

you never have to worry about specifying axioms at the top dimension to seal up the ends, because there is no top dimension. I like to think it's the fact that if we were immortal we could procrastinate forever.

24.10 The moral of the story

Higher-dimensional category theory is currently still developing as a field of research. It is difficult because of the infinite levels of nuance and subtlety that can arise, and because of the difficulties inherent to representing infinite dimensions in our very limited three-dimensional physical world.

The point of doing it is something like going up high to get a better overall view of a situation. I'd like to end by coming back to the analogy I made in Chapter 2 about shining a light. I like the analogy for abstraction in general, and it is relevant to higher-dimensional category theory specifically. As I said, if you're shining a light on something to see it better you can raise the light higher up to illuminate a broader area and see more context, but you might see less detail. If you shine the light very close it will be very bright but you will lose context. This is a trade-off we make whenever we get involved with abstraction and higher dimensions. We gain overview and context. We gain the chance to unify a wider variety of structures. We see more connections between different areas. But we also lose detail. In the case of abstraction we lose specific detail about specific situations, and in the case of higher dimensions we often lose our grasp on detail, or our ability to grasp the detail, because the complications are so extreme. So while I am very drawn to the higher-dimensional approaches it's also important to have ways to operate in lower dimensions, by finding ways to temporarily ignore the extra nuance of the higher dimensions without ruining too much. Basic 1-dimensional category theory has proved itself to be a particularly good level for balancing those aims in much of mathematics, and beyond.

For me there is one final, even more abstract attraction of infinite-dimensional category theory: I think of it as a fixed-point of abstraction. If science is the theory of life, and mathematics is the theory of science, and category theory is the theory of mathematics, I think that 2-category theory is the theory of category theory, and 3-category theory is the theory of 2-category theory, and so on. Finally, however, higher-dimensional category theory is the theory of higher-dimensional category theory. We have reached a pinnacle of abstraction. Or perhaps the heart of abstraction, or its deepest roots.

Perhaps the pinnacle and the heart and the deepest roots are the same, and that, to me, is the joy of abstraction.

Epilogue

Thinking categorically

In case you have become overwhelmed with details and technicalities, we will end by taking a step back and summing up some of the main principles of category theory.

Mathematics is founded on rigor and precision. The discipline relies on a large quantity of formality in the form of notation, terminology, and use of language that is slightly different from the use of language in everyday life. This is one of the crucial things that gives math its clarity and thus its power, but learning this formality can get in the way of feeling its spirit. I think this is like learning the piano and worrying so much about playing the right notes that you don't enjoy the music itself. But at the other extreme if you never worry about playing any right notes there are some things you will never be able to play.

Most math textbooks focus on the formality and have less focus on the spirit, with face-to-face teachers often being the ones to convey the spirit that the text couldn't, or didn't. The thing is that math can then end up seeming like a compilation of a series of very technical processes. The more abstract the math is, the more it can seem like that, which is a particular risk for category theory. And it is then a risk that the *practice* of category theory actually does become a series of very technical processes, if that is how it is conveyed by textbooks, and thus that is how it is learned. In this book I have wanted to convey some of the technicality and a great deal of the joy. Because for me the act of doing category theory feels more like soaring than taking small technical steps, at least at the beginning; the small technical steps come later when you're making sure everything is truly rigorous.

So I wanted to end by summing up what I think it means to "think categorically" rather than just "do category theory". I want to gather together the various principles of category theory that I've mentioned as we're going along, so that we can end with spirit rather than with technicalities.

Motivations

The purpose of abstraction

The purpose of abstraction is to unify more examples (not get further away from them). Sometimes category theory can feel like arbitrary abstractions for the sake of it; in that case I think something is missing. It's true that sometimes abstraction is driven by some internal logic, but if the resulting abstraction doesn't unify anything then I would be suspicious of it.

Structural understanding

I think that category theory is about understanding structural reasons behind things, not just constructing formal justifications. Some proofs in math just proceed step by step and arrive at the conclusion by stringing all those steps together. I think a truly categorical proof does more than that, uncovering some deep aspect of *why* something is happening, structurally.

One example was when we saw that the category of factors of 30 has the same shape as the category of factors of 42. We could show that those categories are isomorphic in several ways. Here they are, sort of in order of what I consider increasing structural depth.

1. Construct a functor from one to the other, and show that it's a bijection on objects and on morphisms.
2. Construct a functor from one to the other, and an inverse for it.
3. Show that each one is the category of factors of abc, where a, b, c are distinct primes.
4. Show that each one is a product category I^3, where I is the quintessential arrow category.

I hope it is clear that none of these proofs is more correct than the others, and they are all expressed in category theory. But they become increasingly categorical in the sense of the deep categorical structure that they uncover.

"Economical" category theory

Working with category theory usually involves categories with specific kinds of structure in them, like products and coproducts, other limits and colimits, and so on. One approach is a sort of "rich" approach where you throw tons of structure into your category because you think you will always have that in any of the categories you're working in: "We might as well assume we have products because all the categories we work in will always have products." It's the kind of approach some people might take with money if they were sure they were always going to have unlimited access to money. The other approach is a more "economical" approach, where you work out exactly what structure is needed in order to do whatever you're trying to do, partly to get a better understanding of it, and partly to leave open the possibility of doing it in a category with very little structure. This more nuanced approach to category theory has greater potential for finding increasingly diverse applications in places that are less obvious, just like an economical solution to a problem is more inclusive, as it is accessible even to people who do not have money to throw around.

I think that economical category theory is more in the spirit of category theory; perhaps I just mean in *my* preferred way of doing category theory.

The process of doing category theory

So much for the motivations behind category theory. Here is what a typical process of *doing* category theory might look like, in steps.

1. We see an interesting structure somewhere.

2. We express it categorically, that is, using only objects and morphisms in some relevant category.

3. We look for that structure in other categories.

4. We use it to prove things that will immediately be true in many different categories.

5. We investigate how it transports between categories via functors.

6. We explore how to generalize it, perhaps by relaxing some conditions.

7. We investigate what universal properties it has.

8. We ask whether it has generalizations to higher dimensions via categorification.

This often involves having a slight hunch about something, following your categorical-gut instinct, and then making it precise in the language of category theory afterwards. This is why I often proposed "things to think about" that were rather vague, because part of the discipline of category theory is making precise sense of vague ideas.

The practice of category theory

In the actual practice of doing category theory here are some things that often come up.

Morphisms

The whole starting point of category theory is to have morphisms between objects, not just objects by themselves. This idea pervades category theory by us always thinking about maps between structures, not just individual structures. So we think about totalities of a particular type of structure, not just one example of the structure. With morphisms to hand, we think about isomorphisms instead of equalities, and more generally the weakest form of sameness for the dimension we're in.

Reasoning

We use diagrammatic reasoning to make use of geometric intuition as well as logical and algebraic intuition. Proofs are often essentially type-checking. We invoke the "obvious axioms" which might seem obscure but once you understand the principle it's less obscure: the principle is that if we have two inherently structural ways of mediating between the same things, they should be the same in order to prevent structural incoherence.

Coherence

We often have two approaches to the same structure, one that starts locally and expands to a global view, and one that starts globally and then pins down a local view that suffices. These two approaches are summarized in the following table.

	Approach 1	Approach 2
definition:	small set of conditions	broad structure
theorem:	broad structure follows	small set of conditions suffices

The question of coherence is the question of how the local view and the global view correspond to one another. Structures in category theory usually rely on having both, and understanding the relationship between them.

Universal properties

We try and characterize things by properties rather than intrinsic characteristics, and best of all we characterize by universal property, where possible. This sometimes involves souping up a category in order to find a way to express something as a simple universal property in a complicated category, and then unraveling it to be a complicated universal property in a more fundamental category.

Structural awareness

We maintain an awareness of our interaction with structure and how we're using it. For example if we are making use of associativity we make sure we're aware of it rather than taking it for granted. This helps for generalizations, especially into higher dimensions where axioms that we used to take for granted may become structural isomorphisms that we now have to keep track of. We also maintain awareness of whether we're invoking the *existence* of some structure, or exhibiting a *specific* piece of structure. For example, we remain aware of whether we're saying things are "isomorphic", or whether we're exhibiting a specific isomorphism between them. Being specific is harder but gives us a more precise understanding.

Higher dimensions for nuance

Morphisms are higher-dimensional than objects and give us more nuance. More generally, every higher dimension we involve gives us even more nuance, along with more complications. But the aim is to learn how to deal with the complications in order to benefit from the added nuance. I think this is important in life as well. Arguments in life have become too black-and-white. I wish we could all become more higher-dimensionally nuanced in our arguments in math and in every part of life.

APPENDICES

Appendix A

Background on alphabets

People sometimes tell me that they were fine with math "until the numbers became letters". We use letters to represent unknown quantities, and in abstract math basically everything is an unknown quantity so we end up needing rather a lot of letters. We also often want to use different types of letter for different things, to help us maintain some intuition about what is going on. Here are some ways in which I typically use roman letters throughout this book. These are not hard-and-fast rules.

- natural numbers: n, m, k
- elements of sets or objects of categories: a, b, c, x, y, z
- functions or morphisms: f, g, p, q, s, t
- sets: A, B, C, X
- functors: F, G, H, K
- specific sets of numbers in standard notation: blackboard bold $\mathbb{N}$, $\mathbb{Z}$, $\mathbb{Q}$, $\mathbb{R}$, $\mathbb{C}$
- small categories: blackboard bold $\mathbb{A}$, $\mathbb{B}$, $\mathbb{C}$, $\mathbb{D}$
- locally small categories: curly $\mathcal{A}$, $\mathcal{B}$, $\mathcal{C}$, $\mathcal{D}$

Once we find ourselves running out of roman letters we branch out into Greek letters. It's not a bad idea to look them up and familiarize yourself with them although in this book really the only ones I use are the first four (shown on the right) for natural transformations, and ω (omega) for countable infinity.

α	alpha
β	beta
γ	gamma
δ	delta

The next letter is epsilon (ϵ or ε) and is classically used in analysis to torture undergraduates in calculus class, or rather, to construct rigorous proofs in calculus using the idea of an unknown distance which can be arbitrarily small but is never 0. Thus proofs in calculus often begin "Let $\varepsilon > 0$", so much so that I have been known to tell students that if all else fails they can write that and I will give them one point. This also gives rise to a joke which I find ludicrously funny: "Let ε be a large negative integer." This is funny because once you're so incredibly used to ε being an arbitrarily small positive real number, the idea of ε representing a large negative integer is, well, inexplicably hilarious.

Appendix B

Background on basic logic

Logical implication

Basic logic begins with statements of logical implication of the form "A implies B". We write it $A \Longrightarrow B$. The converse is $B \Longrightarrow A$ and is logically independent of the original statement, which means they could both be true, both false, or one false and one true.

If a statement $A \Longrightarrow B$ and its converse $B \Longrightarrow A$ are both true then A and B are logically equivalent, and we say "A iff B" (short for "if and only if") or use the double-headed arrow $A \Longleftrightarrow B$.

Another logical statement related to $A \Longrightarrow B$ is the *contrapositive*, which is the statement "not $B \Longrightarrow$ not A". This is logically equivalent to $A \Longrightarrow B$. (Note the reversed direction.)

Quantifiers

The expressions "for all" and "there exists" are called quantifiers, and we notate them by $\forall$ and $\exists$. It is important to understand how to negate them. The negation of a statement is the statement of its untruth. We can always negate a statement A by saying "not A", but sometimes this can be unraveled to something more basic. Here is a table for negating quantifiers involving statements A and B:

original:	$\forall x$	A
negation:	$\exists x$ such that	not A

original:	$\exists x$ such that	B
negation:	$\forall x$	not B

When we use quantifiers in a formal logical statement it might seem that we need some commas, but we often omit the punctuation because it gets in the way without really clarifying anything. Here's an example:

$$\forall b \in B \quad \exists a \in A \text{ such that } f(a) = b.$$

404

Appendix C

Background on set theory

The only thing we really need is to be aware that there's a hierarchy of sets: not every collection of things counts as a set as some are "too big". So we distinguish between *sets* and *collections*. Here "too big" means something very rigorous but quite arcane, and the definition is there to avoid Russell's paradox.

C.1 Russell's Paradox

Russell coined this paradox in 1901 and for a while it seemed like the foundations of math had fallen apart. This is because the paradox seemed to show that there was a logical contradiction at the very start, with the notion of a set. However, it turned out that this only happens if you take a very naïve definition of a set, and it can be avoided if you take a rather subtle and careful axiomatization instead.

The paradox is informally stated as "A barber shaves every man in the town who does not shave himself. Who shaves the barber?" The problem is that if the barber shaves himself then he doesn't shave himself, and if he doesn't then he does. We are stuck in a contradiction.

The paradox stems from some self-reference in the situation, and this is what happens when we state it formally as well: instead of people who may or may not shave themselves, we have sets that may or may not be an element of themselves. Remember we write "is an element of" using the symbol $\in$.

We define the following set of sets: $\boxed{\{\, S = \text{all sets } X \text{ such that } X \notin X \,\}.}$

This definition might sound confusing, and that's sort of the whole point. We then ask: is S an element of S? To untangle this more carefully, first note the following facts about any set X:

- If $X \notin X$ then $X \in S$.
- If $X \in X$ then $X \notin S$.

Now this is true for all sets X, so if we put $X = S$ we get these statements:

- If $S \notin S$ then $S \in S$.
- If $S \in S$ then $S \notin S$.

405

That is, either way we are doomed to a contradiction. The way to avoid it is essentially to declare that this S is not allowed to count as a set. More precisely, we set up some careful axioms for what *does* count as a set, and we are careful to avoid arbitrary collections of "all things", unless they were already elements of a set. So given something we already know is a set A, we can make a new set of "all things in A satisfying something-or-other", but we can't, out of nowhere, produce a set of "all things satisfying something-or-other". Thus, crucially, we can't produce a set of all sets, nor can we produce a set of "all sets satisfying something-or-other". Thus the thing called S in Russell's paradox no longer counts as a set, and so we can't set $X = S$ and get the contradiction.

The collection of all sets is *something*, however. We call it a "collection" and it exists at a different level up in a sort of hierarchy of sizes of collections of things. We prevent self-reference at each level, so that we don't just end up with Russell's paradox at the level above. So the totality of all collections is not a collection, but something bigger.

C.2 Implications for category theory

We have to do something analogous to avoid a Russell-like paradox in category theory. Basically we keep track of the "size" of categories and put them in a hierarchy, disallowing self-reference within any level.

So if the objects form a set, and every collection of morphisms $\mathcal{C}(a, b)$ is a set, then it's called a small category. But the totality of small categories can't be a small category: it's something one level up, which we might call "large". One level up from large we might call "super-large" and so on.

The extra subtlety with categories is that we have objects and morphisms, and even if the objects don't form a set and the *global* collection of morphisms doesn't form a set, it's possible that each individual collection of morphisms $\mathcal{C}(a, b)$ is a set. This is useful to know as we often only consider one hom-collection at a time, which is what we call doing things "locally" in a category. Categories in which every $\mathcal{C}(a, b)$ is a set are called *locally small*. They are larger than small categories but still somewhat tractable.

That's about all you really need to know to get going here. Basically bear in mind that any time we are thinking about collections of sets-with-structure, that's a type of "all sets such that [something]" so is destined to be large; however the morphisms between them are "functions such that [something]" so form a set, so the totality as a category is probably locally small.

Appendix D

Background on topological spaces

For the examples in this book all you really need to know is that topological spaces are sets with some extra structure enabling us to define continuous functions, that continuous functions are ones that do not break things apart, and that a homotopy is a continuous deformation of one continuous map into another. Here is a little more detail.

D.1 Open sets

A topological space is essentially a set equipped with a notion of "closeness", but that sounds like it has something to do with things being near each other, when it really doesn't. The technical definition can be thought of as a generalization of the idea of open intervals of the real line. Open intervals are the ones that do not contain the endpoints. For example, this is the open interval from 0 to 1: $(0, 1) = \{x \in \mathbb{R} \mid 0 < x < 1\}$.

Open intervals are used all over the place in analysis, and if we combine them we get open sets. So we could have a disjoint union of two open intervals, say $(0, 1) \sqcup (10, 11)$ and it still counts as open as it doesn't have any of its endpoints (the formal definition is more precise than that). A non-disjoint union of open intervals is just another open interval; this is important as we often patch together properties like continuity across large open intervals by patching together small ones.

Here's an example of a non-disjoint union $(0, 2) \cup (1, 3) = (0, 3)$. It has an overlap from 1 to 2. It's also rather useful that the overlap (that is, intersection) is itself another open interval $(0, 2) \cap (1, 3) = (1, 2)$.

This idea is then generalized to sets other than the real numbers. It's done in a way that is typical in abstract mathematics: not by finding some characterization of what "open" really means, but by examining some of the relationships that open sets satisfy on the real line, and then defining open sets by those behavioral properties rather than by any intrinsic ones. So the idea is that we have some other set X and we want to say what it means to have a collection of subsets of X called "open". Those key properties turn out to be these:

1. The empty set and the whole set X count as open.
2. Any union of open sets counts as open.
3. Any *finite* intersection of open sets counts as open.[†]

Note that the whole set counts as open, so if our set X is a closed interval, say $[0, 1]$, then although X is closed as a subset of $\mathbb{R}$, it is open as a subset of itself. Other subsets of X might be open in X although they're not open in $\mathbb{R}$, such as the half-open interval $[0, \frac{1}{2})$. This is open in $[0, 1]$ because the only closed end is the boundary (also we can check that the complement is closed) but it is only half-open in $\mathbb{R}$. We will need this in a moment.

A system of open sets on X satisfying these behaviors is called a *topology* on X, and makes X into a *topological space*. It is possible to have different topologies on the same set.

D.2 Continuous functions

Arguably the point of having a topology is to define continuous functions, and these are sort of defined as "preserving closeness", but we have to be careful how we say it. It is tempting to say "open sets are preserved" but that's not quite right.

Consider this example of a function $f \colon X \longrightarrow Y$ which "breaks" the interval $(0, 2)$ in two. We're taking $X = (0, 2)$ and $Y = (0, 1] \sqcup (2, 3)$. A formal definition of f and an intuitive picture are shown below.

$$f(x) = \begin{cases} x & 0 < x \le 1 \\ x + 1 & 1 < x < 2 \end{cases}$$

Intuitively this shouldn't count as continuous because the interval X is broken apart at the point 1. However, this will not be detected if we think about open sets being preserved: an open interval in X might be split apart by f when it is mapped to Y, but the image will still be open in Y even though it's not open in $\mathbb{R}$. Instead we have to look at pre-images: we take open sets in Y and look at the set of all points in X that are mapped there. Now $(0, 1]$ is open in Y (as

---

[†] We have to specify that we only care about finite intersections because infinite ones can turn out to be closed, for example: $\bigcap_n (-\frac{1}{n}, \frac{1}{n}) = [0, 0]$.

the closed end is just a boundary) but its pre-image in X is $(0, 1]$ which is not open in X, because the closed end is *not* a boundary in X. This is not formally stated, but captures the intuition.

So the definition of continuous function is as follows.

Definition D.1 Let X and Y be topological spaces. A function $f\colon X \longrightarrow Y$ is *continuous* if the pre-image of any open set in Y is open in X.

If you have studied formal real analysis you might want to investigate why this is equivalent to the epsilon–delta definition in the reals:

$$\forall a \in X \quad \forall \varepsilon > 0 \quad \exists \delta > 0 \quad \text{such that} \quad \left| x - a \right| < \delta \implies \left| f(x) - f(a) \right| < \varepsilon$$

The advantage of the definition using open sets is that it works in places where we are not using a notion of distance to define our topology. In the reals we are using open intervals to define the topology, and the open intervals are defined using distance: in fact this is the notion of metric space, a very particular type of topology.

D.3 Homotopy

A homotopy is a "continuous transformation" between continuous maps. The definition is very clever in that it doesn't require any extra levels of definition — the next dimension up still just uses continuous maps. The extra dimension is provided by making a sort of cylinder on X in the form of a product $I \times X$, rather than by inventing a new notion of map. Explaining this fully is somewhat beyond the scope of this appendix, but basically a homotopy from f to g is defined as a continuous map $\boxed{\alpha : I \times X \longrightarrow Y}$ such that $\alpha(0, x) = f(x)$ and $\alpha(1, x) = g(x)$. Here I is the (closed) unit interval $[0, 1]$ which we can think of as a unit of time. The product space $I \times X$ is like a cylinder on X where we have thickened it up, but at any time t (or cross-section) we have an entire copy of X. At time 0 we do the function f and at time 1 we do the function g and at every time in between we do something that gradually deforms our function from f into g.

This definition can essentially then be iterated by multiplying by I repeatedly to add more dimensions and get higher and higher dimensions of homotopy.

Glossary

This glossary is not to give formal definitions, but to give reminders to jog your memory about what something is.

2-category 2-dimensional category (with 0-cells, 1-cells and 2-cells).

Abuse of notation When we have used the same notation for two different things. Usually we do this when the things are in some sense the same and it would be somewhat tedious to distinguish between them with notation.

A priori In advance.

Arrow Another word for morphism in a category.

Bicategory Weak 2-dimensional category.

Bijection or bijective function Function that is both injective and surjective, that is, a "perfect matching" between inputs and outputs.

Binary operation A process that takes two inputs and produces one output.

Cat The category of all small categories and functors between them.

CAT The category of locally small categories and functors between them.

Categorical In the spirit of category theory. (Not to be confused with the way "categorical" is used to mean "very clear" in normal English.)

Co- Prefix indicating the dual of something (so all the arrows are reversed).

Commutative diagram A diagram of morphisms in which any paths of morphisms between the same endpoints yields the same composite.

Componentwise When you calculate something component by component.

Compose Do one thing and then another; in category theory we are usually "going along" one morphism and then another.

Cone (over a diagram) The basic data for a limit, consisting of an object (the vertex) and morphisms from the vertex to every object in the base diagram, such that everything in sight commutes.

Contravariant functor A functor from a dual category, that is, it reverses the direction of morphisms.

Covariant functor A functor that doesn't reverse the direction of morphisms. Usually we just say "functor" but sometimes we want to stress that it isn't contravariant, especially if there are some contravariant ones hanging around as is often the case when we're dealing with Yoneda.

Degenerate This is often used to mean that something is the identity that didn't need to be. For example a degenerate commutative square has some sides being the identity.

Degenerate category A category with only one object.

Discrete category A category with no arrows except identity arrows.

Disjoint union A disjoint union of two (or more) sets is taken by putting all the elements of all the sets in, but counting them all as different even if there appears to be an intersection. Informally we can think of it as painting each set's elements a different color to force them to be different.

Dynamically Reading a diagram "dynamically" means we progress through it region by region, reading each commuting region as an equality that takes us from one side to the other.

Dual category The same category but with the arrows considered to be pointing the other way. We write the dual of a category $\mathcal{C}$ as $\mathcal{C}^{op}$.

Epic or epimorphism A categorical version of surjective function.

Equivalence of categories Good notion of sameness for categories, weaker than isomorphism.

Equivalence relation A relation that is reflexive, symmetric, and transitive.

Faithful functor A functor that is "locally injective" (on homsets).

Forgetful functor A functor that just forgets structure.

Free functor A functor that creates structure "freely", without any extra imposed equations.

Full functor A functor that is "locally surjective" (on homsets).

Functor Structure-preserving map between categories.

Functoriality The properties required of a functor: that it preserves identities and composition.

Graph Aside from the usual concept of a graph, in category theory this could refer to the underlying data for a category, consisting of a diagram of sets and functions like this: $C_1 \underset{t}{\overset{s}{\rightrightarrows}} C_0$

Group A monoid in which every element has an inverse. That is, a set with a binary operation that is associative and unital, and with respect to which every element has an inverse.

Groupoid A category in which every arrow has an inverse. A groupoid with one object is a group.

Homomorphism Generic word for structure-preserving map between structures. We can specify the context by saying things like "group homomorphism", "monoid homomorphism" and so on.

Homotopy A notion of map between continuous maps of topological spaces. It can be thought of as one dimension higher, and is essentially a continuous deformation of continuous maps.

Homset or hom-set General term for sets of morphisms in a (locally small) category.

Hypothesis This can either mean something you think is true that you haven't proved yet, or else it means one of the basic assumptions you've taken as the set-up for your theorem. Then when you prove it you say "by hypothesis" when you invoke that assumption.

Identity This can refer to an identity element with respect to a binary operation or composition, which is an element which "does nothing". It can also refer to an axiom, which might be called an identity if it identifies one thing with another in an equation.

iff if and only if

Image Given a function $f : A \longrightarrow B$, we can consider a particular element $a \in A$ and then we call $f(a)$ the image of a (under f). But we also say the image of A for the subset of B consisting of everything "hit" by f. To tell which type of image we mean, you need to type-check.

Injective function "No output is hit more than once."

Integers All the whole numbers, including positive numbers, negative numbers, and 0. Denoted $\mathbb{Z}$.

Inverse An element or morphism which "undoes" another one.

Invertible A morphism with an inverse.

Isomorphism Another word for an invertible morphism, perhaps with more emphasis on the source and target objects which are thus isomorphic, which is a more subtle notion of sameness in categories.

Lemma Small result that we prove, usually as a build-up to a larger theorem. Often lemmas sort out some technical details that we need for the theorem.

Locally small category A category $\mathcal{C}$ in which every collection of morphisms $\mathcal{C}(a, b)$ is in fact a set.

Modular arithmetic Arithmetic on an "n-hour clock".

Monic or monomorphism Categorical version of injective function.

Monoid A category with only one object; that is, a set with a binary operation that is unital and associative.

Monoidal category Category with a binary operation on it (on the objects and also on the morphisms), often denoted $\otimes$.

Natural numbers The positive whole numbers, possibly including 0. Denoted $\mathbb{N}$.

Natural isomorphism An invertible natural transformation, that is, an isomorphism in a functor category.

Natural transformation A notion of morphism between functors.

Naturality Commuting conditions in the definition of natural transformation.

Non-image (Not formal terminology) Given a function $f \colon A \to B$ I said "non-image" for the subset of B consisting of everything *not* hit by f.

Parallel Morphisms in a category are called parallel if they have the same source and the same target as each other.

Partially ordered set (poset) A category in which any pair of objects has at most one arrow between them.

Post-compose Compose a morphism after another one.

Pre-compose Compose a morphism before another one.

Preserves A functor preserves a certain structure if the structure still has the relevant properties after the functor is applied.

Presheaf/presheaves A presheaf on a category $\mathcal{C}$ is a functor $\mathcal{C}^{\mathrm{op}} \to \mathbf{Set}$.

Proposition Medium-sized result that we prove.

Pseudo- General prefix to indicate that something is up to isomorphism.

Putative We use this word when we're studying a particular sort of thing, which I'll call X. A putative X is something we are suggesting could be an X but we don't know yet, because we haven't checked it. Usually we say "putative X" when we're about to check that it really is an X.

Quantifier A logical clause that tells us the scope of a statement. The usual quantifiers are "for all" and "there exists".

Quintessential arrow category Informal terminology to describe a category consisting of just a single arrow. Similarly quintessential commuting square, quintessential isomorphism, and so on.

Rational numbers Fractions $\frac{a}{b}$ with a and b integers and $b \neq 0$. Denoted $\mathbb{Q}$.

Real numbers The rational numbers and irrational numbers together, making up a continuum of numbers on a line, with no gaps. Denoted $\mathbb{R}$.

"Recall" "I said this earlier and I'm hoping you have some recollection of it but if you don't there's a chapter or section for you to refer to."

Russell's paradox Paradox about self-reference in set theory, alerting us to the need for a hierarchy of concepts of "set" to avoid the self-reference.

Serially commutes This is when a diagram has some matching parallel arrows in it, and only the sub-diagrams of corresponding arrows commute.

Small category Category in which the collection of objects and the collection of morphisms are both sets.

Source The object at the "beginning" of a morphism in a category.

Strict Generally used to describe a structure in which axioms hold as equalities rather than isomorphisms.

Structure isomorphisms Structure in a higher-dimensional category coming from weakening something that was an axiom (via an equality) in a lower-dimensional version.

Sub- Prefix indicating a part of the whole (although the part could itself be all of it). Thus a subset of A is a set that is part of A (and could be all of it). Similarly subcategories. I also say sub-diagram informally.

Schematic diagram A diagram showing a general scheme of things rather than a formal diagram in a category.

Surjective function "Every output is hit at least once".

Target The object at the "end" of a morphism in a category.

Theorem A result with a proof.

Totally ordered set (toset) A category in which any pair of objects has exactly one arrow between them.

Type-checking A process of checking that all the concepts in a situation (or symbols in a formula) fit together in a way that makes "grammatical" sense, regardless of whether the result is true logically or not.

Underscore The symbol _ that we often use as a deliberate gap left for us to place a variable.

Unital A binary operation is called unital if it has a unit, that is, an object such that doing the binary operation with it does nothing. A composition is called unital if it has identities for every object, such that composition with them does nothing.

Vacuously satisfied A condition that is satisfied by virtue of it quantifying over the empty set. That is, we say "all such-and-such. . . " but that happens to be true because there aren't actually any such-and-suches.

Walking Another (informal) term for a "quintessential" category containing a certain structure. For example the "walking arrow" category just consists of two distinct objects and an arrow between them.

Weak Generally used to describe a structure in which axioms hold as isomorphisms or equivalences rather than equalities.

Well-defined Defined with good logic, without unresolved ambiguity.

Notation

$\forall$ for all

$\exists$ there exists

!   unique, or factorial when after a natural number (type-checking prevents ambiguity)

$\in$ is an element of (not to be confused with the Greek letter ϵ or ε)

$\implies$ implies

$\iff$ if and only if

$\cong$ Isomorphism, sometimes also denoted $\sim$ when it's a label on an arrow.

1_a The identity morphism on an object a.

$\mathbb{1}$ Often denotes a category with one object and one morphism.

$g \circ f$ or gf The composite of morphisms $\xrightarrow{f}\xrightarrow{g}$.

$\mathcal{C}(a, b)$ The collection (or set) morphisms from a to b in a category $\mathcal{C}$.

$\mathcal{C}^{op}$ The dual category to $\mathcal{C}$.

$\longmapsto$ An arrow indicating the action of a function or functor on an individual element, or object/morphism.

$\hookrightarrow$ We often use this shape of arrow with a hook at the beginning to indicate some sort of inclusion, like the inclusion of a subset into the bigger set.

Further reading

I used the following two books to introduce undergraduate mathematicians to Category Theory for several years:

- Steve Awodey, *Category Theory*, volume 52 of Oxford Logic Guides, Oxford University Press, 2006.
- F. William Lawvere and Stephen H. Schanuel, *Conceptual Mathematics: A First Introduction to Categories*, Cambridge University Press, 1997

The following two excellent texts were published later; my students have found them harder than Awodey:

- Tom Leinster, *Basic Category Theory*, Cambridge Studies in Advanced Mathematics, Cambridge University Press, 2014.
- Emily Riehl, *Category Theory in Context*, Dover, 2016.

The following has an emphasis on applications (obviously):

- Brendan Fong and David I. Spivak, *An Invitation to Applied Category Theory: Seven Sketches in Compositionality*, Cambridge University Press, 2019.

The following are, as indicated by the titles, aimed at specific non-mathematician audiences:

- Bartosz Milewski, *Category Theory for Programmers*, available from github.com/hmemcpy/milewski-ctfp-pdf/ .
- David Spivak, *Category Theory for the Sciences*, MIT Press, 2014.

The following are the standard graduate-level texts on category theory.

- Saunders Mac Lane, *Categories for the Working Mathematician*, volume 5 of Graduate Texts in Mathematics, Springer-Verlag, second edition, 1998.
- Francis Borceux, *Handbook of Categorical Algebra 1 and 2*, volumes 50 and 51 of Encyclopedia of Mathematics and its Applications, Cambridge University Press, 1994.

The following is a semi-formal introduction to higher-dimensional category theory, with many pictures:

- Eugenia Cheng and Aaron Lauda, *Higher-dimensional Categories: An Illustrated Guidebook*, 2004, available from eugeniacheng.com/guidebook/.

The following are other books by me which I referred to at various points (note that some of them have slightly different titles in the UK and the US):

- *How to Bake Pi: An Edible Exploration of the Mathematics of Mathematics*, Profile Books (Basic Books in the US), 2015.
- *Beyond Infinity: An Expedition to the Outer Limits of the Mathematical Universe*, Profile Books (Basic Books in the US), 2017.
- *The Art of Logic: How to Make Sense in a World that Doesn't*, Profile Books (Basic Books in the US), 2018.
- *X+Y: A Mathematician's Manifesto for Re-thinking Gender*, Profile Books (Basic Books in the US), 2020.

Acknowledgements

Personal thanks with British spelling, as that's how I spell in normal life.

The most important person for me to thank is my PhD supervisor, Martin Hyland. Martin taught me Category Theory, but also he taught me about mathematical thinking, education, generosity and wisdom. His influence is so strong, that sometimes even now, when I'm stuck with my research, I think to myself "I wonder what Martin would say?" and thereby get myself unstuck.

Martin's lectures were backed up by supervisions (tutorials) with Tom Leinster, who has been an unquantifiable source of support for all these years, from the first day I submitted my work to him and was rather surprised when he said I had done quite well. As usual I felt I was struggling. (It turns out that you can struggle and do well; in fact you can do better by struggling more.)

I would also like to thank my undergraduate directors of studies Paul Glendinning and Jan Saxl, who encouraged me to keep going when, as a congressive student surrounded by ingressive and often over-confident (and usually male) peers, I very often felt sure I wasn't good enough. I am deeply saddened that Jan Saxl is no longer here to receive my thanks; I owe him a great deal, for always believing in me when I didn't. He not only accepted that I would sometimes neglect my mathematics to prepare for an upcoming concert, but he even came to the concerts. He entrusted me with teaching, much earlier than I expected, and went so smoothly from seeing me as a student to seeing me as a mathematician, that I was encouraged to see it myself.

Peter Johnstone not only taught me logic, which was my favourite undergraduate course, but he gave me my first chance to teach a course on Category Theory, which was in fact my first ever lecture course. Those notes are preserved in files typed up by Richard Garner, who was officially my student but I've definitely learnt more from him than whatever I taught him; similarly Nick Gurski, who has been my most fruitful and hilarious collaborator.

Proceeding sort of chronologically, I next need to thank various people who supported my career in its early fragile days, by inviting me on research visits, urging me to apply for jobs that I would otherwise never had dared apply

418

for, and writing reference letters for me. Thank you to Paul-André Melliès, Pierre-Louis Curien, Peter May, André Hirschowitz, Carlos Simpson, and John Baez.

Thank you to many category theory friends who have patiently explained things to me and helped me feel safe enough to ask them questions that I thought were stupid: along with those I've already named, thank you also to Emily Riehl, Steve Lack, Todd Trimble and Pieter Hofstra.

And thank you also to all those who took any of my courses in category theory, at the Universities of Cambridge, Chicago, and Sheffield.

Moving on to this book more specifically, I'd like to thank those who read and worked through early drafts with me, giving me feedback about intelligibility. The first detailed readers were Bronwynn Wordsworth, Akosua Haynes, and my mother Brenda Cheng. Thanks also to the "Categories Club" at the School of the Art Institute of Chicago, for asking me to teach them category theory, and also working through an early draft of this book. Thanks also to Graciela Torres, and MurphyKate Montee and her students.

Also at SAIC I need to thank several years of students taking my class "The Elegance of Abstraction". Their interest in abstract mathematics despite various levels of past maths-phobia and maths-trauma convinced me that it could be possible to teach category theory in an accessible way, to students without undergraduate mathematics experience. Some of the material in this book was developed through those classes. The material was also developed and tested during several iterations of the high school summer camp I run in Chicago for Math Circles of Chicago; thank you to Doug O'Roark for inviting me to run the camp (and organising it), and to all the students who have participated. I have also tried some of this material with teachers, in mini-courses for Math for America; thank you to that organisation and the teachers who have taken those courses.

Finally, I am very grateful to my editor Kaitlin Leach for approaching me in the first place, about writing a book crossing the gap between technical mathematics and non-technical audiences. If I hadn't been asked, I wouldn't have written it.

Index